National Parks of Argentina

and other natural areas

NATURAL HERITAGE COLLECTION

National Parks of ARGENTINA
and other natural areas

Francisco Erize

Marcelo Canevari
Pablo Canevari
Gustavo Costa
Mauricio Rumboll

Translation: Mauricio Rumboll

LIBRERIA-EDITORIAL
EL ATENEO
BUENOS AIRES

INCAFO

```
55(82):502.4   The national parks of Argentina and its other natural
ARG            areas / Francisco Erize... (et. al.). - 1a. ed. -
               Buenos Aires: El Ateneo, 1995.
               238 p.; 32 x 24 cm.

               ISBN 950-02-6339-4

               I. Erize, Francisco - 1. Parques Nacionales
```

Título original: *Los parques nacionales de la Argentina
y otras de sus áreas naturales*

La presente publicación se ajusta a la cartografía oficial establecida por
el Poder Ejecutivo Nacional de la República Argentina a través del
Instituto Geográfico Militar, Ley 22.963.

(The present publication is adjusted to the official cartography
established by the National Executive Power of the Argentine Republic
through its Military Geographical Institute.)

Acknowledgments

To all National Parks Administration's personnel, without hierarchy
or work location distinctions, who provided cooperation, information or
support, and in particular to the chairman of its board, Dr. Felipe Lariviere,
and to Dr. Arturo Tarak, for his critical revision of the manuscript.
Also to Lic. Jorge Cajal, Dr. José Bonaparte, Lic. Mónica Vibbern,
Mrs. Silvia M. de Ross, and to the authors' wives.

Important warning

All rights reserved.
No part of this work may be reproduced or utilized in any form
by any means, electronic or manual, including photocopying, recording
or by any information storage or retrieval system, without the
prior written permission of the authors and the publisher.
Non complyers will be santioned according to article 172 and its
Penal Code's concordants (articles 2, 9, 10, 71, 72 Legal Act N° 11.723)

Queda hecho el depósito que establece la ley Nº 11.723.
© INCAFO, Madrid.
© 1995, "EL ATENEO" Pedro García S. A.
Librería, Editorial e Inmobiliaria, Florida 340, Buenos Aires.
Fundada en 1912 por don Pedro García.

ISBN: 950-02-6339-4 (edición en inglés)
ISBN: 84-85389-23-9 (publicada por INCAFO y EL ATENEO)
Depósito legal: M. 4.728-1995

Contents

- 6   Prologue
- 11   Map
- 12   Introduction

- 20   **The Subtropical Rainforest**
- 48   **The Yungas - Subtropical Cloudforests**
- 60   **Chaco Woodlands and Savannahs**
- 86   **The Espinal Woodlands**
- 96   **The Pampas Grasslands**
- 112   **The Monte - Brushland Steppe**
- 122   **The Patagonian Steppes - A Semidesert**
- 136   **The Atlantic Seaboard**
- 154   **The roof of America - Puna and high Andes**
- 168   **Subantarctic Forests**
- 210   **The Frozen Antarctic**
- 218   **New expectations**

- 226   Bibliography
- 231   Index of common and scientific names

Prologue

This volume from the series "Nature in Latin America" (La Naturaleza en Iberoamérica) covers the natural heritage of Argentina, a country that has within its borders one of the most varied mosaics of ecosystems. Most of the biomes of the world are represented: rainforest; savanna; hot, temperate and cool forest; scrubland; brushland; grassland; steppes, both scrubby and semi-desert; alpine tundra; sea-coasts; oceanic islands and polar desert.

The continental area is huge - it is the eighth largest country in the world - with some 2,791,810 square kilometres without counting the antarctic sector. It extends for some 3,700 kms from north to south, between tropical latitudes (21°46'S) and the subantarctic (55°03'S). The variety of habitats also answers to the altitudinal gradient east-west, from below sea-level on Valdés peninsula to the 23,000 feet (7,000 m) above sea-level of the highest peaks in the Andes.

Though the huge variety of wildlife (300 mammals, some 1,000 birds and 10,000 plant species) is not as great as in other countries, Argentina is amongst the richest in biodiversity when families and genera are counted.

Though considerably more than a decade has elapsed since the first edition of this book, it is still the most complete volume on the protected areas of Argentina. Demand, and the fact that the information contained herein is still as up to date as in that first edition, have merited that this translation be prepared with the additional chapters on the Antarctic, and on recent projects and developments relating to protected areas.

Essentially the book is an overview of the biological diversity of Argentina, including all the ecosystems and most of the diagnostically important species, and covers all the protected areas which conserve portions of the habitats. The edition appears at a time when most of the world's countries have become aware of the importance of this heritage and have subscribed to the Biodiversity Convention.

This Convention was one of the principal fruits borne by the United Nations' Environment and Development Conference (Rio do Janeiro, 1992). It gives importance above all to the "in situ" conservation of all the components of this diversity, and highlights therefore the enormous importance of protected natural areas. The treaty endeavours to share equally the benefits and the product of such benefits from biodiversity, and implies recognizing the fact that the resources of each country not only cross international borders but are part of the whole world's heritage.

This book aims to describe fully the rich natural patrimony with which Argentina has been blessed, to encourage its conservation; the authors and publishers are convinced that only that which is known can be loved, respected and therefore protected.

AREAS OF THE NATIONAL PARKS SYSTEM

1. Laguna de Pozuelos Natural Monument
2. Baritú National Park
3. Calilegua National Park
4. El Rey National Park
5. Formosa Nature Reserve
6. Pilcomayo National Park
7. Iguazú National Park
8. San Antonio Nature Reserve
9. Colonia Benítez Natural Park
10. Chaco National Park
11. Sierra de las Quijadas National Park
12. El Palmar National Park
13. Diamante National Park
14. Otamendi Nature Reserve
15. Lihue Calel National Park
16. Laguna Blanca National Park
17. Lanín National Park
18. Los Arrayanes National Park
19. Nahuel Huapi National Park
20. Lago Puelo National Park
21. Los Alerces National Park
22. Perito Moreno National Park
23. Bosques Petrificados Natural Monument
24. Los Glaciares National Park
25. Tierra del Fuego National Park
26. Los Cardones National Park (Project)
27. Aconquija National Park (Proyect)
28. Mburucuyá National Park (Proyect)
29. El Leoncito National Park (Proyect)

OTHER PROTECTED AREAS

A. Urugua-í Provincial Park
B. Iberá Provincial Reserve
C. Yala Provincial Park
D. San Javier University Reserve
E. San Guillermo Biosphere Reserve
F. Ischigualasto Provincial Park
G. Talampaya Provincial Park
H. Ñacuñán Provincial Reserve
I. Laguna Llancanelo Provincial Reserve
J. Monte de las Barrancas Provincial Reserve
K. Chancaní Natural Reserve
L. Mar Chiquita Provincial Reserve
M. Laguna La Felipa Natural Reserve
N. Samborombón Provincial Reserve
O. Península Valdés Provincial Reserve
P. Punta Tombo Provincial Reserve
Q. Cabo Blanco Provincial Reserve
R. Río del Deseado Provincial Reserve
S. Mar Chiquita Municipal Reserve
T. Payum Provincial Reserve

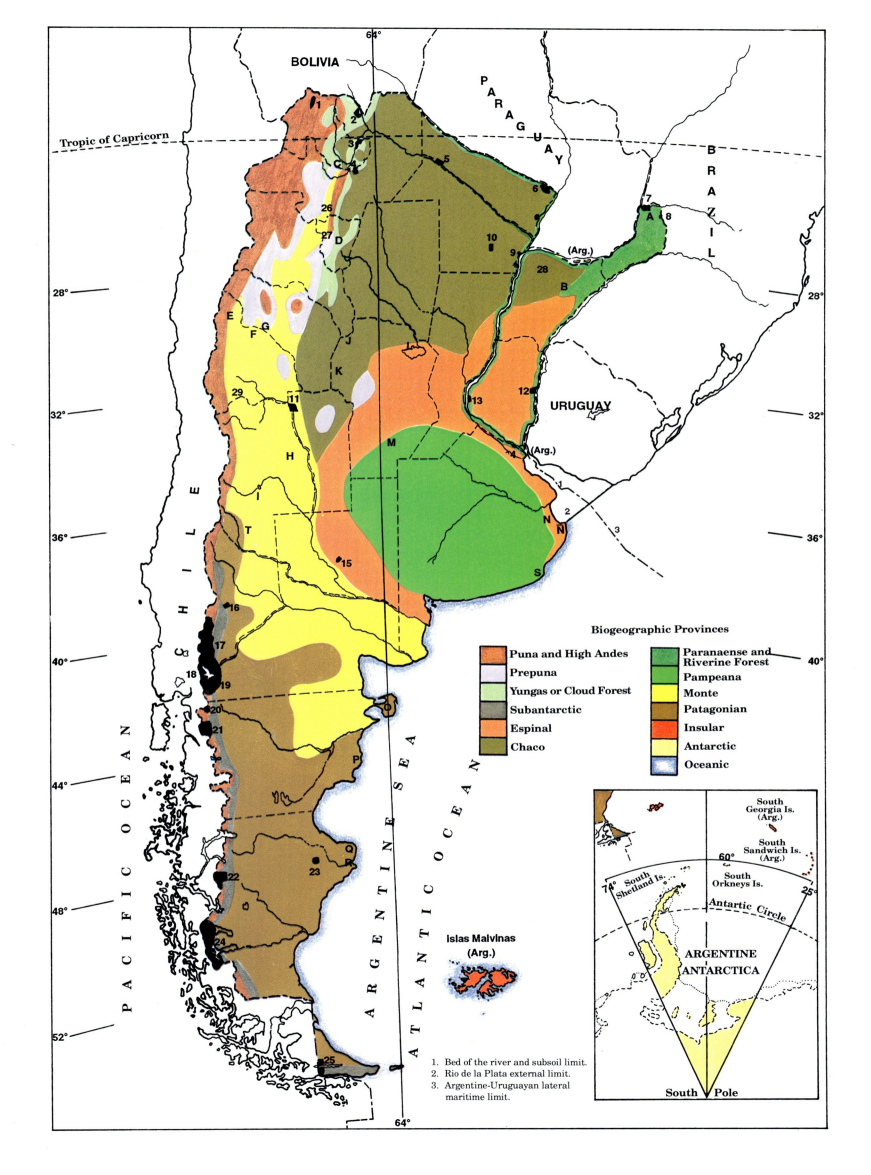

Introduction

In the face of environmental degradation, the continuous destruction of the natural world, the extinction of animal and plant species and the disfigurement of scenery, well-meaning conservationists the world over have, since the middle of the last century, sought to set aside natural reserves to ensure the survival of the most spectacular areas, of the most interesting biological communities.

Historical precedent for these measures had its roots in much earlier times. In order to preserve their game, many feudal lords in the middle ages took to conserving their forests and wildlife in general. Even before this, in China and India there were areas reserved for the protection of the animals, but it is only in 1861 that the first decree officially setting up a nature reserve comes into effect to preserve 624 hectares of the forest of Fontainebleau in France.

The concept of large areas being protected however started in USA where in 1872 Yellowstone, the first National Park was created "as a public park for the recreation, benefit and enjoyment of the people". This held new connotations - that as well as conserving the nature of the place, man could visit these areas and enjoy contact with them. It was also USA which, in 1916, first established a National Parks Service with the mission of administering the areas and ensuring that they performed their function.

Since then a majority of the world's countries have tried to preserve part of their natural heritage through systems of national parks or equivalent reserves. The more civilized countries now recognize as a duty of the first order both for their people and for future generations, to maintain a system of protected natural areas as complete as possible. That is to say to include all major ecosystems (habitats and all their biological communities), all species (especially those under threat), the major scenic features and natural monuments.

If we understand by "nature reserve" any area not noticeably changed by man which for aesthetic, scientific or educational reasons is safeguarded from man's whims to do with as he likes, and where competent authorities adopt pertinent measures to protect the ecological or scenic values which have merited its being set aside, national parks are special nature reserves.

The specific functions of National Parks are: the strict protection and preservation of the renewable natural resources, especially native flora and fauna, of the main physiographic features and scenic values; the maintainance of the natural balance and of the pristine condition of the natural habitats or the restoration of those which have been altered; their use for scientific studies on the species in their natural environment and the provision of recreational activities and educational oportunities not only for this generation but for those yet to come.

The Xth General Assembly of the IUCN (International Union for the Conservation of Nature and Renewable Natural Resources) held in New Delhi in 1969 drew up the definition of this concept which has been universally accepted ever since :

National parks are relatively large areas:

—containing one or more ecosystems, not (or little) altered by exploitation or by human occupation, where plant and animal species, landforms and habitats are of special scientific, educational and recreational interest, or contain natural landscapes of great beauty;

—where the country's highest competent authority has taken adequate measures to prevent or eliminate as soon as possible the occupation or exploitation of the area and to undertake all obligations of strict respect for the ecological, geomorphological or aesthetic features of the park which led to its creation;

—where entry is permitted to the public under certain conditions, for the purpose of inspiration, education, and cultural or recreational activities.

These are areas therefore, under the direct authority of the national government for maximum guarantees of protection, where all economic activities are forbidden except those which are connected with tourism, and even these are subject to regulations dictated in the best interests of conservation. Being considered the patrimony of all the inhabitants of the country, they must be beyond any political or administrative division.

It must be understood that national parks be large enough to make it feasible to protect viable ecological units whose ecosystems must have biogeographic representativity. The size of the area will obviously vary with the characteristics of the ecosystem which is being protected, but it must be remembered that sooner or later parks will end up as islands in a sea of totally man-changed surroundings and so will require buffer zones. It has been determined by research that the variety of beings which islands can support is directly related to their size.

National parks can be important gene banks. As such, they could preserve all the variety of species' diversity, including those species which could have great potential value medicinally or industrially, as yet unrecognized; or wild varieties of those already domesticated species which could boost our stocks through hybrid invigoration. These areas also protect the quality of the water, and the forests, which in turn protect the soils from being washed into rivers and lakes; or the vegetation cover which helps regulate the world's climates.

The national parks of nearly all the world, however, face innumerable problems, and it can be a measure of the success of these areas, just how they have succeeded in overcoming them. From the very creation of the park the borders can be inadequate, either because they are woolly, or because they do not include sufficient habitat for the animal species; they may not contain the whole of a watershed; they still may have within their borders some human settlement or some extractive industry, and dealing with these last two could bring on a socio-economic problem leading to political upheavals; often there is serious pressure on the very existence of the park or on its intangible nature, from the very neighbours who would like to exploit the resources the park contains; pollution in the lakes, poaching, illegal lumbering can be more than the security force can handle - rangers or whatever equivalent - as could even be the case of tourism; sometimes it is found that the native flora or fauna have been altered or affected, making it necessary to reintroduce some species; exotic species often get out of hand and must be controlled.

The solution to these problems usually lies in having an effective law and regulations, in being a strong administration for the management of these parks and in the development of a conservation consciousness, not only among the general public, but very especially in the neighbouring communities.

Argentina was the third country to set aside a national park (Canada had followed the example of USA and created Banff in 1885) when President Julio A. Roca's decree dated 1st of February 1904 accepted a donation made by Francisco P. Moreno. Moreno had been given land in recognition of his explorations of the southern Andes and the border with Chile. He donated it to the country with the stipulation that "the country should keep possession of certain exceptionally beautiful places for the use of present and future generations." On the 6th. November 1903 he gave 7,500 hectares in the lovely Nahuel Huapi area "to be conserved as a natural park for the public", adding his recommendation "that the sorrounding area be not altered, and that no more development be executed than is absolutely necessary for the comfort of the cultured visitor".

While this "Parque del Sur" was taking shape, enlarged to 43,000 has in 1907 and then again to 785,000 in 1922, there were projects under way for the establishment of another national park to protect the Iguazú waterfalls, based on the recommendation of the landscape architect Carlos Thays. So the government bought 75,000 has in Iguazú to be set aside as a national park and a military colony.

These parks did not effectively exist till law #12103 was passed on 9th. October 1934, drawn up by Dr. Ezequiel Bustillo at the instigation of the National Parks Commission presided by Dr. Angel Gallardo; the law declared how these areas were to work and set up the National Parks Administration as an autonomous agency to execute the law.

To the parks at Nahuel Huapi and Iguazú were soon added Lanín, Los Alerces, Perito Moreno and Los Glaciares (law # 13,895 of 11th May, 1937), while the rest of the areas would be added at considerably later dates.

The impetus of the new administration was such that under Bustillo and later under Dr. Luis Ortiz Basualdo, much of the infrastructure which is still extant today - roads, piers, ranger stations, headquarters buildings, hotels - was built in those initial years. Together with all this the Parks Service undertook settlement schemes and planned towns, built the railway stations, the post-offices and schools, hospitals and churches, public buildings in the urban centres of Iguazú, San Martín de los Andes, Villa Angostura, Bariloche (the civic centre of this last is a good example of the period's energy and taste). These towns have since been excluded from the areas of national parks as municipal duties and undertakings are not part of the parks' mandate.

Among the attributes which the law fixed as a function of the administration was to foment tourism. That is why hotels were built for tourist accomodation, and why launch services were started on the lakes. This function, however, led to a certain confusion in priorities .

Criteria for the strict protection of the native flora and fauna and the condition of keeping the areas pristine were, at the beginning, not deeply enough understood, so there were initiatives to introduce, propagate and spread exotic species of animals and plants. Though this today is seen universally as incongruent and incompatible with park management policies, the idea of artificially embellishing nature was at that time accepted, and ignorance reigned as to the ecological imbalance produced as a consequence of this.

The most serious and problematical inconsistency in the philosophy of parks at that time was the survival within its jurisdiction of agricultural, livestock and forestry practices.

At the time that most of these southern parks were created (1937 and before), there were already private land-holdings which were within the boundaries of the parks. With temporary occupation permits a number of pioneers settled in most of the fertile valleys in the hope that the state would eventually cede ownership to them. This possibility was eliminated when the lands were included in the national parks system, but the administration allowed these settlers to stay, exercising certain restraints on their activities - fixing a maximum number of domestic animals, and totally prohibiting logging or hunting.

These permits were intransferable which should have guaranteed an end to the activity with the death of the original settler, and the return of the area to parks and nature conservation. Unfortunately not all the authorities since then have applied this policy coherently - some even revalidated the permits for the descendants, or issued new ones.

It is to be regretted that these settlements, in spite of the restrictions on the use of the natural resources imposed and which they in theory should respect, in practice openly cotravene the law. Overgrazing takes place and there is consequent erosion especially on the slopes. Control is difficult because of the nature of the land. Often the maximum limit of heads of livestock is not respected.

Poaching is often carried out by the farm employees who are encouraged by the employer to supplement their meagre incomes with the product of this activity. Species which could potentially be damaging to livestock, such as the big Red Fox or the Puma are persecuted in flagrant contravention of the regulations which give wildlife absolute priority. The wildlife is constantly harassed by the drovers' dogs. Firewood is over-exploited either for sale or for home consumption which leaves gaps in the forest. Lastly, forest fires caused by carelessness on the part of these settlers in their wanderings throughout the area at all times of the year.

On the other hand several of these settlers have redirected their activities from the agricultural/livestock field to services for tourism and the visitors - camping sites, hostelries, rental of horses etc. - which are in total harmony with the management objectives of the protected area where they live.

In order to accomodate within the system the evolving classification of the different categories of areas which it allows for, and to adjust to the current acceptance on a world basis, as too to include modifications recommended by experience, law #12103 was replaced in 1970 with decree/law # 18945 and this in its turn with 22351 which was passed in 1980 and brings it upto date.

Thus the classification of these areas was established as: National Parks, Natural Monuments, National Reserves, lately supplement the by Strict, Wilderness and

Educational Nature Reserves (see last chapter). In the first two categories any form of economic exploitation is forbidden except if related to regulated tourism. In one category - National Reserve - sporting, commercial, industrial, agricultural and livestock activities are permitted, regulated by the National Parks Administration, but with the conservation of nature taking priority. Hunting and commercial fishing are totally prohibited.

The definition of these National Reserves as "areas of interest for the conservation of ecological systems, for maintaining buffer zones around the contiguous National Parks, or the creation of independant areas for conservation where the existing situation cannot be covered by designation as a National Park" (article 9), relegates these areas to serve as buffer areas around National Parks or be considered as potential National Parks.

In 1970 parks were zoned into "parks" proper, and "reserves", leaving nearly all occupied land in the latter category as the presence of settlers would not permit the application of the regulations of the former, with all its severity. This first division was done on a simplistic basis and it will be necessary at some future date to re-evaluate the zoning on an ecological basis.

The laws cited have been perfecting a National Parks Service, ruled by an autonomous Administration, as flexible and agile as this decentralized system is able to work. It depends on the Executive which is the owner of the lands, ceded to the central government by the respective provinces. This agency which is specialized in the management of natural areas must put special emphasis on scientific research which is indispensible for the task, as too on the development of interpretive techniques - through centres and trails to that end - to teach the visitor about the species of animals and plants found there, their behaviour or needs, their ecological relationships, the geomorphology and climates that affect them.

Control and policing of the parks is done by a force of Rangers who with police attributes must impose the regulations which govern these areas, must help and inform the visitor and must obtain information about the state of the wildlife of the park.

Rangers are professionally trained in the ranger training centre which is part of the National Parks Administration. This is done in conjunction with the National University of Tucumán and takes place at Horco Molle, the reserve just outside the city which will be run under a cooperative agreement. The school is called the Olrog Institute after Dr. Claes C. Olrog, who was instrumental in wakening the vocation of field naturalist in so many young Argentines. Students are trained in biology which they will need for nature interpretation; they are trained in practical disciplines for survival, for doing their tasks in distant situations, fire-fighting, police duties. The para-military aspect of this training is essential as the ranger must often face poachers or other transgressors. In 1968 Bernabé Méndez, a ranger in Iguazú, was shot down by such poachers while performing his duty.

The laws require that the administration evaluate and participate in the planning and execution of all public works within its jurisdiction, which is absolutely essential to avoid their being executed in ignorance of the principles of conservation, or contravening the policies of National Parks. This is of particular relevance in the planning of highways and roads which cross the parks as they can become menaces to wildlife, scars across the countryside or start erosive forces if careful criteria are not taken into account.

Within its nearly 2,800,000 square kilometres and the wide range of latitudes (21 to 55°S), Argentina is one of the best endowed countries as far as variety of habitats goes.

Add to this the elevational factor (below sea-level to about 7,000 metres) to understand just why within Argentine territory can be found such an enormous variety of climates and landscapes.

Subtropical rainforest, wooded savannas, all sorts of marsh and lowlands, grasslands, hot dry subtropical woods, cool, damp subantarctic woodland, brushland, semi-desert, high-mountain tundra, steppe, an extensive coastline, all constitute twelve argentine continental biomes: Paranaense (Misiones) rainforest, Yungas cloudforest, Chaco forest, Espinal woodland, Pampas grassland, Monte brushland, Patagonian steppe, Atlantic coastline, High-andean peaks, Puna and Pre-puna, Subantarctic forests, and the Antarctic wastes - an unusual variety for any country.

The present work is divided into chapters which correspond to each of these biomes, dealing first with the generalities, and later describing the National Parks or other protected areas to be found within each.

The present system of National Parks includes 22 national parks and contiguous national reserves, and three Strict Nature Reserves, but represents only some of the natural biomes of the country, and covers some 28,000 square kilometres, which is a bare 1% of the area of the country. By comparison we can cite other countries with low density of population like the African nations where the proportion of land under national parks is between 3% and 16%, while the USA and Japan have about 5% protected.

It must be a conservation priority then to increase the system to include other biomes and sub-biomes till all are well represented.

As has been stated above, there are also some provincial parks and private reserves which are more than just on paper and which constitute a valuable complement to the national system.

The participation of provincial authorities and even municipal governments in establishing, protecting and managing natural reserves is indispensible. The federal governments must preserve the best samples of the ecosystems in the country, or even those which include exceptional elements of the flora and fauna, or the most spectacular landforms and scenery. The provincial systems must protect other significant areas within their jurisdictions, which are worthy of conservation, to be enjoyed and used by local populations with less possibilities of access to national parks. Even at a municipal level areas of special interest can, and in a few cases do, come under a system of Municipal Nature Reserves, an initiative well worth fomenting.

The Subtropical Rainforest

In the past one vast rainforest covered all southern and eastern Brazil, extending from the coastal ranges - Serra do Mar, de Mantequeira, do Espinhaco - right over to the Paraná river, covering the states of Rio Grande do Sul, Santa Catarina, Paraná, Rio do Janeiro and Espíritu Santo, the eastern portion of Paraguay and the province of Misiones in northeast Argentina. Here, as everywhere in this region, europeans have carried out the most monstrous deforestation.

The part of this forest found in Argentina is known as the *Paranaense* and constitutes without a doubt, together with the Yungas in the northwest of the country, the richest diversity in life forms within the nation.

In truth, no other region can compare in numbers with the 2,000 species of vascular plants, the 400 birds (40% of the species found in the country) and innumerable insects, many of which have still not been discovered or described.

If we compare this area with the Amazon basin it undoubtedly seems impoverished, but all the characteristics of the great rainforests of the world are to be found here, where there is high humidity and temperature, with little temperature variation and no marked dry season.

Taking as an example the forests of Iguazú National Park, with about 2,000 mm of evenly-distributed rainfall annually, (some from the moist Atlantic winds, most from local evaporation), the relative humidity, reinforced by the heavy dew, is usually between 75% and 90%. There is more temperature variation diurnally than seasonally, the mean fluctuating between the average 15°C in June (mid-winter) and 25°C in the high-summer months of December and January.

These conditions of warm temperatures and high humidity make the region feel rather like a green-house, the best conditions for life on land. The essential elements for life - water, sun and warmth - are here always available; no droughts, no cold, those extremes to which life forms must adapt in other areas of the world, life here takes on the most exuberant forms.

The climax forest - that is mature stands which have reached stability - seems to be a compact green mass some 20 to 30 metres high. Wherever a river, a road or a clearing is present, a dense wall of green seems to forbid entry.

The main difference between woodland and forest is that in the latter there are several strata of vegetation which occupy all space from the ground to the tree-tops.

A sort of vegetation ceiling is formed by the crowns of the medium-sized trees which, in competition for the light, squeeze each other and form a continuous layer of foliage between 10 and 20 metres above the ground.

Every now and then, poking up from this canopy, one of the giants of 20 to 30 metres emerges as a silhouette against the sky. These crowns form islands above the green "sea" and are known as the emergent strata. The smaller trees between three and ten metres form another layer called the intermediate storey where the crowns do not touch each other to form a continuous stratum.

These divisions, it must be admitted, are somewhat arbitrary as they often merge one with the other and are furthermore all woven together by a network of lianas and climbers which confuse the issue.

Rainforest trees tend to have straight trunks of upto about one metre diameter in the larger specimens, and branch only at great height, generally in the upper third where the crown seems relatively tiny, but this does not hold for the emergent specimens which, on losing the press of competition for space can expand their crowns at their pleasure.

Though there are some more frequent species like *Balfourodendron reidelianum* (Guatambú Blanco), *Nectandra saligna* (Laurel Negro), *Cabralea canjerana* (Cancharana) and *Lonchocarpus leucanthus* (Rabo Macaco) which, being among the larger trees tend to be canopy species, there is a great variety manifest among these as in all forms of life here. Though all trees tend to look alike because of the shape they are forced to grow into by conditions, and by being similarly dressed in cascades of epiphytes and climbers, in any given area they will all be of different species, and of all the trees growing there, only a few will be of the same more common species, perhaps 5%. In a study on the trees in Iguazú it was found that there were some 300 trees per hectare; of these only 17 were *Balfourodendron*, 15 *Nectandra* and the rest were all of species which were less represented.

Amongst the giants, frequently poking through the canopy of the forest to emerge above - they can top 30 metres - there are many representatives of the Leguminosae family, nearly all shedding their delicate compound leaves in winter, leaves which remind one of the mimosas and acacias; long pods hold the seeds. These are *Myrocarpus frondosus* (Incienso), and *Peltophorum dubium* (Ybirá-Pytá) which produce some of the most sought-after timber; *Apuleia leiocarpa* (Ybirá-Peré), *Lonchocarpus muehlbergianus* (Rabo Itá), *Parapiptadenia rigida* (Anchico Colorado) and *Holocalyx balansae* (Alecrín). This last, as well as being evergreen, possesses a trait common to many amazonian trees - buttress roots which give the lower trunk a peculiar shape.

One of the most spectacular of the giants is another member of the family, the Timbó *(Enterolobium contortisiliquum)* also known as the Black Ear Tree as its pod looks very like an ear. Often thirty metres high or more, its main trunk reaches 1.6 metres in diameter, which when growing on its own, outside the forest, branches low down and gives the idea of a huge umbrella acacia. Other of the larger trees of the upper strata are: *Cedrela fissilis* ("Cedro" Misionero), *Patagonula americana* (Guayaibí), both of which also have valuable wood, *Rapanea lorentziana* (Canelón Guazú), *Luehea divaricata* (Azota-Caballo) and *Diatenopteryx sorbifolia* (María Preta).

Less developed specimens of all these trees are mixed in with the amazing variety of species that make up the canopy where the graceful Pindó-palm *(Syagrus romanzoffianum)* abounds. This palm whose slender trunk reaches over 20 m, like most palms is crowned as if with a feather duster of pinnate fronds. Also very common is *Chrysophyllum gonocarpum* (Aguaí), two more laurels, *Ocotea puberula* (Guaycá) and *Nectandra lanceolata* (Laurel Amarillo), and to a lesser degree the valuable *Cordia trichotoma* (Peteribí). One of the malvaceae, *Bastardiopsis densiflora* (Loro Blanco) has a spidery seed-pod which hooks onto animals and thus achieves its dispersal.

In the intermediate stratum it is the tree-ferns, so typical of tropical forests, which draw ones attention.

Trichopteris atrovirens (Chachí) grows here out of the sun whose rays would be harmful to it. It reaches a height of some 4 to 5 metres, has a single unbranched trunk from the top of which the palm-like crown is made up of fronds of upto 2 m in length. It is a relic of the past - one of the remaining life forms once so dominant in the Carboniferous era, some 250 million years ago. Its use as an ornamental or as the medium on which to grow house orchids has led to its becoming rare through overexploitation.

It is in this intermediate stratum that there are many smaller trees which produce fleshy fruits and berries of great importance in the food-chains. Without a doubt the best-known tree amongst these is the Yerba Mate *(Ilex paraguariensis)*, its dried and ground up leaves are the basis of a favourite infusion in all the southern part of the continent.

Below, in the shady shrub stratum, as well as the seedlings of the large trees and the ferns, there is a huge variety of bushes among which the *Piperaceae, Myrtaceae, Leguminosae* and *Rubiaceae* are well represented. It would be unlikely that the Giant Nettle *(Urtica baccifera)* pass unnoticed, with its huge leaves and ferocious sting.

The understorey seems in places to be dominated by canes which grow everywhere in entwined masses which bar ones passage. Sometimes the stands are 10 to 15 metres high, or more. The canes themselves are of various species: the slender *Chusquea ramosissima* known locally as Tacuarembó which is solid, upto 1.5 cms diameter and leans on the surrounding vegetation; the fragile spotted Tacuapí *(Merostachys clausseni)*; the strong thorny Yatevó *(Guadua trinii)* and the giant of the grass family (which includes canes), the Tacuaruzú *(Guadua angustifolia)* which, growing in a tight clump, reaches 30 m high, a diameter of upto 18 cms and has thorns at the joints.

Like all bamboos these flower but once, and that at the end of their life, every 25 or 30 years. All specimens of the species flower at the same time and die off completely. The seeds however germinate rapidly and new canebrakes soon repopulate the forests.

The lowest storey of this multilayered forest is called herbaceous. Here the plants grow barely a few decimeters off the ground and are decidedly lovers of the shade as very little light can penetrate the vegetation to this level: mainly some broad-leaved grasses like *Pharus glaber, Olyra latifolia,* many ferns like the three lovely *Adiantopsis, (A. radiata* looking like the spokes of a wheel), and non-woody plants like begonias which are so well known as ornamental house-plants. On the floor itself of the forest there is an accumulation of leaf litter and dead plant material - branches and twigs - where fungi, lichens and mosses thrive.

All this varied abundance is growing on a startlingly red soil which contrasts vividly with the dominant green. The abundance might lead one to believe that the soil is marvellously rich. The soils themselves are lateritic and are the result of weathering of igneous rock, rich in iron oxide and aluminium hydroxide but totally lacking in organic matter. In this environment of warmth and high humidity, decomposition is so very fast that there is not time for dead matter to accumulate to form humus, and the heavy rains tend to leech the soils with the result that organic salts and minerals are obtained by the living plants directly from the litter, not from the soil. To this end the root systems are very near the surface and even turn up into rotting logs and such.

With light and part of the rain monopolized by the upper strata several goups of plants have recourse to various techniques to compete for these essential factors. There is a legion of climbers, perhaps next numerically to trees, using fair means or foul to reach the light, twining, using trees as ladders, or germinating on high to start with. Some, like the Yaguápindá *(Pisonia aculeata)* with its hooked thorns which make some parts impassable, only reach a few metres off the ground. Those which are ever present are the lianas.

Holding onto trunks, leaning on branches, hooking into cracks, hugging smaller vegetation, lianas reach the canopy and develop there to hang down in cascades of ropes tying one tree to another, uniting trees to bushes, lying in great coils upon the ground itself. They flower at any height where there is sufficient light; large, showy flowers, which has led to some being used in ornamental gardening. Most species are of the *Bignoniaceae* family, where all the gamut of colours is to be found: red in *Dolichandra cyanthoides,* yellow in *Adenocalymma marginatum* and *Doxantha unguiscati,* lilac in *Arabidea chica,* lilac with white in *Clytostoma callistegioides,* with pink in *Cuspidaria pterocarpa,* orange in *Pyrostegia venusta* which commonly covers areas of gaps in the forest like road-cuttings, flowering in mid-winter. Less varied but more common are the *Sapinadaceae* like *Serjania, Cardiospermum* and *Urvillea* and compositae like *Piptocarpha sellowii* with its great whitish, highly fragrant flowers and white on the underside of the leaves, and *Mutisia campanulata,* a real climbing daisy.

One of the most remarkable characteristics of these forests is the abundance and proliferation of epiphytes, those plants which have sacrificed their contact with the soil and its nutrients for a place in the light - in the crowns of the larger trees. They do not parasitize the trees at all and merely use them as perches, holding strongly to branches or trunks and taking what they need from the air: humidity and sufficient nutrients.

Most conspicuous are the bromeliads, a family which includes a famous member, the pineapple, and many ornamentals. The roots of these epiphytic species are mere anchors, holding fast to trunks and branches. Their long leaves reach 60 cms in certain species of *Aechmea* and are shaped to direct water into the centre of the plant where it accumulates in the rosette formed there at the bases of the leaves. In these "ponds on high" there is a peculiar fauna - mosquitoes and other invertebrates, and even some frogs whose tadpoles develop solely in their own little predator-free tank. Air-plants of this family are common *(Tilandsia meridionalis* and *Billbergia nutans),* especially three: *Aechmea calyculata* with its yellow flowers, *A. disticantha* which has pinkish red blossoms, and *A. bromeliaefolia,* which seem to be on every tree. Their leaves are torn off the plant by capuchin monkeys and the tender bases eaten like we do asparragus.

Among the plants which grow on trees there are some which appear to cover parts of the trunk like upholstery, *Piperaceae* of the genus *Peperomia,* and epiphytic ferns of the genus *Polypodium.* There even are cacti adapted to this way of life, lacking thorns as they are out of reach of the great herbivores. They hang from the branches like green ribbons *(Epiphylum* sp.) or associated with *Aechmea* dangle like bootlaces, looking like aerial roots *(Rhipsalis).*

Of the multitude of epiphytes which as well as those mentioned above, include mosses and lichens, the most spectacular without a doubt are the orchids. The family *Orchidaceae* is considered the most numerous in the plant world - conservative estimates mention some 15,000 species, though the real number is probably far in excess of this - and members of the family are distributed throughout most of the world, including the cold areas like Alaska and Tierra del Fuego. It is only in the warmer climes that they grow epiphytically. Here the family reaches its greatest variety, with the largest plants and the most showy flowers; few are terrestrial.

These epiphytic orchids capture atmospheric humidity through their roots, wrapped as they are in spongy dead tissue which is extremely absorbent but also protects them in time

The Tapir

The largest South American land mammal, weighing upto 300 kgs, the Tapir is strong enough to, and shaped for charging through the undergrowth with ease. Its short trunk helps it to reach leaves, shoots or the fruit on which it feeds. In Argentina it is endangered because it is hunted for its meat which is greatly appreciated.

Epiphytes

The variety and abundance of epiphytic plants that grow here is so great that some of the larger branches look like gardens. On this branch we can see a bromeliad, some ferns and even a thornless cactus, some of the many groups of plants that have adapted to this way of life.

Philodendron *This araceous plant* (Philodendron bipinnatifidum) *with its enormous lobed leaves upto 80 cms long normally grows as an epiphyte on thick trunks and branches, but if its roots reach the ground it receives enough nutrients to develop into a real climber.*

The Jaguar

The largest Argentine predator, the Jaguar has already disappeared from most of its former distribution, and the few that are left are still being hunted by poachers. A small but healthy population of these cats is found in Iguazú National Park.

of drought. Most species have fleshy leaves or thickened parts at the bases of the leaves which serve to store water and nutrients. Whether they produce huge or small flowers on dense flowering heads, all of them have strange shapes which have evolved as strategems to ensure pollination. A large lower "lip", this petal much more developed than the rest, is the landing ground for the insects attracted by the strong fragrance; a single large stamen placed above this powders the head or back of the visitor who has come to sip nectar from the base of the flower; with a good load of pollen, the insect leaves to visit another flower.

To counter the uneven odds against seed landing in just the right place to germinate and later grow well, orchids produce an overwhelming number of tiny seed which are easily wafted about on the slightest breeze. Each pod holds upto one million or more seeds. The odyssey does not end there as, to develop, it must associate with a special fungus (species-specific) with which it will establish a symbiotic relationship, the fungus enjoying part of the sugars produced by the orchid, the orchid using proteins produced by the fungus.

The Paranaense Rainforest is particularly prolific in orchids: *Miltonia flavescens* in October and early November covers whole branches with showers of pale-yellow blooms, *Sophronitis coccinea* produces brilliant orange-red flowers, *Brassavola perrini* throws out pairs of creamy-white blossoms, the "patitos" (genus *Oncidium*), bunches of yellow with brown spots, *Campylocentrum* has no leaves - just roots and flowers, and many other species contribute to the glamour of the tree that hosts them.

This family is of great commercial value, which, coupled with uncontrolled and alarming deforestation, is causing many species to be considered endangered, which in turn requires special conservation measures like the creation of nature reserves.

Another notable epiphyte is the Guembé *(Philodendron bipinnatifidum)* a huge-leaved *Araceae* plant, some of whose roots hang in search of nutrients while others grasp and hug the supporting tree. When the first reach the ground the plant becomes a climber and develops very quickly. It also grows on the ground but with a short thick stem.

The seeds of the Guembé which pass through some fruit-eating bird or mammal often end up in a hollow or crotch of a tree where plant matter accumulates, and there, with humidity and the organic material it develops into a huge epiphyte, with no more relationship than a perch with the tree which acts as its host.

A similar situation can be found with the local strangler-fig *(Ficus enormis)* of the *Moraceae* family, though with a different ending to the story. The young plant starts life as an epiphyte, its seed transported and dropped from some bird or mammal which ate the fruit. The first and main root heads for the ground while lateral roots take hold of the host tree for support. As more roots develop and reach the ground they broaden and slowly cover the trunk completely as they grow together and into each other. This throttles the host over time and the fig's trunk becomes self-supporting against the day the core or trunk of the now-dead host rots away. As a tree the fig reaches a huge size (upward of 20 metres), and is helped in its support by buttress roots - a cunning strategy to have a large crown supported on another's trunk while its own trunk develops and strengthens.

Where the hand of man or fire has produced a gap, a whole community of rapid-growth light-loving plants establishes

Perched on a Pindó Palm (Syagrus romanzoffianum) *this Red-breasted Toucan* (Ramphastos dicolorus) *shows off its colourful plumage and enormous beak with which it can pick berries out of dense thickets, or nestlings out of deep nests.*

itself. Under it soon grow those which prefer the shade and some day will take over from the pioneer community called capueira and establish the second link in the chain towards restoration of the forest. The first, fastest and most abundant species in this cycle is a small tree, Fumo Bravo *(Solanum granuloso-leprosum)* which reaches a height of some 5 metres in a couple of years. Its globular fruit with that of the Palo Pólvora *(Trema micrantha),* of the leaning Tala *(Celtis* spp.), of *Casearia sylvestris* and of Mandioca Brava *(Manihot flavellifolia)* attract a great quantity of birds and mammals to these areas. In wetter parts of these capueiras is *Croton urucurana,* the Sangre de Drago, with foliage of different colours - gray-green, orange and red - which have made it a sought-after ornamental, and whose sap is much used in native medicine.

Another small tree which awaits patiently the oportunity to fill some gap caused by man or even by a falling giant is the Ambaí *(Cecropia peltata),* a moraceous tree whose trunk branches at a fair height above the ground and which develops a parasol shape; its large palmate leaves, silvery underneath, darker green above and its fruit in bunches like dangling fingers. This tree has a stem which is hollow in segments, like a cane, and in one or more of the cavities there usually are nests of the small, red and very aggressive Aztec ants *(Azteca)* which defend the tree ferociously.

Flowering in subtropical rainforests is not synchronized to any season, nor does it coincide with rains or warmth, but is distributed throughout the year, each species having its own time. Some of the trees are spectacular at that moment, like the great Lapacho Negro *(Tabebuia impetiginosa)* which reaches 30 metres and a girth of 1.5, and bursts into bright pink before winter closes and before it has put out leaves, to attract myriads of hummingbirds to its generous offering. The Yellow Lapacho *(Tabebuia alba)* blooms a week or two later in the same way and both these carpet the ground around themselves with fallen blossom. The Ybirá-Pitá *(Peltophorum dubium)* does the same at year's end in yellow, and a forest

Toucans

coral tree *(Erythrina falcata)* in a bright orange-red in November.

Because of the low density of the population of each species due to the extraordinary diversity in this biome, trees in flower will be few and far between at any one time.

However, in spite of the fact that a harvest of flowers could be meagre at any season, it being at all possible in all seasons permits several species of bird and mammal to feed exclusively on flowers.

The same is true of fruits which are here one of the most relevant food sources. Amongst the most generous "fruiters and seeders" are several *Myrtaceae* like Guabiroba *(Campomanesia xanthocarpa)*, the exotic Guava *(Psidium guajava)*, and Cerellas *(Eugenia involucrata* and *E. uniflora)*, as too a Gutifera, Pacurí *(Rheedia brasiliensis)* and a native pawpaw *(Jacaratia dodecaphylla)*.

When fallen fruit accumulates on the ground around one of these trees, most of the vegetarian ground mammals converge on it: tapirs, peccaries, brockets, pacas and agoutis. Because of the paucity of the herbaceous stratum these animals graze or browse preferably in the wetter areas, more open above and therefore with a better bushy or shrub storey, and on the edges of rivers or streams where the vegetation even at ground level is profuse and offers some form of concealment. They take to the forests to get out of the heat, to seek refuge from their predators and from mosquitoes, to move from one fruiting tree to another.

Nearly all these animals share two characteristics dictated by the forest habitat where they live, to wit: their general shape and their familiarity with water. They swim well which is essential for their movements around a habitat criss-crossed with water-courses. Not even the major rivers seem to detain them - troops of 300 or so White-lipped Peccaries have been seen crossing the Iguazú river.

The need to move through the intricate web of vegetation such as cane-brakes, fallen trees, bushes, and all this woven together by lianas and creepers armed with thorns, has led to the animals having compact, robust bodies on short strong legs, veritable torpedos to hurl themselves through the densest vegetation to escape from their predators. The strategy is often to break the obstacles rather than jump them or weave through. In this way a tapir with a jaguar on its back with intent to kill, might be able to scrape the predator off before the fatal bite.

The Tapir *(Tapirus terrestris)* is the largest animal of the Paranaense Rainforest and the largest terrestrial mammal of South America, weighing some 300 kilos. Its solid body, muscular neck and thick hide give it all the characteristics of an armoured battle-waggon, the best adapted lower-storey ground-dweller. The short trunk stretches for leaves, sprouts, tender twigs and fruit which otherwise might be out of its reach. Solitary and fairly sedentary it moves along trails it has opened up through the thicket, and spends long hours wallowing in some backwater or pool to get cool or to keep blood-sucking parasites off. The tapir also feeds on aquatic vegetation and is capable of diving to find it; it often takes to the water to escape when pursued.

Perhaps the most common terrestrial herbivores in this biome are certain large rodents: the Capybara, giants of the order which are pig-sized and weigh upto 70 kgs, Pacas, upto 10 kgs and 60 cms long, and Agoutis all have bullet-shaped bodies like all those which charge off to get away from danger, and, almost lacking tails, remind one more of small antelope of the African forests whose role they seem to duplicate, than of rats and such, the real rodents.

The Paca *(Agouti paca)* is nocturnal, spending the day in some burrow dug into a riverbank, under a log or roots, or some other natural cavity. Alone or in pairs they patrol riverbanks or search for fallen fruit, their favourite food.

The Agouti *(Dasyprocta azarae)* is a little smaller than the Paca - some 50 cm long - faster and more agile on its long, slender legs, and more diurnal. It sometimes sits on its haunches to eat, taking its food in its forepaws. When food is abundant the Agouti buries some to store it.

In this habitat there are also small solid deer, the stags having the minimum possible antlers - just two spikes. These Brockets live in pairs: the Red Brocket *(Mazama americana)*, the Brown Brocket *(M. gouazoubira)* and the Dwarf Brocket *(M. rufina)*.

In South America peccaries take the place of the pigs and boars of the other continents. There are two versions in these forests - the Collared Peccary *(Dicotyles tajacu)* which in

Butterflies

The abundance, variety and extraordinary colouration of the butterflies of the Paranaense Rainforest make them one of the most attractive features of its wildlife. Perhaps the most spectacular are Morpho anaxibia *(top left) because of its size and iridescent colour. The* Brassolidae *(an exclusively South American family) like this* Brassolis sophora vulpeculus *(bottom left) and* Eryphanis polyxena *(bottom right) appear at dusk. They have "eyes" on the undersides of their underwings to suddenly resemble an owl's face when approached by some potential predator.* Metamorpha stelenes *(top right) is of the* Nymphalidae, *an extraordinarily diverse family.*

smallish groups of upto 15 individuals wander in search of their favourite foods - fallen fruit, while the White-lipped *(Tayassu pecari)* forms enormous herds of 200 or more, forever on the move and covering huge distances. They are not exclusively vegetarians as their diet also includes insects and grubs, the larger spiders, snakes and lizards.

The greatest abundance of food however is not found on the floor of the forest but high in the trees. Fleshy fruits, flowers, tender leaves, an abundance of insects are reason enough for many mammals to have taken completely to an arboreal life. To move with agility through the tree-tops, representatives of very different orders, with totally different diets, through a process of what is called convergent evolution have all developed a truly prehensile tail which serves as a fifth limb. This appendage is found in the opossums *(Didelphidae)* of the order of the marsupials, the Collared Anteater or Tamandua *(Tamandua tetradactyla)* an edentate, the Dark Coendú or Tree-porcupine *(Sphiggurus spinosus)* a caviomorph rodent, and the local Brown Capuchin Monkey *(Cebus apella vellerosus)* and Red and Black Howler Monkeys *(Alouata guariba* and *A. caraya).*

The Brown Capuchin is remarkably common in this habitat. Whistles and grunts from the tree-tops above give them away. They move stealthily through their world, except when hurrying off. Troops of upto 200 cover many kilometres per day in search of fruiting trees. Careful inspection of the branches and twigs provides them with moths, other insects and larvae which complete their diet. Water is found in the bases of bromeliads and other epiphytes.

Opossums are also omnivorous but the main part of their diet is small animals instead of fruit: insects, grubs, snails, reptiles and amphibians, birds and their eggs, the smaller mammals and even carrion. The evolutionary success of the *Didelphidae,* the South American marsupials, probably because of the family's adaptive flexibility, can be seen in the many species of which nine are found in these forests.

They vary from the mouse-sized Murine Opossums *(Gracilinanus agilis, Marmosa cinerea)* and the striped Bare-tailed *Monodelphis henseli,* to the cat-sized White-eared and Black-eared Opossums *(Didelphis albiventris* and *D. marsupialis).* In between these extremes are the Four-eyed Opossum *(Philander opossum),* the Rat-tailed Opossum *(Metachirus nudicaudatus)* and the Red Water-possum *(Lutreolina crassicaudata).* Though some are found more in the trees than others, they are all good climbers using their prehensile and naked tails, and are mostly nocturnal.

The most extraordinary of these marsupuals is the Yapok *(Chironectes minimus),* totally adapted to aquatic life, with webbed hind-feet and a watertight pouch; it occupies an ecological niche similar to that of otters. Like these it is an excellent swimmer and diver, feeds on fish and aquatic invertebrates, and its burrows are in riverbanks just above the waterline. It has seldom been seen, probably more because of its habitat and nocturnal habits than because of its rarity.

Here too there are other aquatic carnivores than those just mentioned above as the abundance of rivers and streams provide choice habitat for them, the otters. There are two species of these, the Paraná Otter *(Lutra longicaudis)* and the Giant Otter *(Pteronura brasiliensis)* upto seven foot long, with its laterally compressed tail. This species has been so hunted for its pelt, its diurnal habits making this easier still, that its survival in Argentina is a matter of grave concern.

As herbivorous mammals are found on the ground and in the treetops, any predator must be able to hunt in any or all strata of the forest to take full advantage of the opportunities, because (as in the plants), any single species is too sparsely scattered to be a dependable source of food for a specialist. Even the Jaguar *(Panthera onca),* the supreme predator of the area, though it shows a preference for larger prey like young tapirs, brockets, peccaries, and pacas, also feeds on lizards and snakes, smaller birds, small rodents, catches truly arboreal species like monkeys, fishes in streams and rivers, and even at times eats fruit.

The major group of carnivores in this biome is the cats. Clever, agile, silent, good swimmers and climbers, usually well-camouflaged and equipped to kill efficiently, these are the most successful carnivores as can be seen by the proliferation of species. Of different sizes and shapes, with different colours and patterns, as well as the Jaguar and in order of diminishing size we have the Puma *(Felis concolor),* the Ocelot *(F. pardalis),* the Margay *(F. wiedii),* the Jaguarundi *(F. yaguaroundi),* and the Little Spotted Cat *(F. tigrina)* .

Another remarkable predator is the Tayra *(Eira barbara),* known as Irará locally, which, though much smaller, can follow a single brocket all day, tire it out, and finally start eating it while still alive. They often hunt in pairs, usually take smaller birds or mammals, and climb extremely well when necessary.

The most common mammal is surely the Coati *(Nasua nasua)* which travels around in groups of between eight and over twenty specimens - all females and young, as males are solitary except in the breeding season - combing the forest, exploring every crevice and hole with their flexible snouts in search of anything edible, vegetable or animal. They wander about with their curiously bear-like flat-footed gait, tails on high to display the striking rings, but are equally at home in the trees where they find birds' nests with eggs or young, fruit and a cooler perch during the hotter hours of the day, their tails dangling like counterweights.

Also fairly common is the Crab-eating Raccoon *(Procyon cancrivorous),* usually near water where it carefully washes its food before chewing and swallowing it. The Crab-eating Fox *(Cerdocyon thous)* a widely distributed species, is the most common canid in these forests. The real curiosity however is the Bush Dog *(Speothos venaticus),* a very strange beast with its long body, short legs, small, rounded ears, which was fairly recently discovered in the area.

The mammals which show the greatest diversity in this forest are in the order of bats - Chiroptera. While a large number of Vespertilionid and Molossid bats keep to the insect diet which is the most common in these night-flying hunters, others have widened the spectrum of their diets to include nectar, fruit, blood, fish and even other small vertebrates. The Vampire *(Desmodus rotundus)* is the specialist which feeds exclusively on the blood of mammals and birds. Arriving near its victim on the wing, the final stealthy approach is on all fours - tiny legs and the knuckles of the folded wing. The specially sharp and protruding incisors nick the skin; the saliva has an anaesthetic and an anti-coagulant, and the bat laps up the blood which dribbles out of the wound. The victim sleeps on, unaware of what is happening, so delicate is the whole operation.

Of particular interest are the Phyllostomatid bats. Amongst them the nectar-feeding Long-tongued Bat *(Glossophaga soricina);* while it sips nectar and catches the insects in flowers with its tongue, it also carries out the essential function of pollenation. The Yellow-shouldered Bat *(Sturnira lilium)* and the Neotropical Fruit-bat *(Artibeus lituratus)* with its two white stripes from face to back over the crown are fruit-eaters, while the large Peters' Wooly False Vampire Bat *(Chrotopterus auritus)* is carnivorous, feeding on frogs, lizards, birds, rodents and even on other bats, perhaps supplementing its diet with fruit. A particularly interesting adaptation is that of the Fishing Bats *(Noctilio leporinus* and *N. labialis)* which in flight drag the elongated claws of their feet through the water to catch small fish near the surface.

In spite of the great variety of mammals the Paranaense Rainforest has, including some which are not found in any other habitat in the country, like the Bush Dog, the Dark Tree-porcupine, the Dwarf Brocket, the Giant Otter and the Yapok, one must not commit the mistake of thinking of these forests as heaving with wildlife. In dense vegetation they behave more like ghosts than tangible beings. One can only sense their presence, guess at it, discover the sign and spoor left by them which is so easily visible to a woodsman.

The birdlife of these forests is somewhat more evident because of the calls and songs, their flights and movements, their showiness. But even the most spectacular birds can remain unobserved in the crowns of trees thanks to their use of light and shadow, or of the denser foliage; they keep out of sight hiding in the twilight of the undergrowth.

The amazing variety of birds in this province can be attributed to the enormous possibilities for specialization and the abundance of ecological niches. The quantity, variety and permanent availability of food has led to whole groups or families surviving on a fruity diet (trogons, toucans, parrots), while a world of hummingbirds feeds only on nectar from flowers and others depend entirely on the hordes of flying insects.

As the forests are three-dimensional, the different strata make up different biotopes, each with its own population of bird species. So one can differentiate between those that live on the floor of the forest, the species that are in the shrub storey, those of the intermediate stratum and the birds found in the canopy as well as the species which are to be seen in the emergent giants, or in the air above the forest.

Ground birds include the Tinamous, a family of chicken-like game-birds. Among them is the Solitary Tinamou *(Tinamus solitarius),* the largest of the family, the Brown Tinamou *(Crypturellus obsoletus)* - both searching for seeds, fruit or invertebrates on the ground, though the Solitary Tinamou does roost fairly high in trees. There is also a real quail of the pheasant family, the Spot-winged Wood-quail *(Odontophorus capueira)* which wanders around on the ground in small flocks.

Ground birds tend to be timid and shy, and the large, long-legged Slaty-breasted Wood-rail *(Aramides saracura)* is no exception though it prefers to run from danger rather than count on its camouflage and freeze, and can be glimpsed racing away from one down trails. All these are reluctant fliers but can do so when pushed, with low, rapid and heavy flight.

The bushes and the lower strata of the vegetation are the haunt of many of the Formicarids, ant-birds, an exclusively South American family often associated with the columns of army ants *(Eciton* spp.). These ants are nomadic and settle in temporary nests every so many days - dictated by the period of reproduction. Their advance is a fearsome sight and flushes out all manner of hidden creatures which flee in panic to be snapped up by the attendant Spot-backed *(Hypoedaleus guttatus),* Large-tailed *(Mackenziaena leachii)* and Tufted *(M. severa)* Antshrikes, all markedly sexually dimorphic with striking black or brown plumages boldly barred, spotted or

Concentrations of Butterflies

Pools or mud containing organic salts such as the urine of some larger mammal attract large numbers of butterflies of the most varied species which congregate to sip the dissolved salts and the moisture. In this gathering there are representatives of four families - Papilionidae, Heliconidae, Pieridae and Nemeobidae.

The Coral Snake

Snakes belonging to the genus Micrurus *are feared because of their potent poison which is fatal to humans, but fortunately they are not at all aggressive. Their brilliant colours are an effective warning.*

The Paca

This great rodent (Agouti paca) is one of the herbivores to be found in the rainforests; compact and bullet-shaped it is adapted for escaping at speed. It is evident that some rodents in this continent are occupying niches which in others belong to small antelope or deer.

Cerambicidae *Coleoptera with extremely long antennae, they are among the largest insects to be found in this rainforest. Some, like this* Steirastoma marmoratum *have contrasting colours on the elitra forming curious patterns. They lay their eggs in wood which the grubs bore through while feeding on it.*

plain; also present would be the long-legged ground species like the Variegated Antpitta *(Grallaria varia)* and the Rufous Gnateater *(Conopophaga lineata),* or even the lovely but hard-to-see Short-tailed Antthrush *(Chamaeza campanisoma).*

Sometimes in these nether storeys are found dumpy little birds of another exclusively South American family - the Manakins. Berry-eaters and to a certain extent insectivorous, these creatures draw our attention as much by the brilliant colours of the males as through the elaborate courtship rituals which include dances, strident calls and showing off the best part of the plumage. All this is accompanied by mechanical sounds produced by certain special feather-structures and flashes of yellow, scarlet, electric blue, matte black or blinding white in the various species like the Band-tailed *(Pipra fasciicauda)* or Swallow-tailed Manakins *(Chiroxiphia caudata).* Tree-trunks and branches are the hunting grounds of the woodpeckers *(Picidae).* Inhabited by a multitude of wood- boring insects and their larvae, mostly scarab and long-horned beetles, they provide these birds and the Wood-creepers *(Dendrocolaptidae)* with an excellent diet. Both these families of birds are well-adapted to moving around on trunks and branches, with strong claws to hold on with while chipping away, and rigid tails to serve as the third leg of a tripod. Lacking the chiselling beaks and immensely long tongues of the woodpeckers, the woodcreepers, which are South America's equivalent of the Old World's tree creepers *(Certhidae)* have long bills for poking into crevices and holes, or under bark to search for grubs and insects. All are a reddish-brown suitable for camouflage and the most abundant are the White-throated *(Xiphocolaptes albicollis),* the Planalto *(Dendrocolaptes platyrostris)* and the Plain-brown Woodcreepers *(Dendrocincla fuliginosa),* but the Olivaceous Woodcreeper *(Sittasomus griseicapillus)* seems to be everywhere most noticeable because of its strident descending whistled call.

The canopy, made up of the crowns of the larger trees, is home to the greatest profusion of birds. There live most of the fruit-eaters, many of which have brilliant colours. The toucans use their immense but very light bills to reach the fruit through the densest tangle. The five species here represent all the groups of toucans that exist - the Toco and the Red-breasted Toucans *(Ramphastos toco* and *R. dicolorus* respectively), the Saffron and Spot-billed Toucanets *(Baillonius bailloni* and *Selenidera maculirostris),* and the Chestnut-eared Aracari *(Pteroglossus castanotis).*

The parrot family shows a similar spectrum: the large short-tailed parrots like the Scaly-headed Parrot *(Pionus maximiliani),* the long-tailed White-eyed Parakeet *(Aratinga leucophthalmus)* and the Reddish-bellied Parakeets *(Pyrrhura frontalis),* the small Red-capped Parrot *(Pionopsitta pileata)* and the minute Blue-winged Parrotlet *(Forpus xanthopterygius).* Tanagers *(Thraupidae)* sport bright iridescent plumages like the Green-headed Tanager *(Tangara seledon),* or flash brilliant but generally hidden crests as do the Ruby-crowned and the Black-goggled Tanagers *(Tachyphonus coronatus* and *Trichothraupis melanops);* some wear blue and yellow like the various Euphonias *(Euphonia* spp.); one is simply black and white with a long tail, but in such a pattern as to be most attractive, the Magpie Tanager *(Cissopis leveriana).* A bird which is as brilliant as the tanagers but in a family of its own is the Swallow-tanager *(Tersinia viridis);* the male is electric blue or green - depending on the angle of light - with a black mask and white belly. The Cotingas eat the larger fruit; they include the enormous Red-ruffed Fruit-crow *(Pyroderus scutatus)* some 38 cms long and the Tityras *(T. cayana* and *T. inquisitor),* white, with black caps and tips to tail and wings.

Insectivorous birds are legion in the canopy, the *Tyrannidae* (New World Flycatchers) being the most common and varied. Like their Old World equivalent *(Muscicapidae)* they usually catch their prey on the wing. Most of these are small, olive-brown and yellow and hardly draw ones attention, but the Long-tailed Tyrant *(Colonia colonus),* black with a white crown and two long central tail-feathers, sits high on some prominent leafless branch or twig.

Two Puffbirds, dumpy and big-headed, are of the neotropical family *Bucconidae,* the Eared *(Nystalus chacuru)* and the White-necked *(Notharchus macrorhynchus)* sit on some exposed perch to fly out and snap up some passing insect, or to pounce on some small, unsuspecting lizard or frog.

Such profusion of colours and iridescences might give one the impression that all forest birds are this way. Certainly the trogons *(Trogonidae)* with their metallic green or purple backs and scarlet *(T. surrucura)* or yellow *(T. rufus)* breasts and bellies fit the rule, but they do not stand out; rather they remain inmobile for long periods, and their ventriloqual call of low whistles (two-two-two-two....) is hard to place. They eat fruit which they pluck off on the wing and insects which they capture in tumbling flight. As with the Rufous Motmot *(Baryphthengus ruficapillus),* Trogons are found in the lower strata of the forest. The Rufous-tailed Jacamar *(Galbula ruficauda)* also with metallic green back, could pass for a large hummingbird; it sits at the edge of clearings and snaps up the larger insects like dragonflies and butterflies with its long pointed bill. The usual image of a South American rainforest is one of a profusion of jewelled hummingbirds flitting from orchid to gaudy orchid. There are well over a dozen species of hummingbirds in these forests including the common Violet-capped Woodnymph *(Thalurania glaucopis),* the White-throated Hummingbird *(Leucochloris albicollis),* this last with a white slash interrupting its metallic green at the throat and breast; the Black-breasted Plovercrest *(Stephanoxis lalandi)* where only the male has the characteristics mentioned in its name; the Scale-throated Hermit *(Phaetornis eurynome),* one of a large genus with long decurved bills and the central tail-feathers elongated and whitish, nests at the pendulant tip of a large rolled leaf, lays puce eggs and decorates the lip of the nest with bits of the puce lichen *(Chirodecton sanguineum)* till the chicks hatch, when the decoration is discarded. The hovering flight which is best developed in hummingbirds allows them to visit every flower on the wing and sip the nectar from the depths. Small spiders and insects captured on these visits provide the proteins in the diet.

Many forest species have developed the habit of nesting in cavities in trees or termite nests. The woodpeckers, toucans, parrots, woodcreepers, puffbirds, trogons, jacamars, and tityras all do this; some nest in holes in banks like motmots and kingfishers.

Among the larger birds in this habitat are the Guans *(Cracidae - Penelope obscura, P. supercilliaris* and *Aburria jacutinga),* gallinaceous birds which look like turkeys or pheasants, mostly Neotropical. Highly arboreal, they live in the canopy and the intermediate stratum, nesting there and feeding on fruits and berries. They are hunted for their meat which is much valued by man.

During the summer months the most visible raptors are two lovely birds of prey - the black and white Swallow-tailed Kite *(Elanoides forficatus)* and the Plumbeous Kite *(Ictinia plumbea)* agilely glide and cavort above the canopy to catch the larger insects flying there, or take some smaller vertebrate from the tree-tops, a lizard, a snake or a frog. During migration (arrival and departure from and for the northern climes) they are seen in groups circling on thermals. Harder to see but perhaps more interesting in their adaptations are the Forest-falcons, the Barred *(Micrastur ruficollis)* and the Collared *(M. semitorquatus);* on short, rounded wings and with long tails for steering they can manoeuvre in fast and efficient flight after their prey in the thickets. These characteristics are also found in the booted eagles (Black Hawk-Eagle *Spizaetus tyrannus* and Ornate Hawk-Eagle *S. ornatus)* and the Harpy Eagle *(Harpia harpyja)* all equipped with great strong talons to snatch monkeys and other arboreal mammals as well as guans from the trees. With their beautiful plumages and the adornments of crests which lend them a warrior's aspect they are without doubt the most magnificent birds in the region.

The forests of Misiones are the home to a number of reptiles and amphibians, but it is the snakes which have a reputation. Here there are many species of inoffensive "grass" snakes but it is the seven poisonous species which have caused the false impression of this forest writhing with these much-maligned creatures. The family *Crotalidae* is essentially that of the poisonous snakes of the New world with Rattlesnakes *(Crotalus)* being abundant and varied in North America - only one in Argentina - and *Bothrops* the dominant genus in South America. These Pit-vipers are four in the rainforests *(Bothrops alternata, B. neuwiedii, B. jararaca* and *B. jararacussu)* and are imitated by several look-alikes, all puff and show, no poison. So too are the gentle but lethal Coral Snakes *(Micrurus frontalis* and *M. corallinus)* imitated by others, the most remarkable being *Erythrolamprus aesculapii* which even offers its tail to be attacked instead of the head, just like the shy Coral Snakes. All these "baddies" tend to shy away from contact with any larger being whose presence they can feel through vibrations picked up by the ventral regions on the ground, so the number of cases of snake-bite are few and happen as accidents - the snakes do not go after man but simply defend themselves.

The rainforests are damp places, well-suited to the abundance of frogs found there, in rotting logs, in ponds and puddles or their surroundings, in the leaf-litter, by the streams and on riverbanks. Hylids (tree-frogs) are the most common and numerous, with their fingers and toes ending in "suckers" to allow them to stick to leaves, climb trunks: *Scynax nasica, Scynax fuscovaria, Hyla minuta, Phrynohyas venulosa* amongst others, this last being one of the largest which exudes a foul and sticky, probably poisonous substance in its defence. The most remarkable tree-frog is *H. faber;* the male builds a flooded crater nest by a pond and there emits his striking vocal invitation to any female of his species to join him - his astounding "TOCK" call resembles nothing so much as a hammer-blow on an anvil. One species of enormous edible-frog *(Leptodactylus labirinthicus)* is some 8 inches long and, in Argentina, is only found in this region.

The climate, with its constantly high temperatures and humidity, is particularly favourable to those beings which do not or cannot control or adjust their own internal body temperatures or regulate the water-ballance in their bodies.

Under these conditions the metabolism - internal chemical processes - of the invertebrates, and of insects in particular, can operate efficiently allowing them to carry on and complete their life-cycles in shorter time-periods than similar species in more temperate climates, as well as grow to inordinate sizes. Producing many generations in a single year allows evolution the possibilities of many more genetic variations, and there are so many species of arthropods here that one can truly say that they are the "rulers" of the rainforest. Enormous spiders, huge locusts, giant praying-mantids, great big cockroaches, elephantine beetles, butterflies the size of some birds, an endemic dragon - or damsel-fly some eight inches across the wings only known from the lower-circuit walks, and even a gigantic ant are all here. The large *Phoneutria nigriventer* hairy-legged spider is attacked by a huge wasp of the *Pompilidae* with a dagger sting, paralyzed, and buried with the wasp's egg - soon to hatch into a grub which eats the still living spider in stages. The enormous Tiger-ant *(Dinoponera australis)* is over an inch long. Carnivorous in diet it hunts its prey over the forest floor, kills it with a sting and carries it back to the nest shared with but 20 to 70 others of its kind. This ant also squeaks, a sound it produces with mechanical stridulations. A long-horned beetle *(Acrocinus longimanus - Cerambicidae)* is the largest insect, brick red, black and cream-patterned on its back, has long fore-legs which can spread over 8 inches. It is associated with the strangler fig and even has its own body-fauna - pseudoscorpions live or hitch rides on its "person". Some nocturnal moths are enormous - a hawk-moth *(Xilophanes tersa - Sphingidae)* is some 17 cms across.

The nests of wasps, bees, ants and termites are anywhere and everywhere. There are some 250 kinds of ants in these forests, and during nuptial flights they consitute an important food source for insectivores and even birds of prey.

Cicadas *(Cicadidae,* of the order of the *Homoptera)* sometimes have the brilliant colours one generally associates with other life forms. They suck sap from plants and produce the strident "songs", each specie's differing from the others in rhythm, pulse, pitch, and time of day or season of the year. Tree-hoppers *(Membracidae - Homoptera)* have colours and shapes to blend or imitate the plants on which they live and look like thorns, buds or twigs to escape the notice of their predators.

Many of the multitude of beetles - *(Coleoptera - Buprestidae, Chrysomelidae, Curculionidae, Scarabeidae, Carabidae* and others) have species with metallic-coloured elitra which make them seem more like jewels and precious stones. The rhinoceros beetle *(Enema pan)* whose males carry a long horn on the head, often hatch in such numbers as to cloud the sky and, attracted to street-lights at night, drop and carpet the ground.

There is an infinity of flies, mosquitoes, bush-crickets, true bugs (plant-suckers and predators), multi-coloured dragon-flies - the orders of the class Insecta are nearly all represented in the Paranaense Rainforests. But of all this glut and abundance those which really draw one's attention are the butterflies, as much for their beauty as their variety and numbers. The most spectacular are certainly the *Morphos* whose very generic name means beautiful. The metallic blue wings of *M. achilles, M. anaxibia* and *M. aega* are much sought by souvenir makers. All members of this South American family *(Morphidae),* including *M. catenarius* which is almost white with a string of dark rings at the following-edge of the hind-wing, have this "chain" of dots, rings or "eyes".

The 15 cm *Morpho achilles* flies in floppy flight through all strata of the forests on wings which are relatively huge for the body-weight so that it can even, on occasions, glide. Its colourful hairy caterpillars are gregarious and feed on

Red-rumped Caciques

These birds (Cacicus haemorrhous) *are among the most conspicuous birds of the area because of the brilliant red rump, their noisy but not un-melodius calls and their colonial nesting habits, preferably near humans.*

This aerial view shows the wide "U" formed by the upper Iguazú river, broadening to some 1,500 metres where it splits up into many channels and arms just before falling over the vertical drop to form the waterfalls.

climbing plants. Twilight is the activity time for another whole family of exclusively South American butterflies, the *Brassolidae*, large, dark and fast fliers. One of the largest, *Calligo memnon* has the typical large spots on the underwing which are designed to astonish any predator - when suddenly spreading its wings there is an audible click and two "owl's eyes" suddenly open.

Swallow-tails (fam. *Papilionidae*) are common most of the year; sporting their peculiar "tails" they are fairly large and have complicated designs of yellow, black, white and pink: (*Papilio thoas, P. anchisiades, P. lycophron, P. hectorides*), or white-striped as in *Eurydites stenodesmus* which appears in mass in August and September.

Wherever there forms a puddle and especially if there are salts in solution, from excrement or carrion, butterflies gather to sip up the liquid, often in huge concentrations of many species: clusters of swallow-tails, tight groups of large bright yellow leaf-shaped *Pieridae* (*Phoebis cipris*) which would otherwise be hard to find among hanging dry foliage.

The most varied and numerous family of butterflies is the *Nymphalidae* which includes some fairly large species like the attractive *Historis orion* (red and black) or *Prepona pheridamas* which with its metallic blue looks very like a small *Morpho*. Most are small, like *Anaea morvus* or middle-sized as is *Didonis biblis*, black with red trailing edge, among the more common species. Two favourite species are "80" and "88" (*Diaethria candrena* and *D. clymena*) with patterns incorporating those very same numbers on the underside of the underwings easily seen when the butterfy is perched and at rest presenting this as the outer surface. Almost as common and somewhat similar to these are *Callicore hydaspes* and *C. hystaspes* with the black on yellow instead of white, and with pale blue dots, and red and blue on the top-side. There is a genus (*Hamadryas*) which, being all speckled and spotted, rests head-down on the trunks of trees for protection, looking just like the bark. Some species of this genus also make a clicking noise in flight when scared or when disputing resting territory with another of the same species. As dusk falls the diurnal butterflies are replaced by night-flying moths which

Iguazú Falls

are even more numerous and varied. The Microlepidoptera are legion as can be seen in the vicinity of any light at night, which draws them in like a magnet. The larger, conical-bodied, long (delta) wings, heavy, torpedo or bullet fliers are the Hawk-moths *(Sphingidae)* which can hover in front of flowers that open at night to sip nectar. The silk-moth family *(Saturnidae)* is also represented with species like the large *Rothschildia jacobae*, attractive and colourful. The grubs of this species are huge and so form an important food source for many birds.

This extraordinary aglomeration of plant and animal life-forms held together in a delicate and intricate web of interdependent life, is today, with the Pampas Grasslands, the most threatened of all Argentina's ecosystems. The large trees are much sought for timber, the rest of the forest burned off for reforestation with exotic pulp species like eucalyptus and pines. The larger mammals and game-birds are persecuted for the fur trade or for their meat, or simply as a sport.

Iguazú National Park was, for years, the only preserved area in the biome but happily now there are new provincial and private reserves coming into being.

Iguazú National Park. In 1541, not even 50 years after the arrival of Columbus, Don Alvar Nuñez Cabeza de Vaca made an epic march with 280 men from the Atlantic coast of Santa Catarina in Brazil to the city of Asunción in Paraguay. The things they went though can be deduced from the writings: "Through these lands and provinces we walked for five months without any altercations with the indians nor hostilities; during this time four hundred leagues we walked, nearly 200 opened or cut through cane-brakes and thickets; I went on foot and unshod to encourage my people..." According to Pero Hernández, his scribe and secretary, in places "we had great labour in walking because of the many rivers and difficulties or obstacles encountered; there were days in which, to get people and horses along, eighteen bridges had to be constructed, as well for the rivers as the swamps which were numerous and difficult. And we crossed great ranges of mountains which were rough and steep, thickets of trees and thick canes which had sharp detaining hooks and thorns, and other trees, so that to be able to pass, twenty men were always in the van, cutting and making the trail. And we were many days in progress and many days we did not see the sky". "And as well as the work, we endured, on many days, much hunger".

In the course of this incredible journey they came upon the Iguazú waterfalls which were thus seen by Europeans for the first time: "The river gives a leap over some rocks and boulders very high from below, and the water strikes at the bottom with such force that it can be heard from afar; and the spray from the water as it falls with such force, rises two and more lances high".

He called them the Falls of Santa María though with time the name reverted to the native Guaraní name Iguazú, meaning big water.

Born in the Serra do Mar at over 1,300 m (4,150 feet) elevation and but a short distance from the Atlantic ocean, the Iguazú river flows westwards along a winding course for over 500 kms and reaches the Paraná with barely 90 m (280 feet) to drop and join the sea nearly 1000 miles away. In the upper course, even before reaching Iguazú it plunges over several waterfalls and rapids where the rocks of the bed of the river break the surface (some in Iguazú National Park are San Mateo, El León, Carumbé, Apepú, Las Tacuaras, Irene and Las Hormigas) alternating with still, deep pools known locally as "canchas" where the volume of 1,400 cu.m./sec. doesn't seem to move at all. The upper Iguazú has a variable width of 500 to 1,000 metres, and within the park has several small islands - Cuatro Hermanas and Las Tacuaras - and even one of over a kilometre long, San Agustín.

Below this last the river widens to 1,500 m and takes a swing to the south to immediately return to a northerly course, forming a great "U". Here it plunges into the lower canyon over the lip, forming the famous waterfalls. Just before the falls are a series of channels, islands, sand-banks where the river splits up into many smaller watercourses on the outside of its swing. Some of these islands are fairly large and covered with forest, like Yacú, Ñandú and Carpinchos. It all forms a sort of archipelago. Each of these smaller streams gives rise to a waterfall, some fairly voluminous, others rather timid, the whole forming the great fan of what are called the Iguazú waterfalls.

Rhinoceros Beetle

This big scarab (Enema pan), *in which only the male has a horn, appears in huge numbers at times, becoming most evident on these occasions.*

A World of Insects

Environmental conditions in this region are particularly favourable to the proliferation of arthropods; such are the numbers and species here that one can truly say that they are the real rulers of the rainforests. The photograph shows a couple of copulating Arilus carinatus, *(Hemiptera).*

The Hummingbirds *Hummingbirds (Trochilidae) are exclusive to the New World and specialize in sipping nectar from flowers which they obtain with their long beaks and tongues while hovering in mid-air. In this ecosystem one can find over a dozen species, tiny, having mostly metallic colourations. This is a female Glittering-bellied Emerald* (Chlorostilbon aureoventris) *feeding its young.*

The water falls over the edge of a basalt "sill" of early igneous origin. The area, like most of Misiones is on a mantle of basalt which covered the land in the Jurassic period, flowing gently through rifts and fissures and not explosively through volcanic vents. This is the largest lava- flow known in the world covering about one and a half million square kilometres, and alternates with irregular sandstone lenticula whose origin in turn was from the Crystalline Brazilian Shield during the dry conditions of the Triassic.

The relief of the area was modified during the Quaternary with alternating wet and dry periods. In the wetter periods the dense vegetation protects the soils and avoids massive erosion, but during the dry and lacking the vegetation, even though the precipitations are less, erosion sets in and accentuates the relief. Rivers then establish and incise their courses and waterfalls work their way upstream, Iguazú having started some miles downstream, probably where it joins the Paraná as there is a fault there.

The great line of falls forms a lop-sided and narrow "U". It is some 2,700 m along the top and gives rise to between 160 and 260 seperate cascades according to water- levels. With the vast floods of '83 and again in '93 the whole lot became one, and two when the river virtually dried up during the filling of a dam upstream - just a trickle at San Martín and small volume at Garganta. Where the two sides of the "U" converge is where the main body of water falls. This is known as Garganta (del Diablo) and the focal point is the Salto Unión which divides Argentina from Brazil.

The shorter arm of the "U" constitutes the Brazilian side of the falls, some 600 m long, and the main falls there are the Benjamín Constant and Floriano. The Argentine arc is over 2,000 m long and in the sector which faces the Brazilian falls are Belgrano, Escondido and Mitre. These, together with the waters from Unión, have carved a gorge where the main body of water flows, and all drop the 74 m to the level below, the depths where spray and mist are produced by the pounding of the waters. Downstream, on the Argentine side, the fall is divided in two, there being a ledge almost all the way round, part of which forms San Martín island. Beyond this are San Martín, Adán y Eva, Bossetti, Ramirez falls, amongst others, which join the waters of the lower gorge through the San Martín arm of the river. Below the falls the river tumbles through a 75 m deep gorge which is but 80 m wide in places, right down to the confluence with the Paraná 17 kms downstream.

The Iguazú falls are deservedly considered one of the natural wonders of the world. More imposing, and certainly more natural than Niagara (lower and shorter), these falls compare favourably with Victoria, the latter being higher but shorter and less visible or spectacular.

Their magnificence was convincingly described by the Swiss botanist Robert Chodat: "When we are at the foot of this world of cascades and, raising our eyes see, 269 feet above us, the horizon filled with a line of water, the amazing spectacle of an ocean spilling into an abyss is almost frightening. The waters of the Flood falling straight into the heart of the world, by divine design, in a world of breathtaking beauty, in the midst of exhuberant vegetation, almost tropical, the fronds of great ferns, the giant canes, the graceful trunks of palms and a thousand species of tree with their crowns leaning over the chasm and decorated with mosses, pink begonias, golden orchids, brilliant bromeliads and lianas with trumpet-shaped flowers - all this added to the dizzying, deafening roar of the waters which can still be heard at a great distance, creates an indelible impression, moving beyond words." Remote, in complete wilderness, until the beginning of the XXth. century these falls were only reached by well-mounted scientific expeditions like those of Bové, Holmberg, Ambrosetti, Thays. The enthusiastic reports of the natural wonders observed started to create a consciousness in the National Government of the need to declare a national park around them, the proyect being put in the competent hands of Carlos Thays in 1907.

In 1928 the corresponding lands were bought, but it was not till 1934, with the National Parks Law (# 12,103) that Iguazú National Park came into legal existence. Its area, modified and finally fixed by decrees and laws of 1970, 71 and 72, 67,620 Has includes the National Reserve, protects the Argentine side of the Iguazú falls and contains a sufficiently large area of the Paranaense Rainforest to conserve a good sample of this biome.

The area had been exploited before the purchase, the better specimens of the timber species extracted, but since then natural restoration has been occurring with the satisfactory result that climax forest now reigns over most of the park.

The area can be described as the maximum exponent of the Paranaense Rainforest, but there are within its borders two atypical areas: the community where Palo Rosa *(Aspidosperma polyneuron)* and the Palmito Edible Palm *(Eutrerpe edulis)* grow together, and the extremely high-humidity area immediately around the falls.

In some particular areas of this park the uniformity of the forests is interrupted by giant trees which emerge above the surrounding vegetation. These are the Palo Rosa which reach over 40 m high and whose trunks, some exceeding 1.6 m, are sheathed in a coarse light-gray bark, and only branch at a great height (20 to 30 m) into a few thick, twisting limbs to form a irregular crown. Though the name refers to the colour of the valuable wood, dawn light also tints the light- coloured bark a pinkish hue.

In their shade, as too of other species of large trees are the Palmitos (edible palms), graceful and slender, reaching 15 to 20 m height and seldom over 20 cms diameter. The growing tip, sheathed within the bases of the fronds, is edible and much sought after. Its extraction kills the plant. Poachers have always been a problem in the small area where the remaining Palmitos grow in the National Park.

The shores of the Iguazú river and the maze of island at the falls are especially suited to certain damp-loving plants. So here there are two species of large tree, found nowhere else in the park. Both are *Leguminosae:* the Curupay *(Adenanthera macrocarpa)* with delicate compound leaves like many acacias and mimosas which always grows near or actually in the water, and the Cupa-í *(Copaifera langsdorfi)* also deciduous, which turns a lovely coppery colour with the new leaves in spring. There is also a small tree (5 to 8 m) which in our country is only found at the falls, *Roupala cataractarum*.

The riverine forest (almost a gallery forest within a forest) which grows along a narrow strip with more humid conditions than general, includes most of the plants mentioned so far. It is however enriched by a number of species which are characteristic; some even are exclusive of this area or are particularly abuandant in it. A Coral Tree *(Erythrina cristagalli)* a legume with an unusual trunk whose bright red flower has been declared the national flower of Argentina; the River Laurel *(Nectandra falcifolia)* and White Laurel *(Ocotea acutifolia),* the Aguay *(Pouteria gardneriana)* and the two

Tamandua

The Collared Anteater or Tamandua (Tamandua tetradactyla) *finds its food in termites', bees' and wasps' nests in trees. This climbing anteater is equipped with strong claws to open up these social insects' nests, and a long protractile sticky tongue.*

Ingás with beautiful long pompom flowers (a favourite food of the howler monkeys), white in *Inga marginata* and red in *Inga uruguensis.* This last is the food for the larvae of the white Morpho butterfly *(Morpho catenarius).* Among the typical bushes of the understorey of this riverine forest is Mboreví-Caá *(Faramea cyanea)* with its bright blue flowers and black berries, and a legume of the *Calliandra* genus known as Plumerillo or Borlas de Obispo (bishop's tassels) for the large red or pink and white pompom flowers. Here too are the large stands of the giant cane and some of the pioneer species which, at the water's edge receive all the sunshine they need.

The Iguazú river itself is home to a series of seriously threatened animals, the Giant Otter *(Pteronura brasiliensis)* which was last seen in the early 1980's, the Broad-snouted Cayman *(Caiman latirostris),* already extremely rare in Argentina, and the Brazilian Merganser *(Mergus octosetaceus).* This last, a duck, is seriously in danger of becoming extinct, is endemic to southern Brazil and the adjacent areas of Argentina and Paraguay, and its habits are barely known. Slender, with a long, narrow "toothed" bill typical of the margansers, a long slender crest, this is the last of its genus to exist in the southern hemisphere. It lives and breeds on the rivers and streams in these rainforests and it appears to be sedentary, feeding on small to medium fishes upto 19 cms long. It is evidently a bird of the shade where its gray colouration helps it to pass unnoticed in the presence of its principal predator, the Black-and-white Hawk-Eagle *(Spizastur melanoleucus).*

The Iguazú National Park is faced by the Brazilian counterpart Parque Nacional do Iguacú of some 170,000 has. This is of particular importance as at today's rate of deforestation these will soon become islands, and being contiguous, the effectiveness of a larger island is relatively much more than two smaller islands. The larger mammals and birds should have sufficient of their habitat left to support viable populations of prey and predator species which are all but extinct in the surrounding areas. Further, the magnificent timber trees in the parks will be the last specimens left as they are subjected to tremendous pressure for their wood outside the parks.

The area of the reserve is 6,300 has, but it has no settlers, no exploitation other than in the small area around the falls so there is virtually no difference between it and the park proper.

The purpose of the area for development around the falls is to concentrate all the activity and impact there, the visitors' services, the intensive use. There is an hotel and the start to all the trails; there are bars, restaurants, kiosks sufficient to cover demand, but all the other services are concentrated in the town of Puerto Iguazú some 17 kms away.

The formidable spectacle of the falls themselves added to the possibility of visiting a virgin rainforest make Iguazú National Park one of the leading tourist attractions in the country. Well over half a million visitors arrive per year which includes a high percentage of Brazilians, and many overseas visitors. The Brazilian side of the falls receives an even greater number - upto one million - probably because of the better and cheaper services and higher hotel capacity.

It is to be forseen that numbers will increase which means that measures will have to be taken to regulate visits at the peak seasons (January, February and July as well as Easter week) when the trails are saturated and the natural experience is enjoyed by few if any of the visitors.

Lianas and Climbers

These plants which weave trees, shrubs and trunks together to form a dense thicket, often cheer up the dark forests with their pretty flowers, like this Manettia cordifolia *of the Rubiaceae.*

The Nine-banded Armadillo

*Armadillos are one of the oldest South American Mammal families; there are many species which cover nearly all the habitats on the continent. This Nine-banded Armadillo (*Dasypus novemcinctus*) is one of the mammals to be found early in the morning on the walks around the lower circuit at the falls.*

The Great Dusky Swift   *Great Dusky Swifts (*Cypseloides senex*) feed on insects flying over the forest but return to their waterfalls every evening to roost hanging from the rock-face behind the curtain of falling water, or nest on ledges, nooks and crannies there - surely a predator-free environment.*

The visitor to the falls has wonderful opportunities to view the spectacle in all its splendour walking along a system of cement trails, paths and bridges over the many arms of the river, to get right up to the main waterfalls, all the while contemplating the magnificent vegetation and the birds and animals which are, by now, getting used to the presence of man.

The trail along the top of the falls used to reach the Garganta, with side-trails and lay-bys at most of the falls along the way, but it was severely damaged and washed away by the floods of 1982 and 1983, and the recurrence of similar floods on what appears to be a fairly brief time-span. These are the result of clear-cutting the watershed in Brazil of well-nigh 90% of its forest cover, precisely the factor which retained or slowed the waters and absorbed a fair percentage of them. Bossetti, Ramirez, Dos Hermanas, Adán y Eva and San Martín falls are all viewed from above, as too the splendid arc or amphitheatre into which they all tumble. The continual spray is the medium in which grows the peculiar water-loving grass *Paspallum lilloi*, and *Potostema comata* which looks very like an alga or moss but is in fact a flowering plant of the *Potostemaceae* family, clinging to the rock-face behind the falling water. Flocks of the Reddish-bellied Parakeet *(Pyrrhura frontalis)* flit across the this gorge from the fruiting trees where they feed to the lesser falls where they drink and bathe. They perch to rest on some overhanging climber or on protruding sparsely-leaved branches and twigs. On those parts of the edge devoid of vegetation because of the periodic flooding black vultures congregate to sun themselves with wings spread. Every so often one of these takes flight to rise in circular patterns on the ascending columns of air displaced by the falling water.

From the paths over the islands one can best see the riverine forest in all its splendour. The giant species of tree are not absent, like the Cupa-í and the Curupay, nor are the clumps of the giant cane (Tacuaruzú). Some stretches of the walks at the height of the canopy of the trees growing below permit a view of the bright bromeliads.

Skulking in the vegetation of the bushes by the river are long-tailed heavy-billed glossy-black birds of the cuckoo family, feeding on insects. Their calls are low and gutteral. They are the Greater Anis *(Chrotophaga major)*; the duller, smaller whistling species of these found in the capueiras (communities of pioneer plants) are the Smoothe-billed Ani *(C. ani)*.

Where the walk-ways span the wider stretches of water the ever-present White-winged Swallows *(Tachycinetta albiventer)* and the Blue-and-white Swallows *(Notiochelidon cyanoleuca)* flit and swoop; they even use the underparts of the bridges to nest. Perched low over the water on twigs of the small bushes which grow on the small rocky islands a Kingfisher of any of three species could be seen, Ringed *(Ceryle torquata)*, Amazon *(Chloroceryle amazona)* or Green *(Chloroceryle americana)*, with its bill pointed at, and its eyes fixed on the water below, watching for any small fish on which it will plummet.

From Puerto Canoas an elevated walkway - also destroyed by the floods - gave the visitor access to balconies right on the edge of the Garganta, but boat services ply across the missing gap for an exciting, but extremely safe ride. Garganta is the focal-point of the series of waterfalls, where the various torrents tumble together into the chasm or void, the bottom of which is rather to be imagined than seen, a maelstrom some 245 feet below. From here the spray from the crashing falls is wafted hundreds of feet into the air above, often producing glorious rainbows.

The chasm and spray of Garganta is criss-crossed by dark shapes in a death-defying ballet of swoops and dives, climbs and turns, and every so often one of these seems to shoot through the massive curtains of falling water but in fact is squeezing through some narrow gap with split-second timing and amazing accuracy, to cling to the rock-face behind with long, sharp curved claws.

The Great Dusky Swifts *(Cypseloides senex)* which swarm around the falls at late afternoon and evening, have been feeding on flying insects over the forests perhaps a hundred miles or more from here all day, and return to roost after a frolicsome, cooling bathe in the spray. It is here too that they nest at the appropriate season, on ledges, nooks and crannies in the rock-face behind the tumbling water. This unusual choice of breeding- and roosting-place is surely safe from predators which would be reluctant to enter the swirling chaos to pluck them from their rest or nest.

The lower walks, a series of paths, bridges and steps in a circuit, explore the views from the step half-way between the upper and lower levels of the river, and view the great amphitheatre first alluded to. It should be walked in a clockwise direction as thus one is always facing the views which get better and better, ever more imposing as one proceeds. Opposite San Martín island, facing the fall of the same name or even approaching Bossetti it offers the most lovely panoramas imaginable. The Alvar Nuñez or Bossetti falls are good places to get eye-level views of resting or nesting swifts. This circuit of just over half a mile is perhaps the most fascinating trail on earth, affording, as well as the views of the falls, an intimate sense of the forest through its vegetation; great specimens of the wild Pawpaw *(Jacaratia)*, the Alecrín *(Holocalyx)*, Palo Borracho *(Chorizia)*, Mora Blanca *(Alchornea iricurana)*, Cecropia; here one can see the stages of the Strangler Fig's development, stands of canes *(Chusquea* and *Merostachys, Guadua trinii)*. This walk is also revealing in the numbers of birds to be seen; right at the beginning is a palm with a colony of Red-rumped Caciques *(Cacicus haemorrhous)*, an attractive black bird with a ivory-coloured bill and a bright red patch on the lower back, pale-blue eyes in the male's breeding plumage. The nests are made of woven strands of the Pindó palm's leaflets stuck with a sort of gum from the ripe fruit of the Cancharana *(Cabralea canjerana)*. The chosen palm is like a rowdy neighbourhood: dozens of dangling nests like so many old stockings hung out to dry, and a continual hubbub during the nesting season from September to February, when birds come and go, squabble, hang from the structures to effect repairs, all accompanied by the noise of their many, not displeasing calls. Most often the nesting tree stands alone to avoid the easy access of any tree-climbing predator, and is in the vicinity of homes or other much-frequented places to inhibit visits from toucans. These somewhat shy birds' bills are long enough to reach down into the Caciques' nests to steal eggs and chicks. Caciques nests are parasitized by the Giant Cowbird *(Scaphidura oryzivora)* which like several others of the family lays eggs in other birds' nests to be incubated and the chicks to be raised by the foster-parents. Another curious species associated with these colonies is the Piratic Flycatcher *(Legatus leucophaius)* which steals a Cacique nest for its own use - only one pair per colony as these territorial little tyrants will have it to themselves.

This trail is always best very early in the morning, before the crowds, when one can imagine one is alone. Some busy Nine-banded Armadillo *(Dasypus novemcinctus)* or the troop of coatis may be encountered, almost ignoring the early-riser;

or an Agouti, pretending to be unaware of the discreet human intrusion, sitting on its haunches eating some fallen fruit which it holds in its handy fore-paws.

Among the birds one will probably see there are some spectacular species like the Toco Toucan *(Ramphastos toco)*, the largest of its family with a impeccable white bib and the huge orange bill with a black spot near the tip. This species roosts in the vicinity of the falls all winter when at eventide it can be seen arriving from many directions in small, loose flocks of three or four individuals. They like to roost at the tip of canes for safety, folded into a ball with bill along the back and tail folded over it. Here the Toucan will feel the vibrations of any approach along the cane by a potential predator and have time to "unfold" from its long-held position in which it has undoubtedly become stiff or cramped. At dawn they can be seen exposed on these very tips, slowly "coming-to", silhouetted against a paling sky of gentle dawn hues. There are also Red-breasted Toucans *(R. dicolorus)* so erroneously named as the red starts on the belly in this race, some of the great red-headed woodpeckers like the Robust and Lineated *(Campephilus robustus* and *Dryocopus lineatus)*, unfortunately not now the Helmeted Woodpecker *(D. galeatus)* which is very nearly extinct, the smaller and highly coloured Yellow-fronted Woodpecker *(Melanerpes flavifrons)* so boldly patterned, the Squirrel Cuckoo *(Piaya cayana)* with its immense white-spotted black undertail and generally brown body colouration, or the fairly common and beautiful Plush-crested Jay *(Cyanocorax chrysops)*, cream and purplish-black with an electric-blue "bun" around the nape, usually in flocks calling loudly with metallic voices

During the heat of the day the animals encountered are totally different: it is the time for butterflies which so enliven these walks and trails, and the hour of the lizards. Moving along in front of the visitor, reluctant to leave the warm cement of the path or climbing the hand-rails or bare rock-faces are the *Tropidurus* lizards *(T. catyalanensis)*, or in some clearing under a fruiting tree, or waddling away from the human intrusion, the large land lizard *Tupinambis teguixin*, sometimes over three feet long, which when alarmed vanishes at great speed into the thicket.

The visitor can also use his car to explore some of the roads and vehicular trails for a more extensive contact with the vegetation or in the hopes of finding some exciting mammals. Route 101 is a throughway, some thirty kilometres of it within the park, leading in an easterly direction to the ranger station at Yacu-í on the park's eastern border. Along it one will find the Palo Rosa and edible Palms mentioned above. There is also a track through the area below the falls (in the reserve) called Yacaratiá which can be visited in a guided tour-bus or in ones own car.

For a more intimate contact with the forests, albeit somewhat secondary forest as there were once settlers here decades ago, the Macuco trail, with interpretive stations is some 3 kms long (2 miles) along the lip of the rock-face where the falls were in prehistoric times. After a rain it is exciting to check the edges of mud puddles and spots for tracks left by the mammals. At the end of the trail is a sharp descent to a swimming-hole below the Arrechea water-fall, and the trail joins up with the Yacaratiá which may offer an alternative to returning along the same Macuco.

There are in the forest places where salts occur naturally in the soils, (and sometimes artificially, placed there by poachers). These are the places where mammals concentrate at "licks". On platforms in the trees overlooking these places poachers, the number one scourge of the park, wait for a kill;

Broad-snouted Cayman

it might be feasible to provide such places for the simple observation or photography of wildlife in this national park, but for obvious reasons, there is an understandable reluctance to do this. Tapirs, Peccaries, Brockets and Pacas, as well as a Jaguar could all possibly be seen at such a lick. It is only when the park has sufficient rangers and people are responsible enough not to poach that this idea could be put into effect.

There is however a hide overlooking the marsh on what used to be the airport where some 10 or 12 people can sit to watch for Broad-snouted Cayman *(Caiman latirostris)*, Capybaras *(Hydrochaeris hydrochaeris)*, Least Grebes *(Podiceps dominicus)* and Muscovy Ducks *(Cairina moschata)* as well as the variety of shy rails like the Rufous-sided Crake *(Laterallus melanophaius)*, Blackish Rail *(Rallus nigricans)* and Ash-throated Crake *(Porzana albicollis)* which frequent this area, among the many other species.

Hunting for meat and the additional benefit of skins is firmly entrenched in the minds, culture and traditions of the

One population of this very rare Cayman (Caiman latirostris) *lives in the Iguazú river, protected by the national park. This female is sunning, so too are the young on its tail.*

local agricultural worker and other locals who live or work all around the park. This too exists among certain levels of society in urban Puerto Iguazú, who, like the rest, are known to make sorties into the park and kill anything, including Tapirs and the rare Jaguar. The very existence of route 101 which crosses the park is a threat in that it offers easy, uncontrolled access into the very heart of the park at any time of the day or night. Spot-checks on all vehicles are not able to stop this. There is hope that a new road will bypass the park beyond the southern border, so 101 could be left as an internal road of the park's, easy to control and closed during the hours of darkness. This war against poachers sometimes leads to all-out battles, and it was in just such a encounter that Bernabé Méndez was shot down while on patrol, doing his duty, in 1968. Shoot-outs are not infrequent. It is to be expected then that one of the priorities of the park is soon to establish a sufficiently numerous and well placed net of stations and observation posts, check-points, trails, border posts and what have you, all with sufficient rangers, to effectively eliminate this illegal practice.

The National Park at Iguazú is particularly important, not only because of the numbers of visitors it receives, not even because it contains the most spectacular natural wonder - only the Moreno glacier in the Glaciares National Park can challenge Iguazú for that title - but especially for the ecosystem it protects: subtropical rainforest. As we have seen, the fundamental importance of these forests lies in the maximum variety of life forms thay contain - higher than any other area in the country, and in the vegetation cover which is desperately important as an oxygen factory, in controlling weather patterns and in protecting the soil against the cloud-burst rains typical of these areas.

From the genetic store to be found in any forest man is discovering new substances and products, as much in wild strains for cross-breeding and obtaining the hybrid vigour so needed against plagues and scourges, as for medicinal purposes from plants whose properties till now have been unknown.

These subtropical forests are found mainly in the "developping" Asian, African and South American countries

Below the falls

Here the Iguazú river flows through a narrow gorge towards the Paraná. In this view the falls and the volume of water can be appreciated, and especially the narrow horse-shoe "U" known as the Garganta del Diablo in the upper left-hand corner of the photograph.

where rational uses are not considered and indiscriminate deforestation is the rule, in order to solve today's serious problems and obtain immediate though meagre benefits. The World Conservation Union (IUCN) and the Worldwide Fund for Nature (WWF) gave a figure in 1976 of deforestation taking place at the rate of 50 acres per minute. This has undoubtedly increased to more than double since then. For this reason these institutions started a campaign at that date to save the world's tropical and subtropical forests. The immediate objectives were the setting aside of vast tracts of such habitats for national parks and increasing public awareness at an international level, of the importance of the functions of such ecosystems.

Iguazú National Park is too small to guarantee the survival of all its life-forms. Some of its birds, like the great eagles, and its mammals - wanderers like the Tapir, the Jaguar and the White-lipped Peccary - will possibly go the way of all life. The range of a female Jaguar in the Pantanal in Brazil has been calculated at about 5,000 has.

The possibility of conserving viable populations of this larger fauna will soon be severely limited when the lands contiguous to the park are exploited to the full, cleared, planted to exotics. The shape itself of the park is artificial and inefficient, a large triangle, a bottle-neck and another triangle, so these larger animals are hard-pressed to move around freely.

As it is necessary to increase the size of this park and its effectiveness, it has been proposed to change the category of the Reserve at the western end of the park to that of Park proper as there is no development in that area (always keeping the necessary facilities for attending to the visitors in the area of the falls themselves). This may offset the insufficient area of the park to a certain degree, as might also the creation of further protected areas in the ecosystem (see last chapter).

The Yungas - Subtropical Cloud Forests

In the north of the province of Salta, in the border area bracketed by the Bermejo and Grande de Tarija rivers, from Bolivia to the north, a narrow wedge of forests penetrates the country. These are the Yungas, which in a discontinuous series of patches cross Salta, Jujuy and Tucumán, ending in the province of Catamarca, some 700 kms south.

In Argentina the northern parts of these Yungas grow at the foot and on the eastern slopes of the eastern bolivian range. These mountains, pressing in the west against the pre-puna massif and dipping eastwards under the sedimentary chaco plains, were folded and uplifted during the Tertiary when the movements of the earth's crust gave rise to the Andes. These ranges east of the Puna are contained by faults. They form chains which lie in a north-north-east/south-south-westerly direction, the sinclinal valleys covered by Quaternary deposits, drained by rivers which follow the folds till they find a way out to the east. There the three great rivers are formed - Pilcomayo, Bermejo and Salado del Norte - which virtually without tributaries cross the wide Chaco region. The southern extreme of these forests is in the Aconquija and Calchaquí peaks.

The sudden rising from the Chaco into the mountains in the west causes a noticeable change in the weather patterns. The winds from the north-east, having encountered nothing on which to spill their moisture right across the flat Chaco plains, finally dump their atlantic humidity on these eastern slopes where the forests grow.

Though the wet season is in summer, local precipitation, temperature and humidity are markedly variable according to latitude and the relief factor. The valleys and lowlands are warmer and drier than the slopes, but they also recieve more frosts because the cold air drains towards the valleys.

All these conditions are seen reflected in the vegetation which grows in many different communities according to elevation and the orientation of the slopes.

All along the plains at the foot of the mountains, and up the lower slopes to a height of some 500 metres above sea level is what is called the Transition Forest, what the lumbermen call the timber forests to mark the difference between these and the fire-wood forests of the Chaco. There is a wide ecotone at the interface of the Chaco and the Transition Forests. Rainfall is here about 1,000 mm per annum.

Lumbering in this, the area most suitable for human settlement, has given way to sugar-cane plantations, fruit groves, tobacco, maize, vegetables for market. There are huge areas where no more forest exists, and the clearing is going on today, reducing yet further the area of forest.

In the northern reaches of the district, in the provinces of Salta and Jujuy, the dominant trees of these forests are the Palo Blanco *(Calycophyllum multiflorum)* and Palo Amarillo or Lanza *(Phyllostylon ramnoides)*, which because of their abundance give the forests a certain characteristic with their pale-coloured straight trunks, unbranching upto 20 or 30 metres. Also present, though in less numbers are two legumes: Cebil Colorado *(Adenanthera macrocarpa)* and Horco Cebil *(Parapiptadenia excelsa)* with their delicate compound leaves and long, dark, shiny pods.

Mora Amarilla *(Chlorophora tinctoria)* whose specific name refers to the colour of the wood, the Guayaibí *(Patagonula americana)* with buttress roots at the base of the trunk. The giants in this forest are the Cedro Salteño *(Cedrella angustifolia)* of very valuable timber, Quina *(Myroxylon peruiferum)* and Tipa Blanca *(Tipuana tipu)*, one of the better known native trees of the whole country, growing in city parks and on sidewalks, and whose lovely shape reaches some 40 metres and is covered in epiphytes, especially at the higher elevations of its distribution.

Climbers and epiphytes give these forests an exhuberant aspect, but during the dry winter the leaves fall and the forest turns a sad gray.

With, or even before the first rains of spring many trees bloom prior to putting out their leaves, and this is when the forests are at their most lovely, with several *bignoniaceae* drawing one's attention - the Lapachos, pink *(Tabebuia avallanedae)* and yellow *(T. lapacho)*; the Tarco or Jacaranda *(Jacaranda mimosifolia)* with its blue flowers.

Towards the southern extremity of the region, in Tucumán, the dominant trees are the Timbó here called the Pacará, the Black Ear Tree *(Enterolobium contortisiliquum)* and the Tipa Blanca; many others don't make it this far. But then the forests of these vestigial outcrops seem to have been practically destroyed.

Further up the slopes, above 500 metres the general appearance of the forest starts to change. Here there is a higher rainfall, epiphytes cover whole branches, or hang like curtains while the mesh of lianas and climbers reach for the light. In the understorey, impenetrable and dark, grow ferns and mosses covering the ground and rocks.

This is the montane forest also known as the laurel forest. Here the majestic Laurel de la Falda *(Phoebe porphyria)* towers over the other vegetation on an immensely thick trunk of two metres diameter, some specimens reaching three; branching near the base, its enormous crown tops all the lesser trees round about.

Here too are present the Jacaranda and the Tipa, the Roble *(Amburana cearensis)*, the Southern Wallnut *(Juglans australis)*, two Cedrellas *(C. angustifolia* and *C. lilloi)* and Afata Blanca *(Heliocarpus popayanensis)*.

At greater elevations the humidity increases, not only because of the greater precipitation - some places receiving upto 3,000 mm per year - but also because of the clouds' mist between 800 and 900 metres.

Species give way to others which are hardly present or not present at all at lower elevations. Evergreens increase. This is the Myrtle forest as the *Myrtaceae* are ever-present at all strata. The giant among these is the Horco Molle *(Blepharocalyx gigantea)* also called Palo Barroso because of the colour and texture of its bark - it looks muddy. This huge tree can surpass 40 metres and has brilliant green leaves. It shares the canopy with other myrtles like the Guili *(Pseudocariophyllus guili)*, the Laurel and the Cedrellas.

Intermediate-sized trees are between 10 and 15 m high; two myrtles, Mato *(Eugenia pungens)* with its edible fruit and the Alpamato or Guili Blanco *(Eugenia pseudo-mato)*, two Palos San Antonio *(Rapanea ferruginea* and *R. laetevirens)*, Roble *(Ilex argentina)*, Ramo *(Cupania vernalis)* and others.

Below these are the smaller (5-10 m) trees: the Horco Mato *(Eugenia mato)*, Chal-Chal *(Allophylus edulis)*, Cedrillo or Palo Brillador *(Styrax subargenteus)*, two called Lata de Pobre *(Piper tucumanum* and *P. hieronymi)*, Maitín or Palo Lata *(Myrrhinium rubiflorum)*.

The ground around all the trees and bushes is carpeted with an exuberant layer of green made up of different plant communities according to the light, soil conditions, humidity and so on. Sometimes Ramío Tucumano *(Boehmeria caudata)* dominates, in other places Garrapata Yuyo *(Pteris deflexa)* a beautiful fern upto two metres high, or the solid cane *(Chusquea lorentziana)* some 5 or 6 metres long, usually on the steep banks of a gully or canyon. Others, like the Fuschia *(Fuchsia boliviana)* provide beauty with their flowers so much appreciated by hummingbirds.

Though there are no thorns, the stinging leaves of the Giant Nettle *(Urera caracasana)* are to be avoided.

Perhaps the most noteworthy aspect of these forests is the profusion of epiphytes of many and varied families. In some places growing in mixed communities, in others one single species alone, they cover whole trunks and limbs creating wonderful aerial gardens. The Laurel de la Falda is one of the better trees for colonization by these plants. On one single specimen of the Tucumán forest thirty different ferns, mosses, and flowering plants were counted. Though practically every specimen has epiphytes, the myrtles which shed their bark, only have the smaller air-plants *(Tillandsia* spp) on the twigs in the crown. The larger epiphytes like *Aechmea distichanta* and *Tillandsia maxima* (bromelias both), or the fern *Phlebodium aureum* whose fronds grow to 50 cms grow on the thicker limbs and near the trunk, while some orchids *(Malaxis padilliana, Govenia tinguens, Oncidium viperinum)* and begonias *(Begonia micrantha* and *B. cucullata)* grow in hollow or rotten parts of the trunk and limbs where humus has accumulated.

Dripping, dangling mosses like *Piltotrichella versicolor* and *Meteoropsis onusta* hang in wispy curtains, while the tiny-leafed bromeliad *(Tillandsia usneoides),* which looks like the lichen Old Man's Beard or *Usnea,* waves or dangles from the outer branches of trees.

There are some creeping plants which climb and decorate the trees: ferns of the *Polipodiaceae* family with linear or simple pinna, looking like ribands; flat-stemmed cacti *(Rhipsalis lorentziana).* On other occasions it is the lichens that cover the trunks and limbs, one being *Parmelia cirrhata* with big hemispherical stems.

The *Araceae,* so numerous in the Paranaense rainforests, are here represented only by *Asterostigma vermicida* which is fairly rare and sometimes also grows on the ground.

There is a race for the light between many climbers and lianas. Some lean on and hug trunks and limbs, circling them as they go. Others use tendrils to hang on, or have aerial roots. A Bignonia *(B. unguiscati)* and a *Tropeolaceae* stand out for the beauty of their flowers, this last looking more like an orchid. Also frequent are the Jasmín de Chile *(Mandevilla laxa - Apocinaceae),* Sacha Uva *(Cissus striatus - Vitaceae)* and several *Cucurbitaceae (Sicyos polyacanthus, S. odonelli,* and *Cyclanthera thamnifolia).* There is even an *Ulmaceae (Celtis triflora)* which behaves either like a liana or a tree according to where it is growing. The myrtle forests in Tucumán where they have best been studied, go to an elevation of 1,300 or 1,400 metres; in Salta and Jujuy the number of species is greater and reach 1,700 metres. Where conditions of temperature or soil change, this forest is replaced with another community made up of species which are infrequent at lower elevations or not present at all.

The evergreens are replaced by deciduous species. Many of the genera found here are common to the northern hemisphere or to the southern Andes which seems to indicate some "migration" in the remote past when conditions all along the Andes were different.

These montane forests are patchy; some of the communities are dominated by one species. One such case is the Pino del Cerro *(Podocarpus parlatorei),* the only conifer in north-west Argentina, a genus shared with Misiones and the southern Andes, notable for its dark-green yew-like leaf. It can be found up to 1,700 or 1,800 m - higher in the shelter of some deep valleys, sometimes abundant over large areas while missing in others. Very often the Southern Wallnut *(Juglans australis)* is found in association with these forests. Growing up to 15 to 20 m high its timber is of excellent quality and its nut is the food of many animals. Also at this elevation Alder *(Alnus jorullensis)* is present but this *Betulaceae* generally grows in pure stands where the rainfall is around 500 mm per year though with high humidity from the mist of the often-present clouds. This northern species, found from Mexico to Catamarca grows up to 2,500 m and in some places more. It grows best near water courses and along these down into the myrtle forests.

Some other trees like the Palo Luz *(Prunus tucumanensis),* the Elder *(Sambucus peruviana)* and the Talilla *(Crinodendrum tucumanum)* also grow among these, while in the lower strata White Salvia *(Lepechinia graveolens)* abounds as do many colourful flowers like *Alstromeria* spp, the Amancay.

The tree which grows at the greatest height is the Queñoa *(Polylepis australis)* with a twisted trunk sheathed in paper-thin, multy-layered, loose, peeling bark of an orange colour. Its range starts at 1,500 m. but it can grow at 3,000 though at this latter elevation, conditions dictate its dwarfing.

Where the woods and forests end begins the mountain meadow vegetation, dominated by grasses and herbaceous plants which cover themselves with flowers in the rainy season. These grasslands gradually become the puna steppes though a wide ecotone or transition zone.

The watercourses have patches of woods of the native Willow *(Salix humboldtiana)* and Palo Bobo or Aliso del Rio *(Tessaria integrifolia).*

Though these Yungas forests compare unfavourably with the Paranaense rainforests with which they are closely related, especially those of the southern extreme where they are even less exhuberant, their own characteristics and species give them a special quality.

In bird species alone these forests are uniquely different. The parrots for example are here represented by many different species: a macaw, the Golden-collared *(Ara auricollis)* though ever rarer, still flies in formation over the tree-tops. The Mitred parakeet *(Aratinga mitrata),* with its long tail, the Alder Parrot *(Amazona tucumana)* of the Alder woods, or the small Green-cheeked Parakeet *(Pyrrhura molinae).*

Also the guans: as well as the Dusky-legged *(Penelope obscura)* there is an endemic species, the Red-faced Guan *(P. dabbenei)* only found on certain hillsides at Calilegua in Jujuy, the type locality. Seriously declining because of being hunted for meat, these guans give themselves away at dawn and dusk by calling and "drumming" loudly, though when they are active and feeding during the day they are extremely secretive.

The distribution and fidelity of the various species in the different forest strata is constant.

Tanagers, often highly coloured, especially the males, are mostly birds of the forest canopy, feeding on fruit and insects. The male Blue-hooded Tanager *(Euphonia aureata)* is brightly black and yellow with a blue head, and sings his thin little song from the thicket. Also present are the Hooded Tanager *(Nemosia pileata),* the Orange-headed Tanager *(Thlypopsis sordida),* the Common Bush-tanager *(Chlorospingus ophthalmicus),* and the Hepatic Tanager *(Piranga flava)* whose male is a fine rosy red while the female is yellow.

The Vampire

Bats are most diverse in the subtropical regions of the country. They feed on a variety of diets according to the species. Perhaps the most noteworthy specialist is the Vampire Bat (Desmodus rotundus) *which feeds exclusively on blood. Vampires therefore are potentially dangerous as vectors of certain diseases such as rabies.*

Murine opossums

Mouse-sized marsupials of genera like Marmosa *which includes several species throughout the country are tiny predators. Though they take small vertebrates as opportunity arises, they are basically insectivorous, playing the role of shrews which are absent in South America. They are agile little climbers.*

The Tree Porcupine

Many mammals have evolved to exploit the resources in the crown of the trees and to this end have developed prehensile tails to act as a fifth limb. These Coendús are the only rodents with this characteristic; Coendou prehensilis *is the species of the Yungas, different from the one found in the northeast.*

The Ocelot

In size somewhere between the great cats like the Jaguar and the smaller spotted cats, the Ocelot (Felis pardalis) *preys on fair-sized birds and mammals such as Guans and Coatis which puts it in the range of the forest's superpredators. Its pelt, which is much sought after, has led to this successful predator becoming rare.*

Another bird family of the canopy is the cotingas here represented by the three Becards, the Green-backed *(Pachyramphus viridis)*, the White-winged *(P. polychropterus)* and the Crested *(Platypsaris rufus)*.

Tyrants (the flycatchers of the New World) are many, mostly dull. Some use the crowns of the trees, like the plain-coloured Tropical Pewee *(Contopus cinereus)*, the Dusky-capped Flycatcher *(Myiarchus tuberculifer)*, the Gray Elaenia *(Myiopagis caniceps)*, the Mottle-cheeked Tyrannulet *(Phylloscartes ventralis)*, none of them showy.

The only Toucan in these forests is the Toco Toucan, *(Ramphastos toco)*, very noticeable because of his noisy call, huge beak and contrasting black and white colours. Toucans often flock and move around in search of the fruit, nestlings and the eggs of other birds, on all of which they feed.

In the patchwork of light and shade in the middle strata of the forest the bright-coloured Blue-crowned Trogon *(Trogon curucui)* and the Blue-crowned Motmot *(Momotus momota)* are hard to see in their immobility. The last of these has a strange racquet tail. By contrast the Plush-crested Jays *(Cyanocorax chrysops)* are rowdy, curious, unafraid, and are brightly purple and cream with a bright-blue band around the back of the head and a black top-knot. They travel around in groups of eight or ten individuals, stealing eggs, catching small vertebrates or insects, or picking fruit.

Curiosity in the make-up of several Fringillids renders them conspicuous as they approach the human intruder: the Stripe-headed, the Fulvous-headed and the Yellow-striped Brush Finches *(Atlapetes torquatus, A. fulviceps* and *A. citrinellus)*, as too the Saffron-billed Sparrow *(Arremon flavirostris)*. In the bushy clearings the Red-crested Finch *(Coryphospingus cucullatus)* is at home, a dull plum red with a bright crest which it shows by erecting it at will, and a whitish ring around the eye.

Antbirds *(Formicariidae)*, here with long, heavy tails and robust bodies, seek out their ant prey as well as other insects and larvae in the lower storey of the forest: the Great *(Taraba major)*, the Variable *(Thamnophilus caerulescens)* and the Giant Antshrikes *(Batara cinerea)*, this last huge for the order and sporting a heavy top-knot. The warblers are here represented by the Brown-capped Redstart *(Myioborus brunniceps)*, and the Two-banded Warbler *(Basileuterus bivittatus)*. There is an astounding lack of ground-birds.

Flying over the canopy in search of the carrion on which they feed, detecting it more by their acute sense of smell than by their extraordinary vision, gliding effortlessly on stiff wings, three vultures - Turkey *(Cathartes aura)*, Black *(Coragyps atratus)* and the King, *(Sarcoramphus papa)*, this last all white but for the black flight-feathers and tail, the colourful bare head, and the gray velvet choker. It is not merely the vestments which make it king here, but its enormous size which ensures first pickings at any feast, lesser species stepping aside. The Swallow-tailed Kite *(Elanoides forficatus)* and the Plumbeous Kite *(Ictinia plumbea)* both arrive on migration and spend the summer months, while the Roadside Hawk migrates up from the south to spend its winters in the area. It is an important area for the larger birds of prey - Crested and Solitary Eagles *(Harpyhaliaetus coronatus* and *H. solitarius)*; the largest of all, the Harpy eagle *(Harpia harpyja)* has been recorded here, as also the Black-and-chestnut Eagle *(Spizaetus isidori)*.

The Military Macaw

The Yungas are the home of several parrots which differ from those found in the Paranaense Rainforests further east. Perhaps the most spectacular is the Military Macaw (Ara militaris), among the largest of the family, and now extremely rare.

King Vulture

This spectacular New World vulture of the Cathartidae *family can be found in the north of the country. It is scarce everywhere in the Yungas, Chaco and Paranaense Rainforests. Other vultures give this species first sitting at any carcass.*

The Wattled Jacana

These curious water-birds (Jacana jacana) *have enormously elongated toes to spread their weight over the floating leaves of water-lillies and other such vegetation on top of which they live and nest.*

With nightfall the nocturnal species come into their own - the Spectacled Owl *(Pulsatrix perspicillata)*, the Stygian Owl *(Asio stygius)* and several other species, while the forests echo the mournful calls of the Common Potoo *(Nyctibius griseus)* which passes the daylight hours on the end of some exposed snag pretending to be a stump; or the rythmic call of the Rufous Nightjar *(Caprimulgus rufus)*.

As in the Paranaense rainforests the bats of the Yungas are as varied in their diets as they are in species. The fruit-eaters: Yellow-shouldered Bat *(Sturnira lilium)* and Neotropical Fruit Bat *(Artibeus planirostris)* at all altitudes upto the Alders; the insectivorous Free-tailed Bat *(Tadarida brasiliensis)*, Yellow Bat *(Lasiurus ega)*, Red Bat *(Lasiurus borealis)*; carnivorous Peters' Wooly False Vampire Bat *(Chrotopterus auritus)*; and the blood-licking Vampire Bat *(Desmodus rotundus)*.

The largest herbivore is the Tapir *(Tapirus terrestris)*. There are also two species of Brocket - the Brown *(Mazama gouazoubira)* at the foot of the mountain, and a form of the Red with gray head and shoulders *(M. americana sacrae)* on the slopes right up to the Alder forests. These small deer are to be seen at dawn and dusk grazing and browsing in small clearings.

The peccaries are both here, the Collared *(Dicotyles tajacu)* is the one most hunted, while the White-lipped *(Tayassu pecari)* lives in steep little canyons with very dense bushy vegetation.

Among the smaller herbivores there is an Agouti *(Dasyprocta punctata)* and a small rabbit *(Sylvilagus brasiliensis)* to be seen in the open only at night. There is a large variety of small mice and rodents *(Akodon* spp, *Oryzomys longicaudatus* and others).

There is a special armadillo in the forests, somewhat like the Seven- or Nine-banded, *Dasypus mazzai*, and the Six-banded Armadillo *(Euphractus sexcinctus)* at the foot of the slopes.

Wildlife is always more abundant and varied near the water. Those species which are associated with it are the Paraná Otter *(Lutra longicaudis)*, the Crab-eating Raccoon *(Procyon cancrivorus)* and the Red Water-possum *(Lutreolina crassicaudata)*; they are all found in riverine habitat.

Small Murine Opossums *(Marmosa constantiae* and *Thylamys venusta)* live in thick foliage where they catch the insects on which they live, and there is a mouse which is almost exclusively arboreal, *Rhipidomys leucodactylus*. The rodent which is most noticeable is the Squirrel *(Sciurus ignitus)*, around some Southern Wallnut or other fruiting tree, always alert to the possibility of attack by a Bicoloured Hawk *(Accipiter bicolor)*.

Primates are represented only by the Capuchin Monkey *(Cebus apella)* which in summer ascends to the upper tree-line in search of the fruits on which it feeds.

The Jaguar *(Panthera onca)* still exists in the more distant reaches of these forests though it is extremely rare, and the Margay *(Felis wiedii)* is the most abundant of the cats. This, together with the Crab-eating Fox *(Cerdocyon thous)*, the Tayra *(Eira barbara)* and the Coati *(Nasua nasua)* are the carnivores most likely to be encountered.

Where the forests run out into the Pre-puna there are other mammals which belong more to the neighbouring region, like the Leaf-eared Mouse *(Phyllotis darwinii)*, a Cavy *(Cavia tschudi)*, a Mouse Opossum *(Thylamys elegans)* and some carnivores like the Red Fox *(Dusicyon culpaeus)* and the Pampas Cat *(Felis colocolo)*.

As well as the usual toads, *(Bufo* spp), tree-frogs *(Hyla* spp) and several of the Leptodactylid frogs, here are found two marsupial frogs, recently discovered for the country, in tumbling mountain streams with big rocks and pools. *Gastroteca gracilis* at 1,900 m in the Alder forests near Andalgalá in Catamarca, and *G. christiani* at 1,700 m in Calilegua both carry the ova, and later the tadpoles in a sack, bag or pocket on the female's back.

Both these areas also have *Telmatobius*, a genus which is normally found outside the forests, (*T. barrioi* in Calilegua and *T. ceiorum* in Andalgalá).

Two species of batracian that are of these mountain streams are *Eleutherodactylus discoidalis* and *Melanophryniscus rubriventris*, a brightly coloured small toad, black and yellow on the upper parts, red undersides.

Snakes are many, among them two *Liophis: occipitalis* with a lemon-yellow dorsum, and *miliaris*, near the water.

Lizards include a climbing *Tropidurus, Gymnodactylus* and *Pantodactylus schreibersi*.

The hordes of biting insects which abound, especially at the lower elevations, detract somewhat from any pleasure one might derive from visiting these regions, but as one climbs into the cooler elevations their numbers fortunately decrease. However, butterflies flitting around summer rain- puddles are a delight to the senses. There is an abundance of the *Phycoides* genus (*P. claudina, P. teletusa*) as well as the lovely swallow-tail *(Papilio scamander)*, black with a double line of ivory spots along the edges of the wings.

The hanging nests of the social wasps *(Polystes flavogullatus* and *Mischocyttarus lules)* are made of the cardboard-like material they manufacture with their saliva and scrapings of the bark of certain trees. In the hexagonal cells are the larvae, growing on the food that they are given, chewed up insects which have been caught by the adults.

Under the bark of trees or in the trunks can be found the nests of the ferocious ant *(Solenopsis saevissima)* which delivers a very painful sting; sometimes it nests in the ground and the tailings from digging its galleries are piled in a small dome-shaped heap.

The accumulation of rotting plant material on the floor of the forest feeds and protects a goodly number of mites, springtails, centipedes and beetles as well as several snails *(Scutalus, Radiodiscus, Habroncus, Palaina,* etc.) and slugs *(Vaginulus borellianus)*.

El Rey National Park. In the department of Anta, in the province of Salta, some 80 kms east of the city of Salta is El Rey National Park, created in 1948 by decree 18,800, to preserve a very interesting section of Montane Forest and the ecotone between this and the Chaco.

Other than an area 10 kms wide through which the access road enters the park, the surface of 44,162 has is all of mountain slopes rising to surrounding ridges and peaks of between 1,000 and 1,800 metres above sea level. The slopes from these into the park are gentle, whereas on the outside they fall away in immense cliffs of upto 200 metres drop which turn the park into a sort of impreganble fortress. Inside this ring of crests the park can be compared to an amphitheater rising from 800 to 1,800 metres. Drainage is dendritic, all the small streams from the crests flowing together to form one single drainage river, the Popayán, which after leaving the park with the name of del Valle flows east to dissipate in the Quirquincho swamps in the Salta part of the Chaco. The four main streams in the park which make up the Popayán are the Soco Hondo, Noques, La Sala and Los Puestos.

The park is open to the cooler airs from the south, and being at a higher elevation than the surrounding area is a little fresher.

Winter is the best time to visit, when the average temperature is of 15°C, but there are often frosts and a scattering of snow on the surrounding peaks.

Rainfall is seasonal, the summer being the wet. Measurements are only available for the valley-bottom where 600 to 700 mm fall annually, but obviously this is much greater in the surrounding mountains. Most precipitation is delivered by thunderstorms, so flash floods bring down huge boulders and trees.

The lower parts of the park are alternating patches of dry Chaco Woods on the flat areas, interspersed with grassy meadows though these tend to be closing in and filling with the small acacia *(A. caven)* since cattle and settlers were evicted. The undulating hilly areas are covered with forest which reaches mature stands at middle-elevations. This Cloud Forest is magnificent, with giant Cedrellas, Wallnuts, Black Ear Trees *(Enterolobium)* and many others. Shortly before the park was established a gigantic Cedrella whose girth was over three metres was felled, probably the largest ever of the species. Above grows the Podocarp forest, the Alders, and above 1,600 m, grassy meadows.

The typical Yungas fauna is augmented by the presence of certain elements of the Chaco region such as the Red-legged and Black-legged Seriemas (*Cariama cristata* and *Chunga burmeisteri*), the Chaco Chachalaca *(Ortalis canicollis)* as well as a number of aquatic species which live on an artificial pond by the entrance road.

Access to the park is best in the dry season, the road often being cut in summer at one or more of the fords across streams. Lumbreras, on route 34, is the closest village, over 100 kms away along the road to Salta. There is a hostelry in the park, room for campers and some internal trails and roads for exploring on foot, on horse-back or in your vehicle.

Calilegua National Park. Donated for the purpose of creating a National Park by the company which owned the land, the montane forests on the east-facing slopes of the Calilegua range have, since 1979, become a very important link between Baritú and El Rey National Parks.

The park covers 76,000 has of very steep gullies, canyons, hillsides and valleys of the Calilegua range. The backdrop of peaks reaches some 3,000 m, faced on the east with cliffs of several hundred metres' drop, which, with the dense forest vegetation which covers it all makes access extremely difficult.

The mountains called Hermoso (3,160 m), Amarillo (3,100 m), Morro Alto, Mesías (probably originally Mesilla for its tabular profile) and others, form a most imposing setting for the park as seen from the valley. There are numerous rivers to drain the area, some of which form the borders of the park: the Valle Grande and Tormento in the west, the Ledesma in the south, the San Lorenzo at part of the middle. Others are the river Sora and the streams which flow into it, the Cafetales, del Retiro, del Medio and Canteras. They all join the San Francisco which flows north along the valley to join the Bermejo.

Calilegua National Park

Although for most of the year there is a meandering trickle winding from side to side of the dry riverbeds, during the wet season - summer - they become roaring torrents and have on occasions threatened the town and villages along their courses downstream, carrying trees, rolling huge boulders and bearing an alarming load of silt.

As yet there is little meteorological data except for the lower reaches. 1,000 mm seems to be the precipitation in the valley, but it is most likely doubles up in the mountains.

The vegetation is Yungas transition forests, with some Chaco influence in the lower regions to 700 m, montane forests upto 1,600 m and beyond that the Podocarps and Alders which at 2,500 m give way to the Queñoa and upland pastures right up to the peaks themselves.

Because of the difficulties of access, and mobility within the park, the flora and fauna of the park have been virtually unaffected by man. Here there are important species elsewhere extinct like the Jaguar, Tapir, Paraná Otter, the Guans (*Cracidae* - those turkey-like birds of the trees), and

This park was created on the eastern slopes of the andean foothills and is mantled in Montane Forests. It protects the headwaters of rivers which give life to a large area. Mostly inaccessible, there are basic camping facilities and a road right through the park open for part of the year.

most important of all perhaps, the extremely rare and still poached northern species of the Andean Deer, the Taruca *(Hippocamelus antisensis)* at the upper limit of tree-growth and on the high grasslands.

Access to the park is only from route 34, the main north-south road to eastern Bolivia via Orán, some 150 kms to the north. The neighbouring town is Libertador General San Martín though park headquarters are in the nearby village of Calilegua. There is only one road into the park; it connects the distant village of Valle Grande to route 34 (and the world). Beyond Valle Grande are the tiny settlements of Alto Calilegua and San Fransisco only reached on mule-back, and surviving on an almost totally closed pastoral economy. The road is often impassible in the rainy season and for a time after the rains stop, narrow, winding, steep, climbing 1,000 m in 22 kms - but very beautiful. There is a ranger station near the entrance at Aguas Negras and another half-way up at Mesada de las Colmenas. A simple camping area attends the basic requirements at the entrance, on what may be the only level piece of land in the park. There are Inca ruins at the peak of Cerro Hermoso and elsewhere along the ridge.

Baritú National Park. Baritú National Park was created in 1974 (Law #20,656) and is in the province of Salta some 70 kms from Orán.

It covers about 72,000 has of Montane Forests which have been virtually untouched thanks to their inaccesibility.

The whole is bordered on the west by the Pavas chain of mountains and on the south by the Negro, while other mountainous ridges cross the park, like the Porongal. Their heights are over 2,000 m.

Three major rivers drain the park and all end up in the Bermejo: the Lipeo along the northern border, the Porongal, and the Pescado in the south, while the very Bermejo, the border with Bolivia, is also the border of the park.

This park is still inaccessible, the reason why its vegetation and fauna are as yet untouched, as it is necessary

to cross into Bolivia to get into the park. There is a road which is being built towards the park from Orán - route 50.

The varied nature of the Cloud Forests as described in the main part of the chapter is here even more varied and exhuberant, but hardly any research has been done here yet.

As well as all the rest, there have been records bordering on hearsay but obviously founded on something, of a Sloth *(Bradypus variegatus)* and the Spectacled Bear *(Tremarctos ornatus)* though this last is often referred to as a type of forest yeti, the Ukumari.

Tree-ferns *(Cyathea o'donelliana)* are to be found and a thorny palm called Chunta *(Acrocomia chunta)* which only exists in the park, and a Giant Fig *(Ficus maroma)* which starts life as an epiphyte and ends up strangling and replacing the original host, and dropping roots to the ground from the huge horizontal branches as well as helping support itself on buttress roots.

The first initiatives to protect this area came from scientists of the Lillo Institute in the University of Tucumán, and from the Province of Salta.

El Rey National Park

The heart of this park has Montane Forest vegetation with some influence from the Chaco. Here a Warden on patrol is about to cross the La Quina stream, one of many which drain this enormous amphitheater and form the Popayán river.

Chaco Woodlands and Savannahs

From the mountain ranges of southern Bolivia and northern Argentina a vast and uninterrupted plain extends eastwards, reaching the Paraná and Paraguay rivers. It is mantled with dry woodland and tall-grass savannahs and is known as the Chaco which in the native Quecha language means "hunting grounds".

To the north of this enormous inland region which is bisected by the tropic of Capricorn, lie the vast tropical forests (beyond Santa Cruz de la Sierra); the Chaco ends some 1,500 kms south in northern Córdoba province where a wide transition zone separates it from the Pampas grasslands. Thus it is the only area on earth where a desert does not separate the tropical from the temperate regions.

The area covered is about one million square kilometres, shared between Paraguay and Argentina, with a smaller sector in southeast Bolivia.

Geologically it is made up of a vast sedimentary basin between the Andes and the Pre-cambrian Brazilian shield, as much composed of Cretaceous and Tertiary marine deposits as of material from land erosion, later covered by the loess of the Quaternary pampas deposits.

During the uplifting of the Andes in the Tertiary, the Brazilian shield felt the pressures from the west but unable to fold and rise because of its crystalline nature it faulted and fractured along the lines now occupied by the Paraná and Uruguay rivers, depressed in the Chaco region, uplifted in the ancient Pre-cambrian Pampas mountain ranges.

The sedimentary deposits are of a later date, filling the depression during alternating periods of wet and dry during the Quaternary. These periods probably coincide with the advances and retreats of the sea during glacial periods.

Loess, a deposition containing little clay but much calcium carbonate, is the main material; wind-borne during the long spells of drought, it not only covers the Chaco but also the Pampas and the great riverine plains, alternating with fine clays of alluvial origin during the wetter times, spread over all these extensive plains. These deposits, originating in the mountains in the west, are sandy near the foothills and become finer clays towards the east.

The Chaco basin is part of the vast Chaco-pampean plains which at the south and east of the Pampas reach the River Plate and the Atlantic.

Very few rivers cross the Chaco and these flow down the barely-detectable incline in a south-easterly direction. Their headwaters are not in the Chaco but beyond, from the steep andean slopes and valleys. Hardly any watercourses flow into these rivers on their long traverses, and these are at the downstream ends. In these eastern reaches where rainfall is abundant, there are many swamps and marshes and a richer hydrographic network.

The Pilcomayo, Teuco/Bermejo and Salado rivers rise in the mountains, and like so many silt-laden rivers on almost-level plains, they meander. Their volumes fluctuate considerably between seasons; they are accompanied by series of ox-bows and old river-courses all covered with aquatic plants. They change their courses frequently as their beds become obstructed with sand and silt and vegetation. The rivers flow between levées which are forested with subtropical riverine gallery forests, incursions which penetrate from the west.

There is a pattern in the fluctuations in water-levels: high water levels occur in summer and flood the surrounding plains to form stagnant temporary marshes which dry up and seep away over the dry winter months.

The Chaco's rivers flow into the Paraná or its tributary the Paraguay which drains the Mato Grosso. The very size and proximity of the Paraná and Paraguay rivers affects the vegetation and wildlife of all the eastern Chaco.

The central Chaco is the driest part with less than 500 mm per annum, but this increases eastwards to 1,000 to 1,200 mm near the great rivers. Rainfall also increases in the west where the summer's damp easterly winds come up against the rising relief, to average 800 mm per year. Though there is a great variation in precipitation from year to year, rains are markedly seasonal - upto 80% falling in summer and less than 40 mm in several winter months.

As a defence against this seasonal variation most Chaco trees are deciduous, and even the "evergreens" like the White Quebracho shed many of their leaves in the dry to cut down on water-loss.

Summer rains between November and April are delivered by tremendous storms after days of north wind and increasing humidity. These cloudbursts cause erosive damage and plants benefit but little as the water soon drains away. In the east however the greater vegetation cover slows the drainage so the vegetation gets more benefit from each rain.

Temperatures are subtropical and continental, with very hot summers and dry, temperate winters with the occasional frost, especially in the west. The maximum mean varies according to the area from 25 to 30°C while the average minimum is between 10°C and 17°C. It is in the Chaco that South America's absolute maximum temperatures have been recorded: well over 48°C in eastern Salta, Tucumán and Santiago del Estero provinces, while the minimum recorded was in the central Chaco at -8°C. Nearer to the great rivers extremes are moderated and frosts are rare.

Botanically the Chaco is characterized by different Red Quebracho species (*Schinopsis* spp.) of the *Anacardiaceae* family. The Chaco is divided into four districts: eastern, western, hill and savannahs.

WESTERN AND HILL CHACO

The dry, thorny woodlands which are so characteristic of the Chaco reach their best exponents in the western central portion where rainfall is between 500 and 800 mm per year. Mile after endless mile of these woods extend over western Formosa and Chaco provinces, extreme northwest Santa Fe, nearly all Santiago del Estero, eastern Jujuy and eastern Salta, eastern Tucumán and a tiny bit of Catamarca.

Where the actions of man have not changed the physiognomy of the woods, over and above all else emerge the crowns of the Red and White Quebracho trees, clearly dominating the surrounding vegetation. Though they share the same vernacular generic name they are in fact not related at all. The Santiago Red Quebracho (*Schinopsis quebracho-colorado*), the species whose distribution marks clearly the extent of this district of the Chaco, is one of the most imposing of the native trees of Argentina. The shape of all the trees in the Chaco, rather than being the tall trees of the forests, impress one with their majestic strength and vigor in the face of the adverse conditions which rule the region. The thick, robust trunks of impressively hard bright red wood which gives the tree its local name ("quebrar" means to break, "hacha" is an axe), reaches girths of upto 1.5 m diameter. It grows straight and unbranching in its lower portion, the crown is open, the compound foliage sparse. The clusters of small

The Giant Wood-rail

This bird (Aramides ypecaha) *has an onomatopaeic specific name: its strident calls uttered in chorus at eventide are the most obvious part of this shy, secretive, fast-running ground-loving bird of thickly vegetated marshes and their contiguous woods.*

The Whistling Heron

The lovely Whistling Heron (Syrigma sibilatrix), *with pastel shades of powder-blue and ochre is surely one of the prettiest herons in Argentina. Typical of the Chaco and the Espinal it has of late been spreading into the northern reaches of the Pampas.*

The Yellow Palo Borracho *One of the most characteristic trees of the western Chaco is this peculiar tree* (Chorisia insignis). *Here it grows with a typically bottle-shaped trunk of spongy porous fibre which allows it to store water through the drought.*
"Borracho" means drunken and refers indirectly to the shape of the trunk.

yellow flowers end up as bunches of red woody winged seeds for dispersal by the wind. From mid-summer to autumn they attract large flocks of parrots and parakeets. The White Quebracho *(Aspidosperma quebracho-blanco)* is the other essential component of these woods, but its adaptability allows it to grow outside the Chaco also. Its almost lanceolate elliptical leaves end in a sharp point, and the seeds are flat, enclosed in a woody castanette-like pod, and have a broad tissue-like skirt, also for being pushed around on the breeze.

All somewhat lower in height, other species of tree also grow with the Quebrachos in these open woods, all thorny and small-leaved, growing on the poor soils which turn to talcum-dust in the dry winter. Several *Leguminosae* like the Guayacán *(Caesalpinia paraguariensis)* which grows upto 18 m high, has a large crown and sheds the bark of its trunk to give it a peculiar greeny-gray mottled colour; the Itín *(Prosopis kuntzei)* with its dense globular crown which is entirely composed of long green thorns on which, for a brief period each summer, grow tiny leaves; the Chañar *(Geoffroea decorticans)* with its bark peeling to reveal the yellow-green trunk in younger specimens and which often grows in thick patches from the roots. In spring its golden-yellow flowers cover the tree.

Here too is the Mistol *(Zizyphus mistol)* with its edible fruit, the Peje or Sombra de Toro *(Jodina rhombifolia)* with tough evergreen rhomboid leaves with a thorn at each corner; Brea *(Cercidium australe)* all bright green branches, twigs and trunk in winter, completely covered with bright yellow flowers in spring, and a smallish very spiny palmate palm, *Trithrinax campestris*.

Surely one of the most curious trees in the region is the Yuchán or Yellow Palo Borracho *(Chorizia insignis - Bombacaceae)* which in the rainforests to the east where it also grows, in competition for the canopy and light has a straight, slender trunk. However in the Chaco it acquires a pot-bellied barrel shape with huge, short thorns all over it, and looks most like an armoured chianti bottle (hence the connected name in spanish, "borracho" means drunken). The wood is spongy and fibrous to hold a reservoir of water for the dry season.

On drier ground, often saline, the White and Black Algarrobos *(Prosopis alba* and *P. nigra* respectively) dominate, having better resistence than the Quebrachos to these adverse conditions.

The bushy stratum of the vegetation includes plants of the genera *Acacia, Prosopis, Maytenus, Colletia* and *Capparis* among others, and several cacti. The removal of the larger trees for wood and timber, coupled with overgrazing, favours the growth of bushes to the extent that they develop into a virtually impenetrable mass.

The cacti have specially adapted stems, both to hold volumes of water and for photosynthesis; their leaves have become long piercing spines for defence.

Some Chaco cacti are columnar and grow to great heights like the Ucle *(Cereus validus)* and the Cardoon *(Stetsonia coryne)* which looks more like a monstrous candelabra. Another, the Quimil *(Opuntia quimilo)*, some four to five metres high, produces fruit which is much sought by wildlife, its stems are segmented, flattened fleshy ovals these; with many lesser species, they all give the landscape a singular and exciting beauty.

There is one cactus which grows as an epiphyte *(Rhipsalis aculeata)* with small edible fruit; it sometimes covers branches completely. Other species which grow on trees are to be found among the parasitic *Laurantaceae* with bright flowers, and which are green year-round. As in the subtropical forests bromeliads abound though here the species are smaller, like those of the genus *Tillandsia*.

However, ground bromeliads like the Chaguares *(Bromelia serra* and *B. hieronymii)*, because of the overgrazing of cattle, grow on the virtually grass-free soil into sizeable clumps, sometimes hundreds of yards across which are totally impenetrable because of the thorns and hooks on the edge of each leaf.

These bromeliads form an important refuge for many insects, spiders, reptiles and even frogs like the hylids *Phyllomedusa sauvagii* and *P. hypocondrialis,* both of attractive green colouration and slow movements. They tend to live at the base of the leaves, protected by the thorns and where they find the much sought damp and shade. Other amphibians like the horned frogs *(Chacophrys pierotti, C. cranwelli, Lepidobatrachus laevis* and *L. llanensis)*, aggressive, voracious, feeding on other amphibians, rodents, even birds, solve the dessication problem by burying themselves in the mud of the temporary ponds where they breed, till the next rains.

There are amphibians which simply hide under fallen logs like the Giant Toad *Bufo paracnemis,* and *B. granulosus.* Others use mammal burrows like those of the once-abundant Plains Viscacha: the highly-coloured Coraline Frog *(Leptodactylus laticeps)* red, yellow and black in spotted pattern, an obvious warning of the toxicity of the glands in its skin; and *L. bufonius*. The microhylids *Dermatonotos mulleri* and *Elachistochleis bicolor* get into termite nests and feed off their hosts.

In parts of the Chaco termite nests are very numerous, especially to the west, but the record for diversity goes to the ants. In some areas they are considered the principal croppers of the vegetation. The nests of some species of leaf-cutters like *Atta* spp. consist of extensive underground galleries, the earth removed being dumped above ground in amazing structures and piles. From the nests gigantic columns of foraging workers depart for the trees and bushes to cut off bits of green leaves which are then carried back to the nest. There, other workers which specialize in "gardening", cut the bits of leaf into smaller bits and prepare beds to cultivate fungus upon which both adults and larvae feed. Yet another caste, the soldiers with much larger heads and mandibles than the rest, are the defenders of the colony in the depths of which lives the queen. She is the only fertile female and her sole purpose in life is to lay masses of eggs - several times her own weight in eggs in each warm season. After the first summer rains winged males and females leave the nest on their nuptial flight. The fertilized females drop back to earth, shed their wings and bury themselves to start new colonies, while the males all die off.

There are other leaf-cutting ants like *Acromyrmex,* while some are granivorous and gather seeds *(Pogonomyrmex)* and others are simply predatory *(Ectatomma, Dorymyrmex,* etc.). Some species make their nests in fallen rotting logs *(Camponotus)* while others live in live trees *(Crematogaster).*

The huge biomass of ants is the staple for a large number of their predators. The wasp *Polybia ruficeps* for example stores the live but mutilated bodies of ants and termites which it captures during their nuptial flight, as the food for the larvae when other food is not available.

Ants are also an important source of food for some reptiles and birds, these last keeping in mixed flocks for feeding.

But perhaps the most important role played in cotrolling these social insects was performed, in eras before man's appearance, by the edentate mammals, descended from a great group of neotropical families which had gigantic forbears. Such were the armoured Glyptodons and the Giant Ground Sloth.

At present the most numerous family of these is that of the armadillos which are very well represented in the Chaco, varying in size from the tiny Fairy Armadillo *(Chlamyphorus retusa)* of virtually unknown underground habits, to the largest living species, the Giant Armadillo *(Priodontes maximus)* weighing upto 60 kgs. In spite of its size the Giant Armadillo can run fast and agilely to seek refuge in its huge burrow when threatened by any potential enemy.

The Three-banded Armadillo *(Tolypeutes matacos),* is the only species which can perform the expected trick of rolling up into a virtually unassailable ball, painfully squeezing any finger or snout inserted to explore for a weakness. The Six-banded Armadillo *(Euphractus sexcinctus),* the Wailing Pichi-armadillo *(Chaetophractus vellerosus)* with its pitiful cry of despair when caught, and the virtually unknown Bare-tailed Armadillo *(Cabassous chacoensis)* are other Chaco members of this family. All these live in deep burrows which they dig with the powerful claws on their front feet, claws used also to search for their food, digging up insect larvae, the bulbs and rhyzomes where plants keep their reserves, or breaking into ants' and termites' nests. Undoubtedly the most specialized animal for this diet is the Giant Anteater *(Myrmecophaga tridactyla)* with its enormous and strong claws for opening up these nests, and its long sticky tongue to gather the thousands of ants and termites it needs daily, claws, one might add, that it values greatly, walking on its knuckles to protect them and not wear them down. These claws also serve as a lethal weapon in defence. Generally solitary and peaceful, when threatened, this anteater sits up on its haunches, arms flung wide with claws at the ready to disembowel any dog or larger cat like the Jaguar or Puma that comes within range of the slash.

The Jaguar *(Panthera onca)* which once used to roam all over the north of Argentina as far south as the Rio Negro, is still to be found in the Western Chaco, one of its last redoubts. Its main prey here are the peccaries, herds of which it attends to capture some separated, wandering individual; the troops of White-lipped Peccaries *(Tayassu pecari)* however, offer formidable resistance both with their numbers and sharp tusks which can rip a Jaguar to pieces if caught. Together with the White-lipped and Collared *(Dicotyles tajacu)* Peccaries of the Chaco there is third species, the Taguá *(Catagonus wagneri),* in smaller herds, apparently better adapted to the drier areas, the only place it lives. It was as recently as 1974 that the surprising scientific discovery was made that this species, previously known only as a fossil, was still alive and to be found in the Paraguayan Chaco. Further research also found it in Argentina where settlers had known it for generations - they even had a vernacular name for it: "Quimilero", as it feeds on the fruit of that prickly-pear cactus *Opuntia quimilo.* It is larger than the other two species, has a collar like the smallest, longer legs, and lacks the rudimentary hoof on the side of the hind legs.

The Tapir *(Tapirus terrestris)* and the Brown Brocket *(Mazama gouazoubira)* complete the list of the larger herbivores.

Unlike in the rainforests, tree-climbing mammals are scarce in the Chaco where the vegetation's strata are not very marked, where insufficient protection and food is available; bats and ground-dwelling mammals abound however, many of these being burrowers as these dens offer protection and more stable conditions of temperature.

As happens in so many other parts of South America, rodents in variety and numbers here too comprise a very important link in the food-chains. The Plains Viscacha *(Lagostomus maximus)* and the Woodland Mara *(Pediolagus salinicola)* are everywhere, favoured by man's elimination of their major predators: the Puma *(Felis concolor),* Geoffrey's Cat *(F. geoffroyi),* Pampas Cat *(F. colocolo)* and the Crab-eating Fox *(Cerdocyon thous)* amongst others.

Even the birds have an appreciable list of ground-dwellers (or mainly so) which belong to three exclusively Neotropical families: the Rheas *(Rheidae),* the Tinamous *(Tinamidae),* and the Seriemas *(Cariamidae).* Only one of the Rheas is found here, the Greater Rhea *(Rhea americana),* the great ratite of the open spaces in lowland temperate South America, which lives in flocks. The Brushland Tinamou *(Nothoprocta cinnerascens)* and the Lillo Tinamou *(Eudromia formosa),* both with cryptic colouration, are better adapted to the dry woodlands. They escape by running and take to the wing only when pressed. The Seriemas are long-necked, long-legged birds with strong hooked beaks and big fan tails, in appearance and habits somewhat like the African bird of prey the Secretary Bird, and certain Bustards. The Red-legged Seriema *(Cariama cristata)* and the Black-legged Seriema *(Chunga burmeisteri)* are the only two species in the family; they stride purposefully across the land in search of the insects, reptiles, young birds and rodents on which they feed.

Arboreal birds are easy to see because of the sparser vegetation and the open foliage. Sunrise is the most active moment for most birds, rowdily announced by the Chaco Chachalacas *(Ortalis canicollis)* so typical of these woods, found in flocks feeding on berries and insects.

Woodcreepers *(Dendrocolapidae)* and Woodpeckers *(Picidae)* forage on the trunks and limbs of the trees which they inspect carefully in search of their insect food, the former family in subdued tones of reddish-brown like the Great Rufous Woodcreeper *(Xiphocolaptes major)* and the Narrow-billed Woodcreeper *(Lepidocolaptes angustirostris),* the latter family with the White-fronted Woodpecker *(Melanerpes cactorum),* the Cream-backed *(Campephilus leucopogon)* and the Golden-breasted Woodpecker *(Chrysoptilus melanochlorus).*

The South American family of the Furnariids is particularly well-represented. Their plumage, in browns, chestnut and buffs is hardly stunning but their nests merit our close attention being large and often very odd. The Rufous Hornero's *(Furnarius rufus)* nest is made of mud mixed with bits of grass, about the size of a soccer ball with a curved entrance hall leading to the hole into the nest chamber. This is our national bird. Its proximity to man leads it to build its nest on the cornices of buildings, on lamp-posts in the plazas in the great cities. The Crested Hornero *(Furnarius cristatus)* is similar but with a crest, and considerably smaller.

Many other Furnariids use thorny twigs to make their nests, like the Greater Thornbird *(Phacellodomus ruber)* or the Freckle-breasted Thornbird *(P. striaticollis),* the Firewood Gatherer *(Anumbius annumbi)* or the Short-billed Canastero *(Asthenes baeri).* Some of these nests are inmense and last for years after being abandoned by their first owners, becoming a refuge for other mammals and reptiles (mouse opossums and snakes), and for insects and spiders.

Riverine Gallery Forests

These are found along the banks and on the islands of most rivers. The Paraguay river, in its seasonal floods and surges switches courses, creates new islands, cuts off ox-bows; the result is a chaotic pattern of forested banks interspersed with running or stagnant waterways very rich in wildlife.

Frogs

Amongst the batracians that live in the damper eastern part of the Chaco are the Green Leaf-folding frogs, chameleonlike in their slow measured movements.

The Capybara

The Capybara (Hydrochaeris hydrochaeris) *is the largest rodent in the world, weighing upto 69 kgs. They are found in herds and always near water to which they take for refuge and comfort. They feed on the grass at the edge of the bodies of water where they live, or on the aquatic vegetation. At night they roam some distance from the safety of water.*

The Collared Peccary

Peccaries are the main herbivores of the Chaco where the larger herds of the White-lipped Peccary are the nearest equivalent to the herds of herbivores on other continents. The smallest is the Collared Peccary (Dicotyles tajacu).

Another huge twig nests is built by several pairs of the Monk Parakeet *(Myiopsitta monachus)* nesting colonially in the higher and outer branches of the taller trees or on pylons. Each nest has several individual entrances leading to separate nest chambers. They too later serve as nesting places for such as the Spot-winged Falconet *(Spiziapteryx circumcinctus)* and the Ringed Teal *(Anas leucophrys).* On the other hand most other parrots of the Chaco use holes in trees: the Blue-crowned Parakeet *(Aratinga acuticaudata),* the White-eyed Parakeet *(A. leucophthalma)* and the Turquoise-fronted Parrot *(Amazona aestiva).* In noisy flocks these birds fly over the tree-tops in search of fruiting trees to extract the seeds and break into them with their powerful beaks.

There are other fruit- and seed-eating birds in the Chaco - the glamorous Red-crested Cardinal *(Paroaria coronata)* in its smart, almost military plumage, the Golden-billed and the Grayish Saltators *(Saltator aurantiirostris* and *S. coerulescens),* all with thickish bills to shell seeds; pigeons and doves mostly feed on the ground and swallow seeds and small fruit whole, as is the case with the White-tipped Dove *(Leptotila verreauxi),* the Eared Dove *(Zenaida auriculata)* and the Picazuro Pigeon *(Columba picazuro).*

Insectivores dominate the avian scene by far, and to those families already mentioned must be added the Ant-birds, Warblers, Tapaculos, Cotingas and Tyrants at least.

The Tyrants or New World Flycatchers have the largest number of representatives in Argentina as a whole; they are mostly dull, brownish and olive, though some by their plumage or behaviour stand out in any landscape. Perched on some prominent look-out, as much as a function of territory as to spot flying prey, they fly out in pursuit of some passing insect and return to the same spot. Some of the most outstanding are the brilliant-breasted Vermillion Flycatcher *(Pyrocephalus rubinus),* three Monjitas, the White *(Xolmis irupero),* the Dominican *(X. dominicana)* and the Gray *(X. cinerea),* all with contrasting black, white and gray, and the Fork-tailed Flycatcher *(Tyrannus savana).*

There is also a host of insectivorous reptiles. Nocturnal Gekkos *(Homonota horrida* and *H. borelli)* spend the hours of daylight under bark or logs; like so many lizards they drop their tail when hard-pressed by some predator, and this keeps wiggling crazily to attract the attacker's notice. The tail also serves as a reservoir for laid-up fatty reserves, growing tremendously fat to see them through the lean dry season.

Diurnal lizards are much more numerous, mostly terrestrial like the very large Red Ground-lizard *(Tupinambis rufescens)* or the small *Cnemidophorus leachi* and *Teius cyanogaster,* all of the *Teidae* family which fill the equivalent niches of the *Lacertidae* of the Old World.

The dominant Neotropical family is however the Iguanidae of varying sizes and shapes, having colonized most habitats. Some are speedy runners like *Liolaemus chacoensis,* others are slower but turn and face the pursuer with an intimidating show *(Leiosaurus paronae),* some are climbers like the territorial *Tropidurus spinulosus,* and there even is a species *(Polychrus acutirostris)* with opposable digits and a prehensile tail whose slow movements and latterally compressed body are somewhat chameleonlike.

The snakes include some arboreal species such as *Philodryas baroni,* slender, bright green and amazingly agile, a predator on birds, their nestlings, rodents and frogs. Most however are terrestrial. There is a Boa *(Boa constrictor),* usually around the colonial burrows of the Plains Viscacha;

the poisonous snakes include the only rattlesnake in the country *(Crotalus durisus)* and the pit-vipers of the same family *(Bothrops alternata* and *B. neuwiedii).* In the damper situations, as under fallen logs are two small specialists *(Sybinomorphus turgidus* and *S. ventrimaculatus)* preying on slugs and snails, the latter being sucked out of their shells after these have been pierced or punctured by the snake's teeth.

There is a whole photophobic fauna under logs where conditions are more stable. It is much more varied than all the rest put together and emerges only at night - the beetles of various families (scarabs, weevils, long-horn cerambicids, and so very many others), as well as a host of spiders, scorpions, crickets and centipedes.

Towards the west of the Western Chaco District and rising into the hills to some 1,800 metres above sea level, the composition and appearance of the woods change somewhat. The Red Quebracho is replaced by the Horco Quebracho *(Schinopsis haenkeana)* with a more twisted trunk and grayer foliage, growing with the Coco *(Fagara cocos),* the White Molle *(Lithraea ternifolia)* with its leaves dripping a caustic gum, the thorny Churqui or Tusca *(Acacia caven)* and others. This sub-district is known as the Hill Chaco and is found in eastern Jujuy, central Salta and Tucumán, extreme eastern Catamarca, and reaches the hills of La Rioja, San Luis and Córdoba, in places being a sort of transition ecotone with the Yungas.

The fauna does not here present many differences with that of the greater Western Chaco though there are some endemics like a lizard *(Cupriguanus achalensis)* and frogs, including *Pleurodema kriegi.*

The Formosa Nature Reserve. When in 1968 the Province of Formosa ceded to the National Parks Service these 10,000 has, they already lacked the conditions for being included in the national parks system. In the south-west of the province, between the rivers Teuco and Teuquito this is the region where the semi-arid Chaco was once most productive. Overgrazing by herds of cattle from Salta in the west destroyed the grass cover and allowed the Vinal *(Prosopis ruscifolia)* free rein. The Vinal is an aggressive invader with enormous thorns upto nine inches long. Coupled with this was the virtual destruction of the forests and woods and the almost total elimination of the Red Quebracho. The few specimens left were too small or too distant or too twisted or too sick to merit the axe. They are now the hope of the future as producers of seed for recuperation.

This seemingly hopeless situation was complicated by the presence of settlers, but happily these are now gone, with their destructive goats, and their habit of helping themselves to any wild meat they fancied or which crossed their paths. Again, the remaining Brockets, Giant Anteaters, Peccaries and the several armadillos including and especially the very rare Giant Armadillo now have a chance to bring their populations into line with what should be the normal density.

The birds are more numerous and include a great variety of water-birds which flourish in the old river courses and oxbows. Multitudes of herons, storks, spoonbills, ducks are present in the wet as is the odd specimen of the Cayman.

With all this it has been essential to have permanent personnel in the park and the buildings necessary for this. Till recently the wardens had only visited periodically. A perimeter fence is now erected to stop cattle invading and it is hoped that

The Giant Anteater

This animal has disappeared from most of its former range in Argentina and is now virtually only found in the Chaco. This surprising edentate wanders through the woods and savannahs in search of ants' and termites' nests which it breaks into with its formidable claws, to explore the galleries in search of the insects with its long, sticky tongue.

The Four-eyed Opossum

South America's marsupials are well-adapted omnivores with a preponderance of small animals in their diets. Most have prehensile tails for climbing trees. This Four-eyed Opossum (Philander opossum) is one of the middle size-range species.

The Tapir

Threatened throughout its range, the National Parks of northern Argentina may offer it the last of its refuges in the Paranaense Rainforests, the Chaco and the Yungas. It is important to add new areas to the system of protected habitats, like the "Impenetrable" - in western Chaco Province.

The Maned Wolf

The Maned Wolf (Chrysocyon brachyurus) is without a doubt one of the most threatened species in the country. This curious long-legged canid is solitary and in spite of its size feeds only on smaller vertebrates.

the grasses will return to their former glory, as they have done in experimental plots. It would now be important to increase the size of this reserve by incorporating the public lands all around it; but if this could be coordinated with the setting aside of a similar stretch across the provincial border in the Chaco a goodly area of wetlands could be included with all the arms and old river-beds of the perpetually course-changing Teuco.

The "Impenetrable" - a future National Park? In 1977 the provincial authorities of the Chaco started a campaign of settling fiscal lands, which they called "The Conquest of the Impenetrable". The object is to turn the last and largest block of unaltered native Chaco woodland, and one of the world's most complex semi-arid regions, - some 4 million hectares - into something more traditionally productive.

Though all the plans talk of maintaining the balance of nature there is no mention of a National Park.

The creation of just such a protected area in the region is an almost mandatory priority as there is none of this habitat represented at all in the whole National Parks system. Besides, there is still time for this to be achieved before this natural wildlife reserve is destroyed. Some progress has been made to cover this need but the 40,000 has (1% of the total area) which have been set aside for conservation by the provincial authorities is not nearly enough to effectively preserve a representative portion. A calculation has yielded the figure of nearer 400,000 has or some ten per cent, in order to maintain viable populations of the largest predators like the Jaguar, with their extensive home-ranges. Such a park would also protect other endangered animals like the Tapir, the Giant Armadillo and the Giant Anteater, and become a park of world significance.

Tentatively two different areas are being considered, one in the northwest between the old course of the Bermejo river and the Teuco, bordering on the Formosa National Nature Reserve. Some very interesting and endangered trees are here represented like the Palo Santo *(Bulnesia sarmientoi)* growing in almost pure stands, and also some riverine gallery forests of species characteristic of the Yungas. The rivers and the series of oxbows and lakes in the now isolated old river-beds provide the habitat for an abundant and varied aquatic fauna.

The other area which is being considered and where there is little or no surface water and very dense woodland, backs onto the northeast corner of Santiago del Estero province where there is already a 114,000 has provincial Forest Reserve called Copo, poorly protected by the nearby village's police force. Were this project to come to fruition there would be an extremely valuable area protected.

Córdoba's Protected Areas. Several interesting protected areas come under the jurisdiction of the provincial Department of Natural Areas where a number of warden-rangers work who graduated from the National Parks Ranger-Training Centre.

Monte de las Barrancas is a reserve in the Western Chaco region, in the extreme northwest of the province of Córdoba, near the Great Salt Flats (Salinas Grandes). A patch with less salinity than the surroundings supports a scrubby woodland vegetation where herds of Guanacos *(Lama guanicoe)* live, the last remnants in the province.

Chancaní, at the western foot of the hills, on the Rioja Flats, is a fairly large piece of Quebracho woodland where the Boa *(Boa constrictor)* still abounds and the Woodland Mara *(Pediolagus salinicola)* flourishes.

Mar Chiquita is a huge inland lake of some 100 by 50 kms dimensions but much larger during wetter eras when the salinity is less. It is fed by an extensive endorrheic drainage system from the north through the Dulce river, and several smaller rivers from the south. In Mar Chiquita vast flocks of the Chilean Flamingo *(Phoenicopterus chilensis)* with some Andean Flamingos *(Phoenicoparrus andinus)* concentrate and nest. Black-necked Swans *(Cygnus melancoryphus)* and Coscoroba *(Coscoroba coscoroba),* a strange swan/goose/tree-duck, as well as thirteen other species of waterfowl are but part of the avifauna found here. The sea-lake is also used in summer by large numbers of migratory shorebirds and has been declared a link in the Western Hemisphere Shorebird Reserve Network. Sandpipers, phalaropes and others, including Franklin's Gull *(Larus pipixcan)* visit this highly saline habitat from North America.

The Hill Chaco District is represented in the Cerro Colorado Cultural and Nature Reserve and in Condorito, soon to become a 40,000 has national park surrounded by a much larger provincial reserve. It will include the highest peaks like Champaquí at well over 9,000 feet or 2,800 metres, canyons and a large part of the rolling upland grasslands (Pampa de Achala) with peculiar wooded patches. Here there is an important population of breeding Andean Condors *(Vultur gryphus).* Cerro Colorado is the northern outpost of the Córdoba hills, all sedimentary rock, where the indigenous people who have now disappeared (Comechingones and Sanavirones) left their story in portraits of the natural features of their lives - fauna, flora and landscapes - their achievements such as domesticated llamas and agriculture, and the horsemen and long black-robed, flat-hatted priests who drove them away in single file into what was to be a better life. Condors, rheas, coral snakes, armadillos, cacti, llamas, foxes, brockets, medecine-men are all to be found painted in 35,000 rock-paintings on the walls of their shelters. It is a site worthy of man's most careful protection.

EASTERN AND SAVANNAH CHACO DISTRICTS

The eastern part of the Chaco is of hardwood woodland (known as "strong" woods) occupying the higher ground, alternating with savannahs and palm groves, marshes and swamps in the lower areas. This is the Eastern Chaco District and covers the eastern parts of Formosa and Chaco provinces, northern Santa Fe, spilling over the Paraná river into the northwest corner of Corrientes.

The Santiago Red Quebracho is here replaced with the Chaco Red Quebracho *(Schinopsis balansae),* somewhat similar, but here the vegetation is enriched by the presence of less hardy species of trees: the magnificent Urunday *(Astronium balansae)* upto 20 m tall, the Marmelero or Viraró *(Ruprechtia taxiflora),* the Black Lapacho *(Tabebuia impetiginosa),* the Crown of Thorns *(Gleditsia amorphoides)* whose trunk bears the ring of branching thorns to which the name refers, Zapallo Caspi *(Pisonia zapallo),* the Tataré *(Pithecellobium scalare)* and others. The Yellow Palo Borracho is here replaced with the Pink *(Chorisia speciosa)* with large orchid-like pink flowers with creamy-yellow centres liberally speckled with black, which is everywhere planted as an ornamental in gardens and parks.

Cacti are here less abundant. Epiphytic and ground-delling species of Bromeliads are joined by some climbers and orchids like the White Zarzaparrilla *(Smilax campestris)* and Monkey's Comb *(Pithecoctenium cynanchoides)* with its striking white flowers and rough seed-case which gives the plant its name. All this, together with the denser grass cover, gives this Eastern Chaco a more exhuberant appearance. Where soils are more saline or alkaline because underground water is close to the surface, extensive groves of the White Palm or Caranday *(Copernicia alba)* grow, with glaucous green palmate fronds, upto 14 m tall. In almost pure stands, rising out of a dense herbaceous ground-cover, they are part of a wonderful landscape.

On the lower ground grasses dominate: *Elionurus muticus* is nearly one metre tall and grows in enormous swales dotted with the occasional island of trees. These grasslands stretch southward well into the north of Santa Fe forming the district of the Savannah Chaco.

Where seasonal flooding occurs other grasses thrive, like the Yellow Grass *(Sorghastrum agrostoides)*, Paja Boba *(Paspalum intermedius)* all about one metre tall, growing where Termite mounds abound - concrete-hard earthen structures looking like so many stumps rising above the level of the surrounding vegetation. The grass used for roofing *(Panicum prionites)* upto two metres tall, is found on the regularly flooded banks of rivers, streams and lakes.

The marshes of the Chaco are the most interesting and exciting habitats in the region. They are clay-bottomed lowlands fed by rainfall or flooded by overspilling rivers, in depth between half a metre and 1.5 m which only dry up during severe droughts. Their elongated shape, the fact that they occur in series, and the gallery forests along their banks all seem to indicate that they were once river-beds. The aquatic plant around the edges and in the shallower areas is mainly the Giant Sedge *(Cyperus giganteus)*, with Cattails *(Typha dominguensis)*, the Pehuajó *(Thalia geniculata)* with its large leaves and blue flowers, and some others. On deeper waters the floating plants are at home - Water Hyacinths *(Eichornia crassipes, E. azurea* and *Reussia subovata)* with their lovely blue flowers, and stems and petioles of air-filled spongy tissue for flotation. One of these *(E. crassipes)*, introduced into North America, Europe and Africa has become a pest of major proportions; lacking all natural controls it increases rapidly to choke the waterways. Other floating plants are Water Lettuce *(Pistia stratiotes)* and the tiny floating ferns *Salvinia auriculata* and *Azolla filiculoides*.

It often happens that these floating plants form a dense mass, on and into which grow other rooted species as if in the soil. With time this mat thickens, offering yet more dead plant material, becoming in the end a veritable floating island, supporting bushes and sizeable vertebrates like Capybaras, Caymans and Water Boas. Locally known as embalsados they sometimes completely cover a marsh, other times there is a patch of open water in the middle. It is here that the magnificent water-lilly *Victoria cruziana* grows, its leaves like huge circular dishes with upturned edges, thorn-protected on the outside against herbivores. Many water-birds walk on them in search of their food.

Other bodies of water, like oxbows, concentrate the largest biomass in the area. The abundance of nutrients together with the elevated humidity and temperatures stimulate plant-growth at an amazingly rapid rate. They become one of the most productive ecosystems on earth, and so one of the most valuable in food production.

Huge schools of tiny fish take every advantage of the multiple opportunities offered by the habitat. Some live in open

The Eastern Chaco

water while others take refuge among the stems of rooted vegetation or under floating plants. Some are found near the bottom of the ponds, others near the surface.

Caraciform fish are legion; many sport bright colours like *Aphyocharax rubripinnis* with dorsal, anal and ventral fins all bright red and contrasting with its silvery body, *Moenkhausia sanctae-filomenae* with a red eye and black marks on the tail, *Pyrrulina australis, Hemigrammus caudovittatus* and many more. They live here with several species of Cichlids like the Chanchitas *(Cichlasoma facetum* and *Aequidens portalegrensis)*, interesting little territorial fish and devoted parents.

The greatest predatory fish of these marshes is the Tararira *(Hoplias malabaricus)* looking like a primitive pike, skulking in some dark or muddy corner to dart out on any prey which inadvertently swims by - fish, bird, batracian, reptile or mammal. It is well-adapted to life in these conditions as it can breathe air to a great extent and thus survive in drying puddles till the next downpour. There are other species which are even better adapted, not only breathing air but being able

These lowlands are mostly poorly-drained plains and are covered with swamps and marshes, often temporary. There are also open savannahs and palm groves.

to cross land from one pond to the next with eel-like movements *(Synbranchus marmoratus)*, or using their pectoral fins and tails to "walk" as do the small Cascarudos or Tamboatás *(Callichthys callichthys* and *Hoplosternum thoracatum)*. A further adaptation is found in the "annual" fish like *Cynolebias* spp. where males and females are markedly different. They spawn in the mud of drying puddles and die. The ova are resistant to desiccation and the fry hatch when the puddle fills again to repeat the cycle. They do not spawn if the puddle does not dry up.

Another species which is wonderfully adapted to the cyclical "wets" is the Lung-fish *(Lepidosiren paradoxa)* which has close relatives in Africa and Oceania. When puddles dry up this elongated fish digs itself into the mud of the bottom and, curled up, spends a period of inactivity till the next rains. When these arrive they promptly spawn and it is the male who is left in charge of the fry. These have gills when they hatch but soon develop the lungs which make them so peculiar among fish.

Summer rains also trigger amphibian activity. At dusk the variety of calls emitted by the males, inviting the females to spawn, amply demonstrates the amazing abundance and diversity of these beings. The cacophonous chorus of their many voices is composed of the tinkling and metallic notes of certain Hylid tree-frogs *(Scynax nasica, Hyla pulchella, H. phrynoderma)* which come from males clinging to emergent vegetation; other whistled notes from the Moustached Frog *(Leptodactylus mystacinus)*, the strange mewing of the Weeping Frog *(Physalaemus biligonigerus)* or low-pitched and resonant like an old diesel motor, from the giant Cururú Toad *(Bufo paracnemis)* calling from the edge of the water - and hardly a let-up till dawn.

This abundance of fish and batracians as well as the snails and various other life-forms are food for a world of different predators. Water turtles *(Phrynops hillarii* and *Acanthochelys spixi)*, large water-snakes *(Cyclagras gigas)* and the Water Boa *(Eunectes notaeus)* take their prey under water or on the floating vegetation. But the lords of the marshes are the

The Great Black Hawk

*Some of the larger raptors have specialized in taking fish and other aquatic prey as so much of the land is covered by water, as is the case of the Great Black Hawk (*Buteogallus urubitinga*) which is here seen together with a scavenging Crested Caracara (*Polyborus plancus*).*

inscrutable caymans. Unfortunately both species, the Black *(Caiman yacare)* and the Broad-snouted *(C. latirostris)* are today very rare and shy. The value of their skin has led to this near-extinction of the great herds that used to sun on the banks of rivers and marshes. The few that remain are small and timid, the result of man's greed. Caymans lay their eggs in a mound of vegetation which produces a certain amount of warmth through decomposition. The female waits in fairly close attendance on the nest, helps the young to emerge and to reach the water, and protects them till they reach a certain size.

Birds of the marshes, with few exceptions, are fortunately less persecuted than years ago as plumes and feathers have happily gone out of fashion. Today it is the abundant and varied birds which first catch the eye in any aquatic environment.

Roseate Spoonbills *(Ajaia ajaja)* walk along the edge with the flattened tips of their peculiar bills in the water, scything from side to side, filtering out their food. Great Egrets *(Egretta alba)* and Cocoi or White-necked Herons *(Ardea cocoi)* stalk fishes in slow motion to dart out their necks for capture with the long bill; Snowy Egrets *(Egretta thula)* on the other hand stir up the bottom with their yellow toes to scare up their prey; flocks of Wood-storks *(Mycteria americana)* of one accord advance abreast in puddles and rhythmically dip their beaks and heads underwater to feel for something edible. Majestic Jabiru Storks *(Jabiru mycteria)*, standing much taller than all the rest and showing off their naked black heads and scarlet chokers to contrast with their impeccable white plumage, capture large fish and eels with their heavy, strong bills.

Muddy flats, damp or flooded fields are frequented by White-faced Ibis *(Plegadis chihi)* and Bare-faced Ibis *(Phimosus infuscatus)* in search of small invertebrates and some seeds.

Also on the shores or on the higher ground the three species of Tree-Ducks (Fulvous *Dendrocygna bicolor,* White-faced *D. viduata* and Black-bellied *D. autumnalis*) and the Comb Duck *(Sarkidiornis melanotos)* flock together in hundreds or even thousands, while the enormous Muscovy Duck *(Cairina moschata)* is usually seen perched in a tree; its meat is everywhere much prized by locals. Muscovies were domesticated in pre-historic times but wild birds are now extremely rare.

On top of the floating vegetation, held above the water by spreading its weight over a large area through its immensely long toes, the Wattled Jacana *(Jacana jacana)* seeks insects or crustaceans. Perched on a semi-submerged stump or log the Anhinga *(Anhinga anhinga)* which spears fish under water, impaling them on the sharp-pointed beak, spreads its wings to dry in the sun. The Least, White-tufted and Pied-billed Grebes *(Podiceps dominucus, P. rolland* and *Podilymbus podiceps* respectively) all dive for their food - little fishes, invertebrates, larvae and crustaceans. These and many more species turn these aquatic habitats into wonderful places to watch wildlife.

The mammals on the other hand are shy, rare and many are nocturnal. The largest is the Capybara *(Hydrochaeris hydrochaeris);* weighing upto 69 kgs it is the largest rodent in the world. It lives in herds of often well over 20 individuals which spend most of the day at the water's edge, or during the hottest hours cool off and nibble aquatic plants while mostly submerged. At night they feed at some distance from the water. At the slightest alarm a loud short "bark" drives them all back to the water at a gallop where they dive for safety with a rowdy splash.

In spite of their high reproductive capacity these mammals are now only to be found where they are protected. Their white flesh is delicious and their hides make the finest soft leather, as good as any pigskin, so they are persecuted almost everywhere. The same is true of the Coypu *(Myocastor coypus)*, another aquatic rodent often misnamed "nutria", especially in the fur trade. It is everywhere trapped for its skin. The tracks and trails that Coypu make through and over the islands of floating vegetation are also used by water-rats *(Holochilus chacarius* and *Scapteromys aquaticus)* which feed on Water Hyacinth and other aquatic plants.

The tall, floodable grasslands which grow around the marshes and other bodies of water are the habitat of one of the most threatened of Argentina's mammals, the Marsh Deer *(Blastocerus dichotomus)*, the largest of South America's deer. Stags can exceed 100 kgs in weight, have thick antlers of elegant symmetry, the tines being all about the same length forming a goblet-shape. They live in family groups which spend the day hiding in the islands of woodland, emerging at night to graze. Walking in these boggy situations or on these floating islands, offers no difficulty to these animals with their long splayed hooves. Persecuted by settlers for their meat and by "hunters" after a trophy, they are today uncomfortably close to extinction, a real criminal shame for such a lovely animal. Something similar is happening to the northern form of the Pampas Deer *(Ozotocerus bezoarticus leucogaster)* which lives in the shorter grasslands that do not flood.

The Maned Wolf *(Chrysocyon brachyurus)*, a long-legged canid which in spite of its size feeds on rodents and small vertebrates, sometimes stooping to insects and fruit, is also vanishing from these grasslands as man advances.

A varied bird-fauna is to be found on the wide sedge- and grass-covered savannahs. Long-legged rails like the Giant Wood-rail *(Aramides ypecaha)* whose loud cacophonous duet calls aptly imitate its onomatopaeic specific name; the Plumbeous Rail *(Rallus sanguinolentus)* and others, smaller and even less visible, creep through the grasses in search of larvae, insects, small molluscs. In courtship and territorial flight the Common Snipe *(Gallinago gallinago)* climbs and dives emitting its "call", a rattly drumming produced by the outer tail-feathers vibrating during the fast, steep dives.

Flocks of social icterids: the Yellow-rumped and Brown-and-yellow Marshbirds *(Pseudoleistes guirahuro* and *P. viresens)*, the Yellow-winged Blackbirds *(Agelaius thilius)*, a misnomer as only spots on its shoulders are yellow, the rest all black (male) or dark and streaky (female); the Scarlet-headed Blackbird *(Amblyramphus holosericeus)* with flame-orange head and garters all add colour and brightness to the sedges, reeds and grasses which bend under their weight. The Saffron-cowled Blackbird *(Agelaius flavus)* is now extremely rare for reasons as yet undetermined.

There is a profusion of Furnariids and Fringillids, the last including several Seed-eaters *(Sporophila* spp) and the Great Pampas Finch *(Embernagra platensis)*. Easily visible are several tyrants in which the male trails a peculiar tail one would think might make flight difficult: the Strange-tailed Tyrant *(Alectrurus risorius)* and the Streamer-tailed Tyrant *(Gubernetes yetapa)* in the open grasslands, while the Cock-tailed Tyrant *(Alectrurus tricolor)* is a bird of the higher areas. The fauna in general of the Eastern Chaco is similar to that of the drier Western Chaco but more varied. The mosaic of open marshes in the woods, islands of woods in the seas of grass, savannahs, lakes, rivers and streams offers many interfaces between the habitats and therefore a greater supply of niches (food and refuge). In birds of prey for example, many nest and roost in trees but hunt out over the grasses of the savannahs

Pilcomayo National Park

This is undoubtedly the most interesting of the protected areas in the Eastern Chaco biome; here marshes, lakes and White Palm (Copernicia alba) *savannahs predominate. The large Laguna Blanca which can be seen in the photograph is mostly in the park; it is also a Ramsar wetlands site.*

The Six-banded Armadillo (Euphractus sexcinctus) *is one of the several armadillos found in the Chaco, the region in Argentina which has the greatest diversity of these peculiar animals.*

or marshes. The concentrations and diversity of raptors here is amazing, each specializing in one way or another. The Crowned Eagle *(Harpyhaliaetus coronatus)*, huge, with powerful legs and talons, captures medium-sized mammals like skunks and opossums; the Bay-winged or Harris' Hawk *(Parabuteo unicinctus)* goes after the somewhat smaller cavies while the White-tailed Kite *(Elanus leucurus)* is a mouser hanging on the breeze.

Some raptors, like the Long-winged and Cinereus Harriers *(Circus buffoni* and *C. cinereus)* fly low over the ground - flooded or marshy in the case of the Long-winged - to come upon their prey suddenly, while others like the Bicoloured and Sharp-shinned Hawks *(Accipiter bicolor* and *A. striatus)* pursue smaller birds manouvering agilely and at speed through the foliage, steering with their longish tails, powered by short rounded wings. Species with sharp pointed wings which kill birds or bats over open ground and at great speed, are not missing. Such are the Bat Falcon *(Falco rufigullaris)*, the Aplomado Falcon *(Falco femoralis)* and others. Some are specialists at fishing like the Great Black Hawk *(Buteogallus urubitinga)* or the Black-collared Hawk *(Busarellus nigricollis)*, while the Snail Kite *(Rostrhamus sociabilis)* feeds exclusively on the Apple Snail *(Pomacea* spp); in order to get the body out of the shell the kite is equipped with a specially long, slender hooked beak.

Finally there are the carrion-feeders: the Chimango Caracara *(Milvago chimango)* and the Crested Caracara *(Polyborus plancus)*, aberrant falcons which have almost completely lost their ability to hunt and kill and, especially in the former's case, have become more like scavenging crows. New World vultures include the Black Vulture *(Coragyps atratus)* and the Turkey and Lesser Yellow-headed Vultures *(Cathartes aura* and *C. burrovianus)*; they quarter the area in search of any carrion where they congregate in some numbers and squabble to establish a pecking order. Their feet are weak and almost chicken-like as they are not intended to catch or hold prey. Their beaks are strong for tearing and ripping off bits of flesh, their heads devoid of feathers so the

The Six-banded Armadillo

filth, with which they get covered while poking their heads into abdominal cavities, does not foul feathers and washes off easily.

At dusk, some nocturnal birds of prey like the Short-eared Owl *(Asio flammeus)* and the Rufous-legged Owl *(Strix rufipes)* search out small rodents, and the savannahs are patrolled by nighthawks and nightjars - the Scissor-tailed Nightjar *(Hydropsalis brasilianum)* and the Nacunda Nighthawk *(Podager nacunda)* - in search of their flying insect prey.

Chaco National Park and Provincial Reserves. The difficult terrain, adverse climate and the bellicose tribes of indigenous people kept European man virtually out of the Chaco till the end of the XIXth century. His arrival did however substantially alter the landscape in a short time-span. Tannin extracted from the wood led to the destruction of huge areas of woodland, in many places not a tree was left standing. Overgrazing by cattle prevented the native trees from returning, and this led to an invasion by woody shrubs.

At the height of this irrational advance, and with a view to saving some portion of the native woods, in 1934 the Chaco National Park was first created as a National Forest Reserve some 130 kms northwest of the capital city of Resistencia. In spite of its relatively small size - 15,000 has - in it are represented all the habitats which are characteristic of the Eastern Chaco. The centre, east and south of the park have stands of the natural woods in which some magnificent specimens of Red Quebracho are to be found, in places growing in almost pure communities. Areas which had been lumbered before the creation of the park are today recovering fast and the various stages in this process can be distinguished. The shrub stratum is in places very dense and dominated by vines and bromeliads.

The western part is a typical savannah ecosystem with groves of the White or Caranday Palm, while in the southeast the Panza de Cabra lake which diminishes considerably in the seasonal months of drought, possesses the typically intricate web of aquatic vegetation.

Most of the species of birds corresponding to the district are found in the park, many of them are abundant. Mammals however are scarce and only those not persecuted by man are found in important numbers, such as the Black Howler Monkey *(Alouatta caraya)*. The park was full of squatters when created and they have constituted a serious problem in its management. They run domestic livestock and plough small acreages. Poaching by these settlers was rife as too the illegal cutting of trees. Until recently there was only one warden there, but things are getting better as most of the intruders have been relocated, staff has been increased and there is a local appreciation of the park growing in the neighbouring villages, as well as much more use of it made by visitors and school outings.

There are also two small provincial reserves with similar characteristics as the national park. In the same province the Pampa del Indio Provincial Reserve of some 7,500 has lies between the Nogueira and Asustado streams, while in Santa Fe, in the department of Vera, the headwaters of the Golondrina river are included in the 2,000 has La Loca Reserve.

Pilcomayo National Park. Bordering on the neighbouring republic of Paraguay in the northeast of the province of Formosa, the Pilcomayo National Park backs onto the river of the same name. Because of its size and characteristics it is the most interesting of the Chaco parks though it still has management problems.

It covers an area of some 47,000 has where marshes and groves of Caranday or White Palm dominate. Higher ground has woodland and on the shores of the Pilcomayo river a thick riverine gallery forest grows.

There is a rich and varied fauna: starting with the marsupials, there are the White-eared Opossum *(Didelphis albiventris)*, the Red Water-opossum *(Lutreolina crassicaudata)*, the Four-eyed Opossum *(Philander opossum)* and Murine Opossums *(Thylamys* spp); many bats like the two species of Fishing Bat *(Noctilio labialis* and *N. leporinus)*, the Velvety Free-tailed Bat *(Molossus ater)*, or the Palm Bat *(Lasiurus ega)*; the Tamandua or Collared Anteater *(Tamandua tetradactyla)*; the Night Monkey *(Aotus azarae)*, the Coati, the Crab-eating Raccoon and even a rainforest rabbit.

There are a number of birds which deserve a place here as they have not been mentioned before: the Bare-faced Curassow *(Crax fasciolata)* a huge turkey-like bird of the trees which has a crest of peculiarly forward curling crown feathers;

the Red-winged Tinamou *(Rhynchotus rufescens)*; cuckoos like the Guira Cuckoo *(Guira guira)* and the Smooth-billed Ani *(Crotophaga ani)*, both numerous and belonging to grasslands with some bushes and trees; two toucans, the Toco and the Red-breasted *(Ramphastos toco* and *R. dicolorus)*; the ubiquitous Savannah Hawk *(Heterospizias meridionalis)*, elegant, majestic, bright brown. Water-birds include the Southern Screamer *(Chauna torquata)*, the Buff-necked Ibis *(Theristicus caudatus)*, the Plumbeous Ibis *(Harpiprion caerulescens)*, the Limpkin *(Aramus guarauna)*, the Rufescent Tiger-heron *(Tigrisoma lineatum)* and some shorebirds from North America on migration, the Spotted and the Solitary Sandpipers *(Tringa macularia* and *T. solitaria)*.

Both species of Cayman can be found, especially in the most important body of water in the park, Laguna Blanca. Very unfortunately recent discussions with the provincial authorities have led to a large portion of this lake being left outside the park, though recently it has been declared a Ramsar wetland for better protection. The squatters within the park are being relocated and are now limited to a small portion for their cattle for just a few more years before the dead-line, and nobody actually lives now within the boundaries who is not a ranger or Parks' employee. Fires set by the graziers over the years have caused untold damage to the wildlife like the few remaining Pampas Deer, and Marsh Deer seem to have been exterminated. This park is now destined to be one of the most interesting in the country, and should soon have the necessary trails and roads to allow better access for the visitor.

Iberá - a challenge taken up by Corrientes. The north of the province of Corrientes is almost flat and here a vast marsh and lake with many ramifications exists, splitting the province in two from northeast to southwest. The waters are all from the local subtropical rains. Known as the Iberá which in Guaraní means "brilliant water", the whole system covers an area of over 20,000 square kms. Only small rivers like the Miriñay and the Corrientes drain the huge area of standing water, deep and clear, the lakes Galarza, de Luna, Fernández, Iberá, Tigre, subsidiary marshes, etc. The bottoms of these bodies of water are usually sand and though there is emergent vegetation, in places the depth is upto four or five metres where the waters are free from any vegetation and open, bordered with floating islands of Water Hyacinth and Water Lillies. There are sections of sandy higher ground which are the remains of sand-banks of bygone eras which are known locally as "rincones" (corners). Profuse grasslands are interrupted here and there by groves of Yatay Palms *(Butia yatay)* and islands of woods, where many trees are common to the Paranaense Rainforests. Towards the south there are species which are peculiar to the Espinal vegetation province in the savannahs, like the Ñandubay *(Prosopis affinis)* and the Espinillo *(Acacia caven)*. There are in Iberá several species whose survival depends on immediate protection like the Maned Wolf and the Marsh Deer, both of which are found here in their greatest numbers in the country, the Paraná Otter and both species of Cayman. In spite of the low density of humans in the area, the damage has been great as these all lived on hunting for meat and to sell hides and skins.

Iberá may yet hold secrets and there are legends galore of sunken cities and the likes. The province of Corrientes is setting it aside under a new law of protected areas as a reserve. The locals who did so much damage in bygone years are now the avid guides and custodians of the reserve; neighbouring landowners mostly are in favour as there is a possibility of offering the visitor the required services of lodging, guiding. Since 1992 and 1993 there has been ecouraging progress in this direction and it is to be hoped that it will continue.

THE VAST RIVER PLATE WATERSHED

With an area of 3.1 million square kilometres the watershed of the River Plate is the second largest in South America after the Amazon and sixth in the world. From the Bolivian Altiplano to the Serra do Parecis and the southern Mato Grosso come the rivers which feed the voluminous Paraná river which after a delta of countless islands and innumerable channels joins the Uruguay river to form the River Plate. This last is but 300 kms long and 220 Kms across the mouth, so wide in fact the Juan Díaz de Solís who disovered it in 1516 called it the Mar Dulce (Sweet Sea).

All the north and east of the country, which includes the most heavily populated areas, drains into this watershed.

The various rivers, flowing as they do over such vast plains, are constantly modifying their courses, cutting a new course here, abandoning an old course there, breaking their banks, creating sandbanks, bringing down huge rafts of vegetation and an incalculably large load of sediments. Coming from subtropical climes they also constitute important lines of penetration for the plants and animals from those areas.

A tree-trunk or some other such raft of vegetation can ground and become the nucleus for a new island. Sand settles around it and on the bank pioneer vegetation gets established from seeds brought down by the river or by the wind.

One of the first to grow is Palo Bobo *(Tessaria integrifolia)*, a composit whose seeds germinate in the damp sand and in rapid growth of its root system and development, soon forms a copse of some 6 to 8 metres in height. The local Willow *(Salix humboldtiana)* will join it by wind-borne fluffy seed or some bough or twig washed down will take root. This willow is to be found on most of the rivers of South America. Slowly other species will be added and in time the island will achieve a forested appearance with the addition of canes, creepers and lianas, and epiphytes. Though they never reach the height of the subtropical rainforests, there grow specimens of considerable demeanour. Many of these species are especially constant in these riverine forests and some are almost exclusively found in them. Two Laurels, the White and the River *(Ocotea acutifolia* and *Nectandra falcifolia)* stand out as does also the Inga *(Inga uruguensis)*, the Seibo, our national flower which is in fact a Coral Tree *(Erythrina cristagalli)*, Mataojo *(Pouteria salicifolia)*, the Curupí or Lecherón *(Sapium haematospermum)*, the Sangre de Drago *(Croton urucurana)* and the "Orchid" Tree *(Bahuinia candicans)*. There are also some species from climax forests like the Black Ear Tree *(Enterolobium contortisiliquum)*, the Ybirá-Pytá *(Peltophorum dubium)* or the elegant Pindó Palm *(Syagrus romanzoffianum)*, towering above the rest of the canopy, which once gave the name Paraná de las Palmas to that particular arm of the river in the Delta, but where palms are hardly found today. Easy access and water-transport have speeded the devastation of these forests.

The dense ribbon of vegetation grows only on the edges of the rivers and islands where the sediments brought down by the rivers have formed levées. Though the variety of species diminishes towards the south these forests reach the shores

The Red-legged Seriema

A family of endemic birds peculiar to the Chaco - the Seriemas - has but two species. They feed on small vertebrates and insects but their preferred diet is snake. On their long legs they wander in search of these. Their loud caccophonous calls can be heard for miles. This species (Cariama cristata) *is nearly a metre tall.*

The Ringed Kingfisher

Rivers, streams, marshes and the edges of lakes are the fishing-grounds for three species of Argentine kingfishers. They are very different in size, being large, medium or small, so their prey are divided up according to size also. Perched over the water they await some passing minnow and plunge into the water to capture it.

The Marsh Deer   *In the grassland savannah with termite mounds as a backdrop stands a young Marsh Deer* (Blastocerus dichotomus). *This is the largest South American deer and is today very rare because of poaching and shooting for its venison and trophy heads. They usually graze at night and spend the daylight hours in the islands of woods.*

The Black Cayman

Though the situation of this cayman (Caiman yacare) *is not as serious as that of the other species, the Broad-snouted* (Caiman latirostris), *poaching for skins has led to this once abundant species being hard to find over nearly all its former range.*

of the River Plate and are well-represented on Martín García Island and at Punta Lara near the city of La Plata. On the rivers which descend from the western mountains and cross the Chaco, in the central and drier reaches the vegetation is reduced to stands of Tusca *(Acacia macracantha)* and Chilca *(Baccharis salicifolia)* with other bushes, and becomes richer in the vicinity of the Yungas with species from that biome like the Red Cebil *(Piptadenia macrocarpa),* Horco Cebil *(P. excelsa),* Tipa *(Tipuana tipu),* Jacaranda *(Jacaranda mimosifolia)* and others.

The interior of the islands is lower and floods with high-water. Here there are large areas covered with the Roofing Grass *(Panicum prionitis),* with lakes in the sunken centres and criss-crossed by abandonned river courses called "madrejones" which are covered with floating vegetation. Here Water Hyacinths *(Eichhornia* spp.) predominate with burr-grasses like *Panicum elephantipes* and *Paspalum repens* and the cane-grass *Panicum fasciculatum*. During the big floods the power of the river flows into and opens up these old and cut-off river courses, flushing out the mats of vegetation which cover them. Many of these floating islands reach the River Plate bringing with them a multitude of fauna: insects, batracians, reptiles and even smaller mammals like rodents, occasionally something larger like a Cayman or Capybara.

The isolated and periodically flushed old river courses play a tremendously important role in the life of the river. The islands provide the resources for nearly all the life of the river. In but two or three hectares of aquatic habitat some 50 different species of fish can be found. Fry and fingerlings of many of the great fishes of the river live in these places where they feed on the tiny fresh-water shrimps *(Macrobrachium borelli, Palaemonetes argentinus),* crabs *(Trichodactylus* spp. and *Aeglus* spp.), isopods, amphipods, insects, etc.

In winter, when the waters of the Uruguay, the Paraná and the River Plate grow cooler some species migrate upriver to spawning grounds up the Paraguay, the Salado or the Bermejo to the subtropical high-ground. As this migration takes place during the season of low-water, there are places of barely 20 cms depth and the backs of the Surubí Catfish are out of the water, with smaller fish alongside, using the larger fish to breast the current. Sometimes early rains bring down so much muddy sediment that these fish are asphyxiated.

The abundance of sediment has led certain species to base their diets on the detritus and diatoms found there. Bottom-feeders are best represented by the Sábalo *(Prochilodus platensis),* the most important biomass of the ecosystem, and several species of the armoured Loricarids *(Plecostomus* and *Loricaria* spp.). Many species of fish have an omnivorous diet. These include the two catfish Yellow Bagre *(Pimelodus clarias)* and the White or Moncholo *(P. albicans)* which take molluscs, crustaceans and oligochaets but also plant material, mud and sand.

The Armoured Catfish *(Oxydores kneri)* and Bogas *(Leporinus obtusidens)* feed on snails and fish but often graze or browse on the land vegetation like the climbing plants which trail in the water, and the leaves and stems of water-plants.

The most herbivorous fish is the Pacú *(Colossoma mitrei)* feeding on algae, fruits, flowers, leaves, though it completes its diet with molluscs and small fishes.

The main predatory fish of the open water is the Dorado or Pirayú *(Salminus maxillosus)* upto a metre long and 20 kgs in weight which hunts in rapids, or at the outflows and confluences of lakes and rivers where it takes fish as they move upstream or down. But the largest species live on the bottom, in deep pools from which they emerge to take their prey. The giants of the river are the Manguruyú *(Paulicea lutkeni)* and the Spotted Surubí *(Pseudoplatystoma coruscans)*. There are reports of this last species reaching two metres in length and weighing upto 100 kgs. Next in size is the Striped Surubí *(P. fasciatum)* and the Patí *(Luciopimelodus pati)*. All these siluriform fish are generally known as the "skin" fish as they have no scales.

Also on the bottom but in shallower waters are the large sting-rays known as the Boba *(Potamotrygon brachyurus)* and the Overa *(P. motoro)*. Shaped like a great disk of upto 1.5 metres diameter in the case of the Boba, and armed with a venemous spine on the top of their shortish tails, they defend themselves when stepped on by inflicting a severe and rapidly-festering wound. The Pirañas or Palometas *(Serrasalmus* spp.) with short, wide maxillars and triangular sharp-edged teeth, can remove a fair-sized piece of flesh with one bite. *Serrasalmus nattereri* is the species mainly responsible for taking such a bite out of bathers. It is fortunately not gregarious.

Amongst the birds that are found in the gallery forests there are some species from the Subtropical Rainforests as well as others from the Chaco, the latter mainly of marshes and bushy habitat, though at least one species, the Black Skimmer *(Rynchops nigra)* is of the open water and sand-banks. With its lower mandible projecting beyond the upper is slices the surface of the water, flying low and near the shores, to snap up any fish it should feel.

Mammals are represented as much by the strictly riverine species like the Paraná Otter *(Lutra longicaudis),* and the Fishing Bat *(Noctilio leporinus)*, as by upland species like the Black Howler Monkey, the Coati, skunks and the Hairy Armadillo *(Chaetophractus villosus),* but mammals are scarce in the floodable habitats.

Because floods do not reach the tree-tops, and Howler Monkeys have little value as food or for their skins, they survive. The males' powerful voices can be heard at considerable distance at dawn and dusk to ensure the proper spacing of groups through the forests. Most mammals have been trapped for their meat or skins for a long time and it is when we read travelogues like that of the Jesuit Florián Paucke writing in the XVIIIth century of his sailing up the Paraná, that we get some notion of the destruction man has wreaked on nature in recent times:

"The islands of this river are covered with willows and poplars which also reach heights of ten and six metres (varas) but is impossible to describe adequately just how pleasant the islands look with their tall trees. As soon as one island is passed there immediately appears another and another with most agreable aspect. There is also much wild game to be seen on the islands: suddenly one comes upon sea-wolves (Giant Otters), then quantities of water-hogs (Capybaras)"..."soon we saw a number of deer (Marsh Deer) crossing from one island to another. Every day we saw tigers (Jaguars) running on the islands and hunting; sometimes they were at the shore catching fish. Time cannot possibly drag by for anyone on such a navigation."

Island Parks and Reserves. The greatest concentrations of people along the banks of the rivers have for many years put pressure on the island habitats, destroying at an ever more accelerated pace the native flora and faunas. To this must be added the Pharaonic projects for damming the rivers, like the Paraná Medio which will flood 1.2 million hectares; but not

84

only will this rich habitat disappear; with the lateral closure the periodic flooding on which so much depends will also cease. So it is important to protect island habitat below the dam or at the top end of the lake which will form, especially as the natural corridors for wildlife along the gallery forests will be interrupted.

The province of Santa Fe has some island nature reserves which, though small, have all the characteristics necessary. One of these is the complex of islands with a total surface area of 3,600 has which could be increased to 7,000, known as El Rico, opposite San Lorenzo, north of the city of Rosario. Here there have been recent reports of Marsh Deer with fawns.

Even if it only covers 300 has, the Cayastá Nature Reserve beside and around the remains of the XVIth century settlement of old Santa Fe is a perfect example of the gallery forests.

A third island reserve is the Ibirá-Pitá extending over 3,000 has opposite the city of Reconquista which, because of its latitude has a greater representation of subtropical elements.

However, the size of none of these reserves is enough to adequately ensure the survival of this interesting ecosystem. So the feasibility of declaring the fiscal lands in the delta part of Buenos Aires province, a short distance from the capital city must be considered urgently in the face of the rapid deterioration and threats suffered by this biome.

A large section of the Delta which receives the praise and admiration of those visitors who reach it, and which has not been planted with the exotic pulp species like the rest of the delta, is the last remaining area where such important species as the Dusky-legged Guan *(Penelope obscura)*, the Capybara and the Marsh Deer can be found. There is a long- forgotten project somewhere, but meanwhile Otamendi and Diamante Reserves have recently slightly offset this failing (see last chapter).

Large wading-birds

The large numbers of birds of the Ciconiiform order are what immediately draw one's attention in marshes and lakes, one species fishing beside another but using different techniques and catching different prey. Herons are abundant and there are several species of these. The Wood-stork (Mycteria americana - *lower, left and right) are usually in large flocks and are seen either fishing in a coordinated communal way, resting, or nesting in colonies in trees. The Jabiru* (Jabiru mycteria - *lower, centre) is one of the largest storks in the world and usually found in pairs or alone. The Roseate Spoonbill* (Ajaia ajaja - *lower left) are also often gregarious and sometimes associate with storks to nest. They are of the family* Threskiornithidae *which has other representatives in the Chaco, like the Bare-faced Ibis* (Phimosus infuscatus - *top).*

The Espinal Woodlands

A wide belt of arboreal vegetation which surrounds the vast Pampas Grasslands, is known botanically as the Espinal. In places it is interspersed with patches of open savannah.

Flat or rolling plains and low hills are the physical characteristics of this region which occupies the greater part of Mesopotamia, central Santa Fe and Córdoba, southern San Luis, western La Pampa, southern Buenos Aires and a strip of the northeast of this same province.

Mesopotamia, rising to broad uplands of between 100 and 200 metres elevation locally known as "cuchillas", drained by numerous streams and rivers which flow into the Uruguay and Paraná, is the damper part of the Espinal. Rainfall which in wet years can surpass the 1,000 mm mark, is mainly in summer. Towards the west and south climatic conditions become drier and in certain areas of La Pampa province they do not exceed 400 mm per annum. Except for the Colorado river which crosses it in the south, the Espinal lacks any permanent water-courses, and many of the drainage systems are endorrheic, forming large salt-flats. Large areas of fixed dunes, in some places brought back to life and actively mobile because of overgrazing, are almost uninterrupted from Bahía Blanca to Córdoba while another chain stretches from San Luis via La Pampa to the south of Buenos Aires province. Botanically as well as zoologically the Espinal can be considered an impoverished Chaco with southern input from the Monte and Patagonia.

The vegetation is characterized by a predominance of *Leguminosae* of the genera *Prosopis* and *Acacia* and differs from all the Chaco by the absence of the Red Quebrachos (*Schinopsis* spp.). There are less cacti and small-leafed deciduous vegetation. It can be considered an open woodland with but one or two arboreal strata not higher than ten metres, and bush and herbaceous strata below.

The limits and internal divisions of this botanical province are not easily defined as action by man has noticeaby altered the area by destroying woods or permitting their encroachment with changing exploitations. It has been divided into three districts: Ñandubay, Algarrobo and Caldén. The first of these covers the provinces of Entre Rios, part of Corrientes and the centre of Santa Fe, and coincides with what some authorities call the Mesopotamian Parkland. The most representative species is indeed the Ñandubay (*Prosopis affinis*), a legume of some ten to twelve metres height which possesses a wide "umbrella" crown. Its small pompom yellow flowers grow in clusters; the pods are long and curly and much sought by wildlife, rather like the sweet ones of the Black Algarrobo (*Prosopis nigra*) which even man uses as a food and to make an alcoholic beverage. In places these are accompanied by the White Quebracho (*Aspidosperma quebracho-blanco*). At a lower level grows the Molle or Incienso (*Schinus longifolia*), the Tala (*Celtis spinosa*) on damper soils, and the Espinillo (*Acacia caven*) which is so attractive when in flower, its branches and twigs covered with sweet-scented yellow buttons.

Some areas like one called the "Selva de Montiel" in the middle of Entre Rios, have much denser woods and include the small spiny Caranday Palm (*Trithrinax campestris*) whose long-fingered pointed-tipped fronds die and hang down covering the trunk.

However in most other parts the trees are spaced and the dominant vegetation is grass or herbaceous. On sandy soils and in almost pure stands grow the Yatay Palms (*Butia yatay*) upto 18 metres tall with lovely arched pale-green fronds.

West of this Ñandubay District, in western Santa Fe, Córdoba and San Luis provinces comes the Algarrobo District which is the one which has been most destroyed by intensive agricultural and livestock practices. The dominant species here were the White and Black Algarrobos (*Prosopis alba* and *P. nigra*) which grew in communities with many other Chaco species. These woods reached the province of Buenos Aires in the extreme south of the distribution of Algarrobos, and there are extremely few savable vestiges, such as the woods in Baradero and Zárate, or the few isolated specimens still to be found on the old river-bank at San Isidro. Here too the Ombú (*Phytolacca dioica*) grows, erroneously said to be a tree typical of the Pampas since it is characteristic of the northeast. The "tree" is of the same genus as the Pokeweed of North America. The trunk in old specimens stands on a base of immense surface roots and gives off several secondary trunks which form a beautiful rounded crown of dense foliage. From here the strip of woods splits into two, the eastern arm following the coast of the River Plate and the Atlantic as far as Mar del Plata, and inland another grows in patches on the ancient quaternary shell-beaches, also as far south as Mar del Plata. In these easternmost woods of the Espinal the Tala (*Celtis spinosa*) is the dominant tree. It is of the Ulmaceae family and exists as far north as the Chaco but here in Buenos Aires, where there are no other native woods, they are particularly relevant. The trunk is twisted, branches and twigs zig-zag and are thorny, leaves are small and have serrate edges; they fall in winter. The fruit, which is yellowish-orange, ripens in summer and is avidly eaten by birds, which later, perched on the wire fences, void the seed in their droppings so that the tala now grows in rows. There are a few other trees grow with the Tala: the Molle, the Coronillo (*Scutia buxifolia*) and the Sombra de Toro (*Jodina rhombifolia*) which both keep their leaves in winter and so stand out. But the best time for these woods is in spring when most plants are in flower and the Tala puts out its fresh green foliage. In these woods and at this season abundant and showy red flowers are to be seen on the vine *Tropaeolum pentaphyllum*, white on the Elders (*Sambucus australis*) and the highly perfumed Espino Negro (*Colletia spinosissima*), yellow on *Cestrum parqui* and two delicate ground orchids (*Chloraea membranacea* and *Cyclopogon elatus*) which live in the half-shade of the woodland's floor - all these give the Tala woods a very lovely appearance.

The southern part of the Espinal which occupies most of La Pampa, the south of San Luis and the south of Buenos Aires, forms a well-defined district known as that of the Caldén where this species (*Prosopis caldenia*) is the dominant tree and often forms wonderful open woods which some botanists call the Pampa's Parkland; with its great crown lending it a certain similarity to the Algarrobos (*P. nigra* and *P. flexousa*) sometimes growing with it, and the Chañar (*Geoffroea decorticans*), the Sombra de Toro, the Incienso (*Schinus fasciculatus*) and others. The ground is covered with a mantle of tough grasses while in the bush stratum Creosote Bush (*Larrea divaricata*), the Piquillín (*Condalia microphylla*), the Alpataco (*Prosopis alpataco*) and others flourish which, when the woods are removed become invaders.

The Caldén woods which are sparsely populated and sometimes hard to cross, are a good refuge for some species which have already vanished from other districts of the Espinal. Such is the case of the Puma (*Felis concolor*), the most adaptable of the American predators, as much at home in tropical forests as on rocky mountain-tops and everywhere in

The Plumbeous Rail

Secretive among the aquatic vegetation, rails like this one (Rallus sanguinolentus) *only emerge very timidly which makes them hard to see. Their strident voices, on the other hand, are very obvious and easily heard.*

The Wailing Pichi

Chaetophractus vellerosus *is a small armadillo of the hairy group of a very omnivorous diet which includes plants as well as carrion, invertebrates and even small vertebrates. Its water requirement is minimal so it can survive well even in the drier Monte biome.*

Burrowing Owls

This small owl (Athene cunicularia) *of diurnal habits and open ground uses the abandoned burrows of digging mammals to live and nest. They feed on larger insects and small vertebrates.*

between from Canada to the extreme south of Patagonia. A solitary wanderer, the Puma is mainly nocturnal and hunts by waiting and dropping on its prey. When its hunger is satisfied it covers the carcass with earth and vegetation to return to it later.

Rheas, Guanacos and Viscachas as well as other lesser animals used to be the diet of the Puma in these woods, but things changed with the introduction of exotics which are now the main part of its diet.

As well as the Puma there are other lesser cats in the region like Geoffroy's Cat *(Felis geoffroyi)*, the Jaguarundi *(F. yaguaroundi)* with its long, slender body and colouration black or reddish-brown, some specimens being mid-gray, the Pampas Cat *(F. colocolo)* speckled salt and pepper with wide black bands around the tops of the fore-legs, a grassland cat unlikely to be found in trees. These are nocturnal and feed on smaller mammals and birds.

The Pampas Gray Fox *(Dusicyon gymnocercus)* is more of an oportunist and takes what circumstances offer, not turning its nose up at carrion, insects and fruit, though it does take its own prey also.

There are many and abundant rodents like the Leaf-eared Mouse *(Graomys griseoflavus)* which lives in hollow logs or in the nests abandoned by birds and feeds on the seeds of the Caldén and Algarrobo, and *Reithrodon auritus* living underground, sometimes in abandoned ants' nests or burrows of the Tucotuco *(Ctenomys mendocinus)*. Tucotucos are a family of many species which live in arid lands from Peru to Tierra del Fuego, preferring loose sandy well-drained soils where they dig extensive tunnels not far below the surface, with ramifications and nest chambers. They are diurnal but in spite of this they are seldom seen, being so well adapted to their underground existence taking roots from below the surface and rarely emerging. Their thumping sonorous calls are very audible - the name is onomatopaeic. They are territorial so it is presumed that this call helps them keep their distance.

Quite the contrary, the Plains Viscacha *(Lagostomus maximus)* is gregarious and sociable, living in many-mouthed communal burrows. The tunnels are increased by each succeeding generation but some specimens evidently start new colonies. They are nocturnal and come out to graze at sunset feeding around their burrows. Shortening the grass improves their field of view to watch for predators. When danger appears grunts and penetrating whistled "key-ah's" send them all scuttling for the safety of the communal den. Many other species like lizards, amphibians, opossums, skunks and even snakes and tortoises use the burrows as do certain birds - swallows and ground-living furnariids, even, on cold winter nights, the Elegant-crested Tinamou *(Eudromia elegans)*.

Together with the Viscacha, the Guanaco *(Lama guanicoe)* and the Mara or Patagonian Cavy *(Dolichotis patagonum)* are the chief native herbivores but the present balance of mammals in the Caldén woods has been very upset. In 1906 the first specimens of Red Deer *(Cervus elaphus)* and Wild Boar *(Sus scrofa)* were introduced to a private estancia in the province of La Pampa. These European species adapted wonderfully and years later escaped and invaded the whole region, together with the European Hare *(Lepus c. europaeus)* which is now in almost every corner of the country since its release at the end of the XIXth century.

The birds of the Caldén woods are basically that of the Western Chaco region, with some species added from the Monte and some influence from Patagonia.

Though never very evident, Furnariids are the most numerous family and are found everywhere. The Stripe-crowned Spinetail *(Cranioleuca pyrrhophia),* small and never still like the Plain-mantled Tit-Spinetail *(Leptasthenura aegithaloides)* which has a long tail ending in two points, search the branches and twigs for insects and spiders. On the ground the Common Miner *(Geositta cunicularia)* keeps to the open spaces. Very remarkable are the Brown and White-throated Cacholotes *(Pseudoseisura lophotes* and *P. gutturalis),* robust birds with very loud duet vocalizations.

The call of the White-tipped Plantcutter *(Phytotoma rutila)* sounds like a creaking door. It is one of three species in this exclusively southern South American family of birds.

Pigeons, parrots and parakeets, wood-creepers, tyrants and seed-eating Fringillids are others of the many families represented in these woods. To the north and east of the Espinal the number of species increases, augmented by the aquatic species already dealt with in the chapter on the Western Chaco.

El Palmar National Park. "A distant bluish vegetation which was new to me presented itself before us, immense groves of Yatay Palms. As we approached we made out first the separate specimens, then the small crowns on the slender trunks; at last we reached the first of them... I was delighted with this new genus of plants. Everywhere were palms with spherical bluish-green crowns composed of long fronds curving like the jets of a fountain; the marks left by leaves already fallen made patterns of zig-zag lines on the trunks. As we advanced the groves got thicker and no other species of tree was mixed in with the palms which I did not weary of observing. Any lovely object to which one is not accustomed produces a sensation which is difficult to express but which is none-the-less very real: soon admiration is added, and a deeper respect for all nature is involuntarily felt." This description from the pen of the French naturalist Alcides D'Orbigny during the XIXth century, of his first contact with a grove of Yatay Palms in Corrientes, is equally valid for the Great Palmar then already known on the banks of the Uruguay river in central Entre Rios province. The groves of Yatay, dense and extensive, here form the best representation of the species. They grow on rolling sandy soils, interrupted every now and then by surfacing sandstone or traversed by sloughs with tall aquatic vegetation, draining towards the streams which are hidden in the riverine gallery forests. Created in 1966 to protect these palm from the grazing of cattle which eat the sprouts of new palms, the recovery of the flora and fauna of this park is very evident.

On the way into the park one can see Red-winged Tinamous *(Rhynchotus rufescens)* crossing the road. Their mournful whistles can be heard from the tall grasses which are today their refuge.

On the trunks of the palms one can find the several woodpeckers found in the park: the Field Flicker *(Colaptes campestris),* the Golden-breasted Woodpecker *(Chrysoptilus melanolaimus)* with its red nape, or the striking White Woodpecker *(Melanerpes candidus)* contrasting black and white with just a touch of yellow. There are also the Scimitar-billed Woodcreeper *(Drymornis bridgesii),* large and with a very long curved bill, and the Brown Cacholote *(Pseudoseisura lophotes),* its crest and size making it easily recognizable. The noisy Monk Parakeets *(Myiopsitta monachus)* build their huge communal stick nests in the crowns of the palms, among the fronds. A group of Greater Rheas *(Rhea americana)* race though the grass to avoid the vehicle.

The Mataojos

Pouteria salicifolia *of the Sapotaceae family - soap-plants - always grows along the stream-banks as here, on the Palmar Stream, and it is one of the species typical of the riverine gallery forests.*

Pit-vipers

Snakes of the rattlesnake family these (like this Bothrops alternata) *are widespread.*

The Caranday Palm

To protect groves of this small palm (Trithrinax campestris) *and all the life forms which live with it there are no Reserves as yet, though some projects are being considered.*

In the small creeks which cross the road Whistling Herons *(Syrigma sibilatrix)* are common, elegant birds of pastel colours, powder-blue and yellowish ochre, pink bill with a black tip. Here too are the Brazilian Duck, *(Amazonetta brasiliensis)* extraordinarily bright in flight when the brilliant green-sheened wings contrast sharply with a white patch.

Sometimes one can find a swimming Coypu *(Myocastor coypus)* or a Painted Turtle *(Tachymenes scripta d'orbigny)* sunning on an islet or a half-submerged log.

Paths or tracks lead also to the two major streams through the park, Palmar and Los Loros where Capybaras *(Hydrochaeris hydrochaeris)* dwell among the dense vegetation on their banks. The population of these rodents has increased enormously since protection started. Though only occasionally, pairs of the Paraná Otter *(Lutra longicaudis)* have been seen fishing and cavorting in the streams.

As one approaches the Park Headquarters high on the banks overlooking the Uruguay river, the scenery changes and the palm groves give way to the drier vegetation typical of the Espinal: Ñandubay, Coronillos, Molles, Talas and the scarce Algarrobo and White Quebracho. Here too, by the river are the comodities the park offers, such as the camping area where wildlife is easy to observe. The Viscacha colonies are flourishing, perhaps because of man's presence which surely keeps certain predators like the two foxes *(Credocyon thous* and *Dusicyon gymnocercus)* away. The Viscachas have become tame, beg for food and also tidy up the campsite - any gear left outside the tent at night is removed to the burrows and might be found outside the dens as decoration, amid sticks and the other treasures they accumulate.

During the day other species like the large Ground- lizard *(Tupinambis teguixin)* emerge from the colonies; even the Grison *(Galictis furax)* uses these burrows and surely finds in them certain prey.

Around Viscacha colonies the grass is kept extremely short - like a well-tended lawn, and many species of birds frequent the area. Two doves, the Eared and the White-tipped, *(Zenaida auriculata* and *Leptotila verreauxi)*, Rufous Horneros *(Furnarius rufus)*, or the Brown Cacholote *(Pseudoseisura lophotes)*, Red-crested Cardinals *(Paroaria coronata)*, Chalk-browed Mockingbirds *(Mimus saturninus)*, and two thrushes, the Rufous-bellied and the Creamy-bellied *(Turdus rufiventris* and *T. amaurochalinus)* all search for seed, small fruit like the orangey berries of the Tala or the red of the Chalchal *(Allophyllus edulis)*.

The banks of the great Uruguay river are covered with gallery forest in which *Myrtaceae* dominate *(Myrcia ramulosa, Blepharocalyx tweediei, Eugenia uniflora* and many others). Many willows *(Salix humboldtiana)* and Laurels *(Nectandra falcifolia)* are also found on the banks as well as the larger species like Crown of Thorns Tree *(Gleditsia amorphoides)*, Inga *(Inga uruguensis)*, Azota Caballo or Francisco Alvarez *(Luehea divaricata)* etc. Some of the birds here are more typical of the rainforests further north - the Squirrel Cuckoo *(Piaya cayana)* and the Plush-crested Jay *(Cyanocorax chrysops)*.

Being situated not too far from several great cities - Buenos Aires, Rosario, Paraná and Santa Fe - this park receives many visitors, at times exceeding the park's capacity to handle, so in the near future plans for coping with such contingencies must be drawn up and put into effect.

The Field Flicker

Colaptes campestroides *is a bird of the open grasslands and terrestrial in habits. It feeds mainly on ants and is now provided with plantation trees, fence-posts and telegraph-poles for its nests.*

There is a Visitors' Centre and Interpretive display which informs the visitor as to the different habitats and what to expect. One of the major problems faced by the park is the large population of feral Wild Boars *(Sus scrofa)* which uproot vegetation and predate nests and many species of animal.

The largest native mammal in the park today is the Capybara, but in historic times there were Pampas Deer *(Ozotoceros bezoarticus)* and Brown Brockets *(Mazama gouazoubira)* which ought to be reintroduced.

A Project in the Comechingones Area (San Luis). In the Espinal there is another species of palm, also found in a swathe across the north of Córdoba and San Luis provinces. An area of special interest is in the northeast of the latter where this Caranday Palm *(Trithrinax campestris)* grows in terribly dense stands on the flats, and more openly in other parts, mixed with Talas, Espinillos, Chañares, Sombra de Toro and others. The palm groves thin out as the ground rises, and cut off at 1,100 metres above sea level. Here too are the Hill Woodlands with the dominant Molle de Beber *(Lithraea molleoides)* and the Coco *(Fagara cocos)*. The streams are crystalline and their banks covered with Pampas Grass *(Cortadeira selloana)* and the native Willow. This woodland goes from 850 m to 1,300 m where the upland grasslands begin. There are no animal species any different from those already mentioned for the Espinal though some forms are peculiar to the area such as the Red Fox subspecies, large and very red, *(Dusicyon culpaeus smithersi)*. The common thrush is the Chiguanco Thrush *(Turdus chiguanco)* all dark with yellow-orange feet and bill, and the most spectacular of hummingbirds, the Red-tailed Comet *(Sappho sparganura)* with a long red (or orange, or yellow or green according to the incidence of the light) and black forked ladder tail.

One of the inconveniences with this area as a future National Park is the long history of settlers who habitually set fire to the palms to clear them for more grazing land. The palms may not die but their skirts of dead fronds which cover the trunks burn off; they are a valuable refuge for many species of wildlife. The blackened palms trunks are an awful sight, but the real problem is that with the removal of the vegetation cover erosion accelerates - it is severe enough simply from overgrazing. Near Papagayos erosion is particularly bad.

Some 20,000 has have been proposed for a National Park, but they will of necessity have to be considered part of a major restoration project. There are other much more pristine areas which can be considered to protect this important species of small palm and the special habitat it creates, valuable as much for wildlife as for recreational and educational activities.

The Plains Viscacha

The Viscacha (Lagostomus maximus) *is a large rodent which lives communally in colonial burrows with many entrances and ramifications. They emerge at night to feed on the surrounding grass.*

The Pampas Grasslands

The Pampas - an endless sea of grass, horizons set at infinity, uninterruptedly flat, and vast skies - images of the Pampas from bygone years, when yet to be settled by Europeans, where the wildlife roams at will. These scenes have been altered by the plough, fences which cut it all up into squares and rectangles, copses of introduced trees dotting the landscape, roads and railways ruled arrow- straight between villages and towns. The land, which was considered a desert till the last third of the XIXth century is today the most productive area that the country possesses and the mainstay of the agricultural and livestock potential. Some 60% of the population lives here with over 9 million in the capital city of Buenos Aires and its "greater" suburbs. Other nuclei are Mar del Plata, Rosario, Bahía Blanca. This development has led to all manner of alterations, to the extent that today it would be difficult to find, anywhere in the 630,000 square kilometres that the Pampas cover, any fraction which has not been affected in some way.

These immense grasslands occupy virtually all the province of Buenos Aires, southern parts of Entre Rios, Córdoba and Santa Fe, and eastern San Luis and La Pampa. In remarkably few places are they at all undulating, but at the southern end there are ranges of hills - Tandilia and Ventania with a highest peak in the latter in excess of 4,000 feet - separated by 140 kms of flat grassland and which run parallel in a southeast-northwest direction. They are outcrops of the crystalline base-rock which in places is upto 5,000 metres below the surface, covered by the thick deposits of loess, the sediment which constitutes the vast Chaco- pampean flats.

Rivers and streams are few, and typical of flat country, they are slow-flowing and meander. Those in the northern pampas flow into the Paraná and River Plate while those in the south do so into the Atlantic. There are moreover endorrheic drainage systems like that of the Vallimanca stream. The Salado river which has its headwaters in the Chañar lake in the northwest of the province of Buenos Aires and flows out into the Samborombón bay forms and drains a huge depression of the same name of some 80,000 sq kms which is subject to periodic floods of catastrophic consequences for agriculture and livestock production.

The most characteristic and the richest bodies of water in the region are the lakes, marshes and lagoons, as well as the occasional sea-water coastal lagoon. Lakes and lagoons, either temporary or permanent occupy shallow depressions, they have well-defined shores and their own sediments which are different to those of the surrounding area. The most common are those that owe their origin to former ancient courses of rivers and so are found in series, like the links of a chain, as in Chascomús. Others are due to tectonic movements, erosive factors, or drainage interrupted by such as sand-dunes or raised shell-beaches. The water of these comes from rainfall, from streams or from subterranean sources, and if these last have been through underground salt deposits, the lakes are saline, as is the case at Carhué and Guaminí.

Marshes are temporary or semi-permanent and are fed by rainfall; they lack deffinite edges (these come and go with fluctuating water-levels) and occur where the subsoil is impermeable. Marshes come in three types: the spreading "pans" of level open ground, those enclosed between barely noticeable banks called "cañadas", and "cañadones" which are large versions of these last. Being shallow they all have much aquatic vegetation, both submerged and emergent which is the habitat for a rich and varied animal life.

The climate of the Pampas is warm-temperate with an oceanic influence, and though it rains evenly throughout the year there is slightly less precipitation in summer and in winter, there being a negative balance in summer because of the heat. In the northeast the yearly average is around 1,100 mm per annum which decreases westward and southward to 600 mm where the soils are more granular and there is less plant cover. The average daily maximum temperature is 23°C and mean minimum 9°C. Frosts are usual and frequent during winter months but snow is rare and only falls in the south.

Botanists have had to study the region basing their research on the scarce data of chroniclers of yore and on the small less-modified relics which in most cases are the edges of roads and railbeds, as there is mighty little of the pristine pampa left. This reconstruction of the original vegetation is what will be described without much refference to man's subsequent actions.

The phytogeographic province of the Pampas is part of the great Chaco domain whence come the principal contributions to its flora and fauna, and in it can be identified four districts: the uruguayan, the eastern, the western and the southern.

Though there are some characteristics particular to each, all four districts are dominated by grassy steppes, wide seas of grasses waving to the rhythm of the winds, a vegetation perfectly adapted through millennia of being subjected to fires and frosts, droughts and floods, to the animals which made it their home, living, breeding and dying there, trampling and feeding on it, digging up the soil. It is termed "steppe" because of the two deffinite periods of vegetative repose - winter with its cold, and summer with its heat and droughts. In this true kingdom of the grasses there have been 190 native species of *Gramineae* identified, some indication of the diversity of possibilities in what would seem at first sight to be a homogenous plain. If one considers that the main crops as well as the forage grown here are all grasses - wheat, sorghum, maize, and so many others - it will be evident how important it is to protect this ecosystem with all its grass species which could contribute to man's future needs in unsuspected ways.

There have been several and varied theories put forward to explain the total lack of trees on the pampas plains, such as the action of fire, of droughts. Though several factors could influence this concurrently, it is surely important to take into account the action of the dense grass cover and the extensive root-mass which inhibit the germination and growth of tree seedlings. The settler knows that it is only by controlling the grass that he gets the trees he has planted to grow, which, once started, need no further such assistance.

The climax vegetation is the arrow-grass (*Stipa* spp.) steppe which prospers on higher ground and is made up of grass growing in clumps, dominated by the genera *Stipa*, *Piptochaetium* and *Aristida*, all with the awn-shafts attached to the pointed seed which give them their general name. To the north the pampas are warmer and damper with only one period of vegetative repose and that in the winter, and to the genera already mentioned must be added some subtropical forms. In winter grasses are short which permits growth of some annuals like *Glandularia peruviana* with lovely red flowers, several species of *Oxalis*, pink or yellow, orange, violet or white which at the beginning of spring flower to produce one of the most noteworthy spectacles of these arrow-grass prairies. Then the whole area turns green with the sprouting of the grasses which, when they seed, produce hues of violet, pink, reddish, whitish or silvery, according to the dominant species. With the heat of summer the grasses which have grown and seeded die off, bending under the weight of seed. With the autumn rains and the relief from the heat and drought, the vegetation turns green yet again but never with the vigor of spring. Finally the rigours of winter impose a further hiatus and the grasslands turn yellow again.

Mostly associated with the better-developed communities of the pampas plains there is a rich and varied fauna, though nothing like the spectacular mammals which roamed the area and had disappeared before European man's arrival. The only grazer of any importance was the Pampas Deer (*Ozotoceros bezoarticus*), today almost extinct but which even as late as the beginning of the XXth century was extremely common. Others merely touched the region tangentially - the Guanaco, the Mara and the Capybara.

The Greater Rhea (*Rhea americana*), a giant among the pampas' birds, is another herbivore of considerable apetite though it also takes some food of animal origin. Typical of the pampas, its long neck serves it as a periscope - some 5 feet above the ground it offers a vantage point for spotting any danger.

During the winter rheas congregate in flocks of upto 40 or more birds from which the adult males separate in early spring to form their harem which might number upto 15 females. These lay communally for the male's nest which can contain a clutch of upto 60 eggs. It is the male who is in charge of incubating and later rearing the chicks. As soon as the male sits, the females are off to lay for the second most dominant male and soon fill his nest, and so on. This species has been reduced alarmingly in numbers by hunting for its feathers and meat, by dividing up the pampas with fences and the alteration of the habitat, so that it is now absent or rare in many places where it was once abundant.

There were two major predators - the Jaguar and the Puma, but both have paid for their size and carnivorous habits by being erradicated from the plains. Only of the latter might there be a remnant population, in some secluded or inaccessible corner of the hills. Both, being opportunists would have taken an important proportion of their diet from the important biomass of large to middle-sized rodents which surely partly fulfilled the role of the missing ungulates in keeping the grass down. The Plains Viscacha (*Lagostomus maximus*) was evidently the most important and classic pampas rodent, now reduced to vestigial populations in distant corners or in national parks, exaggeratedly persecuted for the damage they did to agriculture. Like all the remaining animals of the pampas it lives in burrows, the only possibility for shelter on these shelterless prairies.

But there are other rodents which are still abundant and common such as the Tucotucos (*Ctenomys* spp.) found in sandy areas which are admirably adapted to their underground existence. Cavies there are also: the Pampas Cavy (*Cavia aperea*) is the larger; the other: *Galea musteloides*. The former, or something very like it gave rise to the domesticated and well-loved guinea-pig. A gamut of lesser genera (*Akodon, Calomys* and *Reithrodon*) make their runs or nests under or in the dense grasses. They have proliferated recently and certain species are links in the epidemiological chain of the virus of haemorrhagic fever, a terrible endemic disease with fatal results for man. With the disappearance of mammalian and avian predators, as much objects of an absurd persecution as by the effects of abuse of pesticides which ultimately affects predators most, the small rodents' population explosion has facilitated life for the mites which are the vectors of the disease.

Geoffroy's Cat (*Felis geoffroyi*) and the Pampas Cat (*F. colocolo*) are becoming increasingly rare as are the Pampas Gray Fox (*Dusicyon gymnocercus*) and the Grison (*Galictis*

The Long-winged Harrier

This raptor (Circus buffoni) *flies low over grasslands and marshes on permanent patrol in search of prey, which it comes upon suddenly. It is a typical sight of the Pampas. The bird in the photograph has caught a legless lizard* (Ophioides spp.).

A good bluff

An imitation of the poisonous pit-vipers is a valid defence for this harmless hog-nosed Lystrophis dorbignyi.

The Pampas Gray Fox

Since the disappearance of the Jaguar and the Puma from the region, the Gray Fox (Dusicyon gymnocercus) *is the largest predator left. Over-trapping for the fur trade has contributed to upsetting the ecological balance as this is one of the important controls of rodent populations.*

Coots

These are the most abundant waterbirds on lakes and marshes in the Pampas. The Red-gartered Coot (Fulica armillata), nearest the camera, has a red line between the yellow bill and shield. The White-winged Coot (F. leucoptera) has no such dividing line, and the Red-fronted (F. rufifrons) has a dark red shield.

cuja) with its elongated weasel-like body for pursuing prey along narrow burrows.

Other mammals of the Arrow-grass are the Hog-nosed Skunk *(Conepatus chinga)* and two species of armadillo - the Hairy *(Chaetophractus villosus)* which, as its name implies is covered with sparse though rough and long hair, and the Seven-banded *(Dasypus hybridus)* neater, taller, with long erect ears. When danger threatens they run much faster than their appearance would lead one to believe possible, and dive into some burrow where they disappear. Finally there are three opossums, the White-eared *(Didelphis albiventris)* everywhere, the grassland Short-tailed *(Monodelphis dimidiata)* and the Red Water-possum *(Lutreolina crassicaudata)* usually in aquatic situations.

Birds of these grasslands there are aplenty, starting with the tinamous. The Red-winged Tinamou *(Rynchotus rufescens)* is large and consequently found in the taller grasses. It was an important game species, but overhunting together with changing grassland management policies where grasses are shorter now, have led to its becoming rare. The Spotted Tinamou *(Nothura maculosa)* however has been favoured by this change. Both species have loud whistled calls, the first mournful, the second either a trill or a long-drawn-out series of peeps. Whistles are evidently the most effective quality of sound to communicate in the vast grasslands.

The smaller passerines of the grasslands have turned to another method of communicating their territorial claims. As the region originally lacked trees or indeed any elevated perch, aerial displays have become the most effective means. Rising on the wing, in some species almost out of sight, they call in different ways, and drop on stiff wings, announcing all the time their species and their claim. All pampas pipits, *(Anthus* spp.), the grassland Yellowfinch *(Sicalis luteola)* and the White-browed Blackbird *(Sturnella supercilliaris)* all do this, this last with the added warning of a brilliant scarlet breast. Other smaller denizens of the deep grass are the Bay-capped Wren-spinetail *(Spartonoica maluroides)* and the small and sociable Double-collared Seedeater *(Sporophila caerulescens)*.

The shorter grass which is the dominant habitat today because of livestock management techniques, is preferred by other species of birds, as it is here possible to see a predator from afar, and walk about with ease. Such are the Southern Lapwing *(Vanellus chilensis)* which is never missing from such habitat, also on the lawn-grasses near the water. Strident calls announce to all the world that there is courtship (often in threes), or some approaching danger such as a dog or a man. The Guira Cuckoo *(Guira guira)* seeks its insect and small vertebrate food in these shortish-grass areas. It is found in flocks and bears little resemblance to typical cuckoos.

The Rufous Hornero *(Furnarius rufus)* also needs short grass, and today it rarely builds its round, mud nest on the ground having millions of fence- and telegraph-posts, trees and buildings to put it on. It is believed that it is a recent arrival in the Pampas from the neighbouring Espinal.

Birds of prey are abundant. The Burrowing Owl *(Athene cunicularia)*, a diunal insectivorous hole-in-the-ground dweller once associated with viscachas, is fairly common away from the main roads - car lights attract insects which attract owls which in turn get dazzled and killed; the Short-eared Owl *(Asio flammeus)*; the ubiquitous Chimango Caracara *(Milvago chimango)* which is an oportunist, a scavenger and carrion-eater occupying the niche of crows; the Aplomado Falcon *(Falco femoralis)* which takes smaller birds.

Grasslands such as these cannot lack migratory species and there are those from the north in summer and others from

The Horned Frog

The striking colouration and the strange shape of this peculiar frog (Ceratophrys ornata), *together with a great show of puffing up and hissing, lunging with mouth agape at any danger, has led to the mistaken belief that this species is poisonous; though it does bite it has no teeth.*

The Common Toad

This Bufo arenarum *of the Pampas spawns in temporary puddles which are everywhere after big rains. Only the males call - croaking to attract the females to the spawning-grounds.*

the south in winter. From nesting-grounds in Patagonia three species of sheld-goose (*Chloephaga* spp.) arrive in autumn to graze on the short grass: the Tawny-throated Dotterel (*Oreopholus ruficollis*) and the Least Seedsnipe (*Thinocorus rumicivorus*) among others, also arrive from the south. From the north several *Tyrannidae* return to their Pampas breeding- grounds in spring, as is the case of the Fork-tailed Flycatcher (*Tyrannus savana*), the Vermillion Flycatcher (*Pyrocephalus rubinus*), and the Tropical Kingbird (*Tyrannus melancholicus*); Swainson's Hawk (*Buteo swainsoni*) comes south from North America in large flocks to spend its winter (our summer) feeding on the abundant grasshoppers, locusts having been exterminated in the forties and fifties.

There are several batracians adapted to spawning in the temporary ponds which abound after heavy rains, such as the two toads *Bufo aranarum* and *B. granulosus*. Others have special techniques to which they recur after a succession of light rains; these belong to a group of burrowing frogs (*Leptodactylus latinasus, L. gracilis* and *L. mystacinus*). After the first rain the males of these species dig small burrows in the mud and there call to the females to join them. Spawning takes place within the burrow and the ova remain enveloped in a gelatinous foam where the tadpoles start to develop, feeding on this substance. When the puddles fill, the tadpoles emerge and start their aquatic life, completing their metamorphosis in but 18 days. Another amphibian which is to be found in the Pampas Grasslands is the striking "Horned" frog *Ceratophrys ornata* so mistakenly held to be a poisonous species. There are a number of "grass" snakes, all inoffensive like *Liophis anomalus,* or *Lystrophis dorbignyi* which in a theatrical bluff tries to intimidate the intruder, while some are slightly venemous like the slug- eating *Tomodon oscellatus*.

The most noticeable group of insects is ants in a great variety of forms like fire-ants (*Solenopsis richtieri*) with a painful sting to paralyze their small victims, *Camponotus mus* with a pale abdomen or *Acromyrmex lundi* with a highly evolved social structure.

There are climax communities unto themselves within this biome where soil and water conditions are such that they are inappropriate for human exploitation so they have better survived as refuges for peculiar floras, and faunas which are generally similar to the surrounding areas already described.

The hills are ancient outcrops, veritable islands in the sea of grass which is so different, and the only such communities which possess endemisms within the Pampas region.

Their age and isolation precisely as island has favoured the evolution of new species. Curromamuel (*Colletia paradoxa*), Chilca (*Dodonaea viscosa*) and Brusquilla (*Discaria longispina*) all woody bushes, form thickets. Above 500 m elevation the windy steppes are composed of endemic grass species like *Festuca ventanicola, F. pampeana, Stipa pampeana* and *S. juncoides* as well as others. Even in bushes there are endemics: *Plantago bismarckii* and *Senecio ventanensis*.

A recently rediscovered lizard *Pristidactylus casuhatiensis* lives only here above the 1,000 m mark and on only one of these ranges. The attractive small black toad with yellow and red belly (*Melanophryniscus stelzneri*) whose distribution is curiously disjunct, being found in several other parts of the country, is also common here.

Another patch or island is formed by the Paja Colorada (*Paspalum quadrifarium*) in extensive areas in the Salado river depression. Stands grow upto 1.5 m tall and offer splendid habitat and refuges for certain species of wildlife.

The White-eared Opossum

This marsupial (Didelphis albiventris) *is highly adaptable to most situations as proven by its being found in close proximity to man in towns and suburbs where man inadvertently provides it with food - rubbish, chicken-runs and so on.*

Sand-dunes are present in the western reaches of the Pampas, and there is one range all along the Atlantic seaboard. Man today, in search of recreation and holidays by the sea, puts ever more pressure on the area with mushrooming towns, forestry with exotic species and his massive hordes in summer.

These corridors, as well as isolated dunes, whether fixed or shifting, support a special plant community which varies with the size and with proximity of the sea. Grasses with rhizomes are typical - *Poa lanuginosa, Panicum racemosum, P. urvilleanum,* and the similar *Spartina ciliata*. There are also composits with rhizomes (*Hyalis argentea*), creeping legumes (*Adesmia incana*) or succulent-leaved plants (*Senecio crassiflorus*). Lizards abound such as *Liolaemus wiegmanni, L. multimaculatus* and *Proctotretus pectinatus* which all perform the clever trick of vanishing before one's very eyes, rapidly wriggling under the sand.

On sandy soils where ground-water is close to the surface the queen of grasses grows - the Pampas Grass (*Cortadeira selloana*) - in tall clumps over which, when in flower, large feathery heads reach three metres in height.

This type of soil is especially common between dunes near the beaches where sand hills have obstructed the drainage creating lagoons and shallow sloughs.

Bodies of standing water like lakes and marshes are to be found dotting the pampas, some permanent, some temporary. These form the most spectacular ecosystems of the Pampas region and constitute an ornithological paradise. Here, the highest bioproductivity is found, starting with the primary producers - phytoplankton, aquatic plants - through the zooplankton to invertebrates and fishes which will in their turn support the amphibians, reptiles and larger fishes which in their own time feed birds and mammals, all in a complex and intricately balanced web of life in which the larger eats the smaller and will be devoured by some even greater predator or, eventually, by carrion-feeders. Saline bodies of water are fringed by wide salt-grass (*Distichlis* spp.) flats, White Duraznillo (*Solanum glaucophyllum*) prospers in the floodable bottomlands and some marsh edges where the twisted stems emerge from the water in well-spaced patterns. The dominant plant here however is surely the Tule (*Scirpus californicus*) a slender *Cyperaceae* reaching over two metres in height, growing in dense stands at the edges, and in the shallower waters of marshes.

In the vast tule beds many species of birds nest, using the very stems as building materials, as they do also in Cat-tail (*Typha* spp.) which has a more restricted habitat and distribution.

Floating on the open water there grow two minute ferns - *Azolla* spp. and *Salvinia* spp. - together with *Lemna* spp. and *Wolfia* spp. Here grows the curious *Utricularia platensis* which catches small aquatic zooplancton with its modified leaves, real underwater traps.

Anseriform birds - ducks, geese, swans, and also screamers - are very well represented on the Pampas. Foremost are the Black-necked Swan (*Cygnus melancoryphus*) and the Coscoroba, a swan-sized peculiarity (*Coscoroba coscoroba*), both of which reach their food - aquatic vegetation - from the surface, dipping their long necks under, or directly up-ending, pointing tails at the sky. Most of the ducks are dabbling ducks which feed in similar fashion: the Brown Pintail (*Anas georgica*) is possibly the most abundant; the Red Shoveller (*Anas platalea*) with its specialized broad bill and upper wing-coverts of a pale blue like the Cinnamon Teal (*Anas cyanoptera*) in which the male is bright chestnut-red, or the smart Silver Teal (*Anas versicolor*) with its dark cap. All these

Campos del Tuyú

nest on the ground as does the Speckled Teal (*Anas flavirostris*) though this last frequently takes over Monk Parakeet nests or any hollow in a tree when these are available. Diving ducks here include the Rosy-billed Pochard, (*Netta peposaca*) the male shiny black with silver flanks and a bright puce bill with a knobby caruncle.

Tree-ducks - the White-faced (*Dendrocygna viduata*) and the Fulvous (*D. bicolor*) also dive for their food. They live in mono-specific flocks according to distribution or season.

Perhaps the most surprising member of this multifacetted family is the Black-headed Duck (*Heteronetta atricapilla*), the only parasitic duck, which completely obviates its parental duties, laying its eggs in the nests of the most diverse marsh-nesting birds - other ducks, coots, swans, gulls, ibis and even on occasions raptors like the Chimango Caracara. The foster-nest merely provides the incubatory facilities and the duckling must fend for itself from day one.

The Southern Screamer (*Chauna torquata*) is a member of a three-species Neotropical family of waterfowl related to

This is the wildlife reserve where several pampas ecosystems meet, some very localized. Tala woods, floodable grasslands, tidal mud-flats and salt-marsh can be seen in the photograph.

those mentioned above though in appearance very different. It has large pink legs and feet but no webs between the toes. The beak is strong and sharp but not pointed, the body huge and gray, with a black choker and a splendid crest. The call, loud and heard at a great distance, sounds like its spanish onomatopaeic name "Chajá", the J pronounced like a loud H. They stand in pairs at the edges of marshes, in the appropriate season accompanied by their five fluffy dark-golden chicks, but they are also found in large flocks of upto 500. They are exclusively vegetarians and feed on the short grasses at the edges of marshes nesting on huge floating nests in the dense reed-beds; they can swim but very rarely do so.

Three species of coot amaze one by their numbers: the Red-Gartered Coot *(Fulica armillata)* is the largest and most aggressive, living out on the open water while the Red- fronted *(F. rufifrons)* is a bird of the dense reeds and the White-winged *(F. leucoptera)* is the most gregarious and usually on temporary water or clearings. There are several species of heron, either abundant or easily seen because of their size like the Cocoi or White-necked Heron *(Ardea cocoi),* or because of their entirely white plumage, the Common or Great Egret and the Snowy Egret, (respectively *Egretta alba* and *E. thula*), being the most noticeable.

In the reed-beds the tiny Stripe-backed Bittern *(Ixobrychus involuchris)* waits immobile, and at dusk the first signs of life come from the Black-crowned Night-herons *(Nycticorax nycticorax).* The long-legged Maguari Stork *(Euxenura maguari)* is usually seen on the uplands stalking insects and vertebrates - reptiles, batracians, small mammals or even young birds - but roosts and nests in the marshes on platforms of floating vegetation. The Chilean Flamingo *(Phoenicopterus chilensis)* frequents more brackish waters where it filters out the planktonic organisms which constitute its diet with its strange bill. Equally strange is the bill of the Roseate Spoonbill *(Ajaia ajaja)* which is flattened and long with a broadened tip and serves a similar function. This bird belongs to the family which includes the ibises, and the ibis

White-faced Ibis

The ever-present flocks of these birds (Plegadis chihi) *are one of the most common sights in wet situations in the Pampas. They are very evident in flight where long strings succeed each other across the sky on their way to roost or on migration.*

The Roseate Spoonbill

This bird nests colonially along with the other wading birds in the reed-beds of Pampas marshes. The brilliant carmine on rump and shoulders, and the yellow around the face are signs of breeding plumage.

The Black-necked Stilt

Himantopus mexicanus *wades at the edge of marshes to capture the invertebrates on which it feeds. In one or two areas they have learned to follow the plough for larvae and insects turned up with the soil.*

which is extremely common on the Pampas is the White-faced Ibis - the white face is seldom seen in this race - *(Plegadis chihi)* flying in long strings across the wide sky. Europeans, whose ibis are ever rarer, are amazed at the numbers here. Something similar happens with the Snail Kite *(Rostrhamus sociabilis)* so numerous here, so rare in the southeast U.S.A.

There is one bird though which had not been found for years and is now occasionally seen in certain areas where bird-watchers are alert - the Eskimo Curlew *(Numenius borealis)* was feared extinct. At the beginning of this century it still migrated in numbers to the Argentine Pampas where it "wintered". Other shorebirds still perform the spectacular long-distance flights from the northern latitudes where they breed, to these southern prairies: the Golden Plover *(Pluvialis dominicus)* a text-book example of a migratory species, the Greater and Lesser Yellowlegs *(Tringa melanoleuca* and *T. flavipes)*, Buff-breasted Sandpipers *(Tryngites subruficollis)*, Stilt Sandpipers *(Micropalama himantopus)* and several others.

Not many perching birds are adapted to marsh situations: the fidgety Wren-like Rushbird *(Phleocryptes melanops)* hunts invertebrates at the base of the reeds and builds an amazing egg-shaped nest of wet algae and vegetation which dries into a cardboard-like consistency with an entrance at the side, with a little overhang and all; the Many-coloured Rush-tyrant *(Tachuris rubigastra)*, awarded "bird-of-the-trip" status by many bird-watching visitors, also nests in the tules but the little cup is built holding on to one or two reeds. The Brown-headed and Yellow-winged Blackbirds *(Agelaius ruficapillus* and *A. thilius)* are - one species only at any one place - visible and vociferous, as is the Brown-and-yellow Marshbird *(Pseudoleistes virescens)*. The Spectacled Tyrant *(Hymenops perspicillata)* perches on lookouts to find its flying-insect prey and to watch for other males' advances into its territory, when fierce disputes occur.

In this kingdom of birds mammals are fairly rare and certainly less visible: the Coypu *(Myocastor coypus)* and the red water-rat *(Holochilus brasiliensis)* as well as the Red Water-possum *(Lutreolina crassicaudata)* which is often found in the marshes and flooded grasslands.

As darkness descends innumerable frogs and toads take up their chorus. *Lysapsus mantidactylus* lives a purely aquatic life never coming to land but rather inhabiting the floating and submerged water-weeds in the deeper parts of marshes, where it captures the invertebrates and certain small vertebrates like fish and immature batracians on which it lives. The tree-frog *(Hyla pulchella)* is everywhere abundant as is the Common Lesser Escuerzo *(Odontophrynus americanus)*; they all need permanent or semi-permanent water bodies to breed in, as their larval development is slow and some species go through the winter as tadpoles.

One of their chief predators is the Tararira *(Hoplias malabaricus)* a primitive fish which can sometimes be seen sunning in shallow water. The Pejerrey *(Odontesthes bonariensis)* is the fish most sought by coarse fishermen; the local eel *(Synbranchus marmoratus)* can spend time under the mud during droughts; the Madrecita *(Cnesterodon decemmaculatus)*, an amazingly abundant minnow, is the food for many species of wildlife.

There is very little of the Pampas left as it was known by the indians, the Spanish conquistadors or the first gauchos. Of this area which has been so very important historically, culturally and economically, there are no representative portions preserved in their natural state as National Parks and the only effective reserve that exists for the protection of the native flora and fauna - Campos del Tuyú - is the work of a private foundation.

The Greater Rhea

This ratite was once one of the important grazers of the Pampas, but is today growing ever rarer as it is hunted for its feathers and meat. Disturbance of the nests is surely another factor. The male incubates and rears the brood, while flocks of females merely lay eggs for the nests; clutches of upto fifty are common though 15 to 20 or more are the norm.

It is tacitly implied in this situation that the prime effort must be made to preserve precisely this habitat before it is too late, and not for aesthetic or ethical reasons alone, nor only for recreational purposes for huge numbers of the population, but for economic convenience, preserving the diversity of plant and animal life which just might come in useful to mankind in the future, and as a yard-stick to measure the extent of the modifications which the rest of the land has undergone.

Campos del Tuyú Wildlife Reserve. The shores of Samborombón Bay are part of the Atlantic coast of Buenos Aires province and are among the least affected by man in the Pampas Grasslands. Here there is an intricate network of temporary and permanent water-courses, and the tidal influence which causes periodic flooding with salt water, which has led to an estuarine salt-marsh ecosystem with the vegetation well- adapted to these saline conditions, and which cannot be used for agricultural exploitation. So this habitat and other more typical neighbouring habitats have been preserved as natural refuges for the interesting and abundant wildlife and vegetation.

At the southern end of this huge bay Campos del Tuyú has a surface area of 3,500 has and was turned into a wildlife refuge by Fundación Vida Silvestre Argentina in March 1979. This Argentine Wildlife Foundation is a private organization, working here through an agreement with the former owners of the estancia. Complemented with a "buffer" zone of a further 4,000 has it covers an interesting area between the sea and National Route 11 and includes a varied succession of habitats.

Here, at the mouth of the River Plate the waters are a mix - muddy and salt. Tides are about one metre though spring tides sometimes exceed two and a half. The extreme flatness of the land allows penetration of salt water over mud-flats bordered by sedge, known locally for the mud-crabs that live there - cangrejales. Thousand upon thousands of crab-burrows honey-comb the mud, and at low tide these denizens of the mud are very visible, the species being *Chasmagnatus granulatus*. The mud seems bottomless to anyone bold enough to venture out onto it. When the highest tides coincide with the strong southeast winds, the river Plate's waters back up and flood to renew the water in many salt lagoons where the vegetation is mostly *Salicornia*. There too there are communities of grasses like *Spartina* spp. and Espadaña (*Zizaniopsis bonariensis*) with cutting edges, which grows to 1.8 m on slightly elevated banks. The lovely Pampas Grass (*Cortadeira* spp.) is also present in clumps. It is usually surrounded by *Juncus acutus*, a viciously pointed sedge, dark green in colour, growing along the edges of these salt sloughs in places upto 300 m across.

On raised beaches of sub-fossil shells, or on fixed dunes, small patches of Tala woods grow, sometimes with the odd Sombra de Toro or Coronillo, a timid reappearance of the Espinal which has its own bird fauna - the Monk Parakeets, Red-crested Cardinals, Vermilion and Fork-tailed Flycatchers, the Yellow-browed Tyrant or the Blue-and-yellow Tanager, all adding their touch of colour.

As one moves inland from the coast the land becomes higher and the Arrow-grass starts to predominate though there are a number of fresh-water marshes which cross the reserve.

The main reason for the creation of the reserve was to protect the last vestiges of the once ubiquitous Pampas Deer (*Ozotocertos bezoarticus celer*). Stags are about 70 cms at the withers and weigh 30 or so kgs. The head bears symmetrical antlers with three points per tine which rarely exceed 30 cms length. The does are smaller. At the base and between the

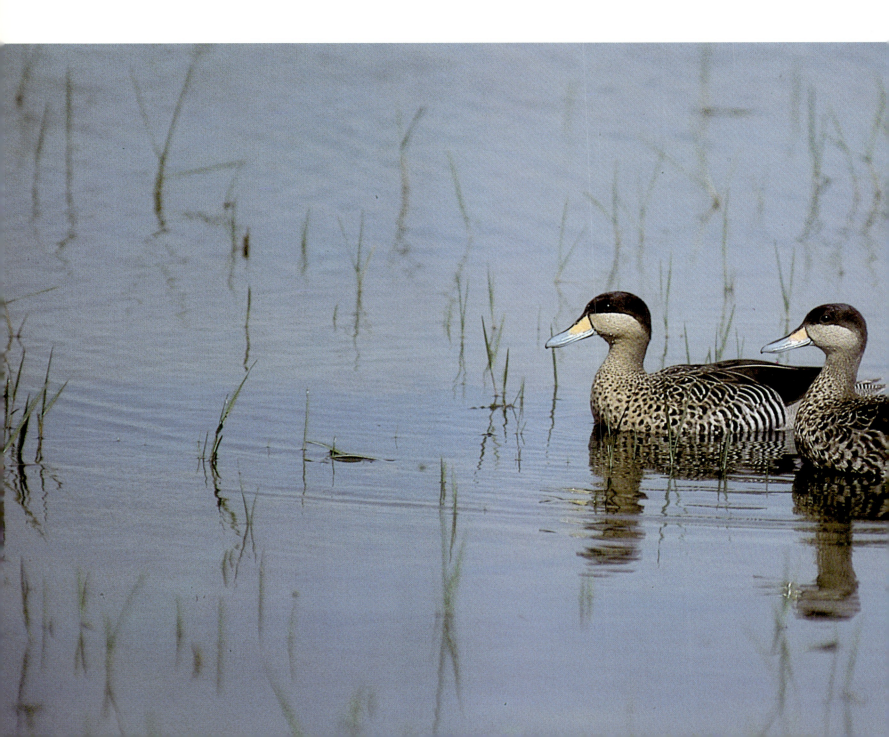

cloven hooves of the hind feet there is a gland which produces musk which marks the trail and can be detected for a considerable time after the deer has passed by.

According to the writer Justo P. Saenz at the end of the XIXth century in some places this deer was so abundant that they jumped up in the espartillo just as today grasshoppers do in the fields of alfalfa. The species was the victim of uncontrolled hunting to such an extent that it is calculated that in the decade from 1860 to 70 some two million hides were exported legally. This, together with the changes in habitat and the diseases of domestic cattle have led to its present critical situation, where barely over two hundred survive in the province of Buenos Aires and between two and three hundred in San Luis.

In an effort to breed it in captivity in 1968 a number of these deer were taken from the wild at the northern end of Samborombón. Unfortunately many specimens died and, though in the long-term fair success was achieved and numbers built up, there was no follow-up, no policy and the effort was wasted as there is no plan for this valuable herd.

In Campos del Tuyú they are protected from poachers by trained rangers so that there is some hope for the future of this reserve.

That area is also the home of populations of other threatened species like the Red-winged Tinamou, the Greater Rhea, Capybara and Geoffroy's Cat. With its varied habitats - salt-marsh, lagoons, marshes, upland grass and Tala woods - and with all the species associated with these, it offers a permanent refuge for the Pampas Deer so that it may interact naturally with the habitat that saw it evolve. The reserve is proof of the valuable contribution that can be made by a private organization to conservation when the national or provincial authorities are reluctant or slow to act.

The Southern Screamer

These grazing birds (Chauna torquata) *akin to the geese, are of a peculiar South American family* (Anhimidae) *with only three species. They are particularly abundant in the damper northeast of the province of Buenos Aires. Flocks of unpaired birds can number upto 500 or more, while pairs breed on the edges of marshes where the nest is floating among the reeds and where the chicks take refuge before any danger. The name is appropriate as the call can be heard at a considerable distance.*

Migratory shorebirds

In large numbers migratory shorebirds come from their breeding-grounds in Arctic North America and Alaska to spend their winters on the Pampas or Patagonia. The Campos del Tuyú area is a jumping-off point on the return migration such as for these Hudsonian Godwits (Limosa haemastica).

The Silver Teal

This is but one of the many species of waterfowl to be found on the Pampas' wetter areas, veritable paradises for observing birds both because of the excellent uninterrupted visibility as also for the variety and abundance of birds.

The Monte - Brushland Steppe

Monte is a term which in this country generally means copse or wood so that to see it applied to brushland and steppe sounds wrong. The meaning in Spain is of course mount and in the flat pampas any rise above the horizon, as in the case of a distant wood, could become a hill - but seldom did.

The German botanist P. G. Lorentz first coined or applied the term to the largest botanical region in the country - dry and dominated by scrubby bush, only supporting trees where conditions permit. The whole region is within Argentina, runs east of the Andes from southern Salta province south to the province of Neuquén where it swings east to end at the coast between southern Buenos Aires and central Chubut.

Its precise limits are subject for much debate as there is a wide ecotone to the south with Patagonia, with the Espinal and the Chaco to the east, with the Pre-puna north-westward. The cut-off westward is sharp because the Andes mountains rise abruptly from the plains.

Although there is such a great latitudinal gradient through the region - 27°S to 44°S - there is uniformity in the flora and fauna because of the climatic similarities which dominate throughout. The aridity is due to the great pressure systems on the planet and the resulting wind patterns which produce desert conditions between 15° and 30° in both hemispheres, here accentuated in a southerly extension by the effect of the Andes which cut off the damp winds from the Pacific, as do the Pampas ranges to a small extent with the rain-bearing Atlantic winds. Annual precipitation is in the range of 80 to 250 mm, mostly falling in a few torrential downpours, but in spite of this there have never been recorded periods exceeding nine months without some form of rainfall. The lack of cloud cover makes solar radiation intense and the yearly mean temperature for the northern parts is 17.5°C while in the south it is 13.4°C.

With the Colorado river as a divider, there are two identifiable types of climate. In the northern part the few rains are in summer and the climate is termed subtropical, while in the south the rains are better distributed throughout the year though the summer is drier because of intense solar radiation, and it is called mediterranean. This aridity dictates that the rivers that cross the Monte all come from somewhere beyond, mostly the Andes, and are fed by snow-melt.

The landscape is rolling without mountains or great heights, alternating with valleys, plains, dunes, plateaux and mountain slopes. The valleys are locally known as "bolsones" and are of three types - "playa" with sandy bottom, "salar" with salt or "barrial" with earthy mud.

The lack of water and the elevated temperatures are not conducive to abundant life forms so that those species found have adopted stratagems to survive in this inhospitable region, and it is amazing that in the Sonora desert some 8,000 miles to the north, mainly across tropical rainforests, there should exist similarities, even in the species of plant encountered. Both regions are populated by ancient floras of the tropics which were widely distributed at the beginning of the Tertiary, but they also respond to what is termed convergent evolution starting with that flora. Other plants which are common to these two areas are well-distributed in other areas also, and Creosote Bush - *Larrea tridentata* of the north is so very similar to *L. divaricata* of the south that until recently it was thought to be the same species. They were probably even more widely distributed until recently.

In the Monte two climax vegetations can be recognized, the most characteristic being the Jarillal (dominated by creosote bush). This Brushland Steppe with bushes upto 1.5 to 2.5 m high, is dominated by the *Zigophilaceae* of evergreen foliage, with a clear dominance of the three Creosote bushes *(Larrea divaricata, L. cuneifolia* and *L. nitida)*. These all branch from the base, have small evergreen resinous leaves which are hard to digest so are not eaten by herbivores.

They all have yellow flowers, and being so dominant dictate the whole aspect of the region. One particularly interesting adaptation to conditions is demonstrated by *L. cuneifolia* which has all its leaves oriented east-west and vertical to the ground, so that the incidence of solar rays is less at the hottest hours of the day. Other bushes that are frequent in the Jarillal are Monte Negro *(Bougainvillea spinosa),* Lata *(Prosopis torquata)* which is endemic to the Monte, and the Ardegras *(Chuquiraga erinacea)* which burns (arde) like fat (grasa). There are plants which, to reduce surface area of leaves and retain their moisture have developed the photosynthetic function in the stems and branches, these being green as in the Pichana *(Cassia aphylla)* which in spring is totally covered with yellow flowers, a patch of colour amongst so much gray, or Mata Sebo *(Monttea aphylla)* with its thick green stems on which grow small violet flowers. These perennial shrubs have developed different answers to the problem of the drought, while annuals have yet other solutions as they spend most of the year as latent seeds, grow and bloom in a brief period after rains, thus escaping the chronic lack of water. They possess minimal adaptation to dry conditions other than the rapid life-cycle with surface roots which take up water before it percolates through to the underground systems or is evaporated. They also produce abundant seed to ensure survival through to the next season's cycle. Several grasses are typical like *Bouteloua aristidoides, B. barbata* and *Eragrostis argentina,* or the composit *Pectis sessiliflora.*

The other climax community is the Thorny Steppe which grows on the coarse detritic soils of the north. The species found in the Jarillal are present here too as there is no definite division, but in different proportions, the thorny *Zygophilaceae* dominating here like Rodajillo *(Plectocarpa rougesii)* which is deciduous, or *P. tetracantha.* There is an abundance of cacti, especially of the genus *Opuntia* which defend the water they store in their tissue with sharp spines, and also unarmed or defenseless shrubs like the legume Jarilla Macho *(Zuccagnia punctata).*

There are also communities existing in isolated pockets living under certain conditions. Dry woodland is the most diffuse and prospers in the northern part, the Algarrobal being the most developed. The trees which grow in these woods are dependant on the water from the underground water-tables, so they don't grow where this is not fairly accessible and near the surface. Most of the elements are of Chaco origin, though numerically and in variety impoverished. They are dominated by species like the White Algarrobo *(Prosopis alba),* the Algarrobo Dulce *(P. flexuosa)* of edible seeds, the Sombra de Toro *(Jodina rhombifolia)* with its strange-shaped rhomboid leaves, the Tala *(Celtis spinosa)* with its small orange fruit so sought by the birds, and the Chañar *(Geoffroea decorticans).* Present also is the Corpus *(Phrygilanthus acutifolius)* with a biology similar to the Guapo-í Strangler Fig of the Paranaense which intertwines and cross-connects to tie up the host, and which in many places forms almost pure copses eliminating the other species. Unfortunately these woods have been irrationally destroyed for the mining industry and vinyards, for furniture, buildings or simply as fuel, so today their distribution is very restricted.

Only along the rivers are there Willow woods of *Salix humboldtiana*, the only tree which is found at the southern extreme of this biogeographical province. The dune areas have characteristic plants with most of the plant underground like the cactus *Pterocactus tuberosus* in which species the part above ground dries up every year, and which survives in the much larger tuberlike roots, or the pretty Amancay (*Amaryllis tucumana*), or another of the same Amaryllidaceae (*Habranthus* sp.) which flowers a bare two days after the rain. In the salt-flats, meanwhile, grow succulents mainly, surviving here thanks to their very high internal osmotic pressure; *Salicornia*, Jume (*Suaeda divaricata*) with fleshy leaves in the wetter periods, smaller leaves in the dry, or the whitish-green Zampa (*Atriplex lampa*).

The resources accumulated by all these plants are the sustenance of a varied fauna which also has to overcome the adverse conditions of this arid zone. Mammals must sacrifice water to perform two basic functions - the elimination of salts through solution in urine, and sweating or panting for cooling-off. So most species are nocturnal, some live in particular microhabitats and underground where at just a little depth, temperature extremes are tempered.

The armadillos (fam. *Dasypodidae*) do not perspire to cool off; they are considered incomplete homeotherms as their body temperatures are dictated in part by the ambient conditions. The Fairy Armadillo (*Chlamyphorus truncatus*) is the smallest armadillo, endemic to the Monte where it has a restricted distribution, lives almost totally underground emerging only very rarely. So adapted is its organism to stable temperature conditions that any brusque change can kill it. The Wailing Pichi (*Chaetophractus vellerosus*), more abundant and better known, is the other species typically found in the Monte, one of the most common mammals. It does not feel the lack of water, feeds omnivorously on plants, insects, other invertebrates, small vertebrates and carrion and like the other armadillos from arid areas is covered with hairs, albeit sparse, which might keep it cooler. The Three-banded Armadillo (*Tolypeutes matacus*) and the Patagonian Pichi (*Zaedyus pichyi*) are also found. The largest herbivore is the Guanaco which is found in herds called "piñas" with one male, his females and "chulengos", as baby Guanacos are called. Other herds are formed entirely of bachelor males. Another is the Mara or Patagonian Cavy (*Dolichotis patagonum*), a large rodent which looks more like a small antelope, a role it seems to fill ecologically. Maras weigh upto 15 kgs and move at speed with their own version of "pronking", a quadrupedal skitter, bouncing on all four limbs at the same time, high into the air, displaying an evident alarm signal in the white "bloomers"; a sight to behold. The Plains Viscacha (*Lagostomus maximus*) is more sedentary and stays around its vast colonial burrows in which it spends the day, emerging only at night to feed.

Smaller yet are the two species of Cavy (*Galea musteloides* and *Microcavia australis*) as well as the underground dwellers, the Tucotucos (*Ctenomys* spp.). Of the small rodents which are well-represented, the most characteristic is *Eligmodontia typus* in the more arid areas, common in the *Larrea*, in dunes and in salt flats; specially adapted to drink water with four times the concentration of salts that sea-water has, it can feed on salt-bush.

The largest predator is the Puma (*Felis concolor*) which is accused of many transgressions. It is said to kill several sheep in a single night for the pleasure of the act, and that foals are the favourite food. Whether this is true or not it has served livestock ranchers as an excuse to declare all-out war on it without considering its necessary and basic role as controller of other competitive fauna.

Other lesser cats include Geoffroy's Cat (*Felis geoffroyi*), the Pampas Cat (*F. colocolo*) and the Jaguarundi (*F. yaguaroundi*).

The most common canid is the small Gray Fox (*Dusicyon griseus*) which omnivorously feeds on insects, fruit, small mammals and vertebrates, and carrion, while the larger Red Fox (*D. culpaeus*) is more of a mountain species. There are three mustelids - the Grison (*Galictis cuja*), the tiny slender Dwarf Grison (*Lyncodon patagonicus*) both agile and fidgety, ferocious little predators alone or in packs (the first), and the Hog-nosed Skunk (*Conepatus chinga*) which leads a more peaceful and sedate life digging up grubs or bulbs, or taking small vertebrates.

Two marsupials are found here: the ubiquitous White-eared Opossum (*Didelphis albiventris*) especially where man lives, and the Murine Opossum (*Thylamys pusilla*) the smallest of our predators, mouse-sized and insectivorous, storing up reserves for winter in its fat tail.

Some bats occupy the aerial space vacated by diurnal birds. *Myotis* and *Tadarida* are both insect-eaters.

Ground birds are most common, running through the bushy vegetation, the largest being the Greater Rhea (*Rhea americana*), followed by the Tinamous - the Elegant-crested (*Eudromia elegans*), the Brushland (*Nothoprocta cinerascens*) and Darwin's Nothura or Pale-spotted Tinamou (*Nothura darwinii*). All tinamous, like the Rheas, breed in a system where several females lay in one nest at a time; the male incubates and raises the brood. The eggs are characteristic monochromes - browns, greens, blues, pinks, lavender - with a high-gloss finish. These birds are often erroneously called "partridges" for their somewhat similar appearance and habits.

Birds, like reptiles but unlike mammals, use uric acid to disolve the nitrogenous salts for elimination from the body, allowing them to be concentrated with little loss of moisture. They are homeotherms with an efficient insulating layer of feathers, can move around freely by flying and only lose water on very hot days through respiration. All in all they are better adapted to arid conditions than mammals.

It is unlikely that anyone should miss either of the two species of vultures gliding overhead. The Black Vulture (*Coragyps atratus*) is found in great numbers where a dead animal lies. Completely black, only in flight can one see the whitish patches towards the end of the spread wings, at the base of the primaries. The tail too is shorter than that of the Turkey Vulture (*Cathartes aura*). Both seem to have a very keen sense of smell by which they detect the odours which lead them to food, wafted on the rising airs. Two Caracaras, the Crested (*Polyborus plancus*) and the Chimango (*Milvago chimango*) are of the falcon family and complete the spectrum of winged scavengers, though neither of these two would refuse any oportunity to prey on something easy.

Among birds of prey however there are the true raptors such as the Red-backed Hawk (*Buteo polyosoma*) which preys on the smaller mammals, the Aplomado Falcon (*Falco femoralis*) which takes birds, the Black-chested Buzzard-Eagle (*Geranoaetus melanoleucus*) taking larger prey like viscachas, skunks and even young foxes.

Smaller raptors like the American Kestrel (*Falco sparverius*) are mostly insectivorous, though the tiny and

The Burrowing Parrot

Like most parrots, Cyanoliseus patagonus *travels in flocks such as this one which is intent on feeding on the fruit of the Molle* (Schinus polygamus). *It nests colonially in banks and cliffs, digging its burrow with the beak.*

The Jarillal

The Jarillal is the most typical of the plant communities of the Monte. A view of the great depression in Sierra de las Quijadas, a future national park.

endemic Spot-winged Falconet *(Spiziapteryx circumcinctus)*, almost exclusively a Monte bird, is a terror on little birds. It nests in hollow cacti among other sites.

The Burrowing Parrot *(Cyanolisseus patagonus)* of medium size and with its long tail, nests in banks and cliffs where it digs its burrows and lays two eggs. In rowdy flocks of upto hundreds of birds it travels around in search of the seeds on which it mainly feeds.

Perching birds here are mostly drab and brown-plumaged, among which the Furnariids are well-represented. This Neotropical family of which several nest in burrows, includes the lesser predator, the White-throated Cacholote *(Pseudoseisura gutturalis)* with a strong beak and robust body, feeding on insects and small reptiles. The Scale-throated Earth-creeper *(Upucerthia dumetaria)* easily recognized by its long, decurved bill, is one of the burrow-nesting species, the entrance to it burrow being either in a bank or simply on level ground.

The Crested Gallito *(Rhinocrypta lanceolata)* is of another South American family - the *Rhinocryptidae* - as is the Sandy Gallito *(Teledromas fuscus);* both species prefer to run along the ground rather than fly and it is common to see them dart across the road, tail held high, like miniature road-runners. In winter the flocks of Fringillids are evident, often composed of several species, more abundant in the areas with trees, the species varying from north to south. In the south flocks are mostly the Mourning Sierra-Finches *(Phrygillus fruticeti)* while in the centre and north it is the Common Diuca-Finch *(Diuca diuca)*. This family has some species with bright colours like the Many-coloured Chaco-Finch *(Saltatricula multicolor)* or the Yellow Cardinal *(Gubernatrix cristata)* today almost extinct in the wild as a result of its popularity for the cage-bird trade, the Ringed Warbling-Finch *(Poospiza torquata)* or the Golden-billed Saltator *(Saltator aurantiirostris)*.

Some tyrants search the bushes for the insects on which they feed, like the Greater Wagtail-Tyrant *(Stigmatura budytoides)* with its attractive duetted call and looking slightly like a diminutive mocking-bird, or the Tufted Tit-Tyrant *(Anairetes parulus)* so pretty with its yellowish breast streaked dark, and its remarkable curling crest. Other tyrants catch their food in flight like the well-named White-winged Black-Tyrant *(Knipolegus aterrimus)*, or the Rufous-backed Negrito *(Lessonia rufa)* - equally well-named - which runs along open ground and the shores of bodies of water, darting and flitting to capture insects. Important bodies of water are habitat for waterfowl of Pampa or Chaco origin, but poorer in species diversity.

Some batracians can cope well with temporary puddles to spawn, thus demonstrating perfect adaptations to what for them is an extreme habitat, the Monte. *Pleurodema nebulosa* is just such a species and has suppressed periodicity in its reproductive cycle to adjust to the intermittent rains of the region. Eggs hatch in a bare 12 hours and the tadpoles complete their metamorphosis in a record seven days, in a race against certain death which they do not always win, the implacable sun drying up the puddle before development is complete. It is the price to pay to ensure survival, one spawning after each rain. Other amphibians of the Monte are the Sand Toad *(Bufo arenarum)* and the Lesser Escuerzo *(Odontophrynus occidentalis)* which spends most of its life buried underground, emerging only when conditions are favourable.

Reptiles are unable to regulate their internal temperature metabolically. They do so through their behaviour and with

Sierra de las Quijadas

This area in the northwest of San Luis province is a monument to the whims of erosive powers, but it also contains a large area of flats with the typical fauna and flora of the Monte.

The Guanaco

Most of the temperate and cool habitats in South America have been home to the Guanaco, a member of the camel family. In Argentina they are found in the High Andes, the Patagonian Steppes, Subantarctic Woods, the Pampas and in the Monte. Here they are in thorny woodland.

Lihué Calel National Park *The Monte vegetation of this Park is enriched with a special element from the Espinal, the Caldén* (Prosopis caldenia) *which grows at the foot of the slopes where streams carry water on occasions.*

the aid of the ambient conditions. When it is warm, as is usual in the Monte, reptiles are more active which is beneficial as they can move faster in the competitive world, to catch their prey or to flee from their predators. The most abundant reptiles are a genus of small lizard (*Liolaemus*) which is composed of several species found among the vegetation, in sandy areas, among rocks, feeding basically on insects. The Lion-lizards (*Leiosaurus* spp.) unjustifiably believed to be poisonous because of their courageous bluff facing most enemies, gekkos which are active at dusk and during the night, like *Homonota horrida* which bears little relationship to its name as it is a rather dainty and charming gekko with big lidless eyes with the vertical pupil. Undoubtedly the most noticeable reptile of the Monte is the boa (*Boa constrictor*) known to frequent the area of viscacha colonies where it finds food. Seven or more feet long (2.2 m) it has been subjected to intense trapping for its skin which is valued in the trade, so today is very rare. Of poisonous snakes the Monte has its share, some being dangerous like the rattlesnake (*Crotalus durissus*) with its neurotoxic venom as is that of the coral snake (*Micrurus frontalis*) which is fortunately shy and non-aggressive, and like all coral snakes is ringed in red, black and whitish-yellow. *Bothrops neuwiedii*, on the other hand, is a pit-viper and more irascible in temperament, likely to strike if bothered. Finally let us mention the Tortoise (*Chelonoidis chilensis*), slow and peace-loving denizens of the region, also victims of greedy man who captures them in mass for sale in pet-shops.

The insect fauna also has some endemism: two neuropteran genera - *Veurice* and *Pastrania*; two species of social wasp - *Polystes buyssoni* and *Misochocyttarus lilae;* and a genus of Bruchid beetle - *Pectinibruchus*. Ants and grasshoppers are everywhere and abundant, fulfilling their important role in the system as defoliators and as food for other species. As in all arid lands, there are several scorpions.

This biogeographical province has suffered numerous alterations at the hand of man - overgrazing, deforestation, indiscriminate hunting and trapping, and much else, so it is important to set aside as soon as possible new protected areas. At the moment there are but one small national park and two provincial reserves which are staffed, in all too little for a region which, apart from being exclusively Argentine, because of its very low productivity is likely to suffer erosion and serious modifications through bad management, damage which will take ages to repair if indeed it can ever be achieved. There have been several initiatives for the creation of national parks such as at Pipanaco in Catamarca, San Blas del Pantano (La Rioja), and now Talampaya is on the books (La Rioja). These would indeed contribute to lessening the defficiency in nationally protected areas of the Monte.

Lihué Calel National Park. This is at the time of writing, the only national park of the Monte province, and is found at the southern end of the region. It includes the hilly outcrop of Lihué Calel, some 15 kms long north-south, and seven wide, reaching a height of 600 m above sea level. These hills rise above a monotonous plain and are of Pre-cambrian igneous rock, gently sloping on the north side, abrupt on the south.

The vegetation is similar to what has been described for the Monte with Creosote Bush and Chañar dominating, added to the Goat's Beard (*Caesalpinia gilliesi*) whose name does it no justice, Brea (*Cercidium australe*) all green-barked and twiggy, covered with wonderful yellow blossom in spring, Piquillín *(Condalia microphylla)* with its little sweet red fruit an important resource as food for many vertebrates. At the foot of the slopes, along the temporary water-courses grows the Caldén *(Prosopis caldenia)* which is typical of the Espinal and other trees like the Sombra de Toro *(Jodina rhombifolia)*. There are three species of plants endemic to these hills which are worthy of mention: the Treacherous Cactus *(Opuntia puelchiana)* all silver-spined and growing at the foot of the north-facing slopes, *Gaillardia cabrerae* a yellow daisy growing out of the crevices in the rocks, flowering in spring and whose leaves have a pleasant resinous odour, and the legume *Adesmia lihuelensis* which grows where humus has accumulated between rocks. There are a number of ferns and pretty-flowering plants, favoured by the special conditions the hills offer which drew commentaries from the first white men in the area, who aptly described them as a green island in the vast, flat, arid steppe.

The fauna of this park is typically that of the Monte as described above, with but a few additions: the Brown Cacholote *(Pseudoseisura lophotes),* like the other species already mentioned of this genus, but richer, redder brown with no contrast on the throat, builds its enormous stick nest on top of a hefty horizontal branch. Its strident duet call can be heard throughout the year. In winter the Fire-eyed Diucon *(Pyrope pyrope)* arrives from Patagonia's andean woods, and others as conspicuous like the White Monjita *(Xolmis irupero)* completely white with black eyes, beak, wing-tips and tail-tip, always perched high to dominate the territory or in search of insects. The fundamentally terrestrial Field Flicker *(Colaptes campestris)* feeds on ants mainly while the Golden-Breasted Woodpecker *(Colaptes melanolaimus)* which sometimes shares this diet is more often seen searching for food in trees. Ground birds must be ever alert for the ubiquitous cats; all the species mentioned are present in the park. Jaguarundis here have been seen sporting a gray pelage as against the more usual red or black coats. Out on any walk one could come across thr Red Ground Lizard *(Tupinambis rufescens)* upto 1.5 m long, a terrible predator for smaller animals. Exotic species include the European Wild Boar *(Sus scrofa)*, a species which seems impossible to control, while feral goats were finally eliminated in the late 1980's.

Lihué Calel is on National Route Nº 152, an alternative route between Buenos Aires and Bariloche, only 200 kms from Santa Rosa, La Pampa's provincial capital. There is an Automobile Club motel almost at the gate, a good place to take a break on the long drive from or to anywhere. Even though this recently-created national park is a very good example of the Monte, its 10,000 has are not really sufficient. There could be much to benefit by if the park were enlarged to include one of the salt lakes close by, where abundant and varied waterfowl are found. The main danger lies in the frequent fires which affect the whole region yearly.

Sierra de las Quijadas. The isolated range of hills which gives its name to this projected national park poke out of an extensive plain in the northwest of the province of San Luis.

To be precise, this area is on the ecotone between the Chaco and the Monte so plant and animal species increase with input from both biogeographical provinces.

One of the reasons that this project is being considered is a strange geological formation in the north centre of the hills known as Potrero de la Aguada. It is a huge amphitheatre of around 4,000 has completely surrounded on all sides by vertical walls of soft red sandstone and conglomerates where erosion has sculpted the most extraordinary shapes: cliffs, cornices, buttresses, columns, all of which, when seen from an overlook, offer a splendid spectacle. The only drainage is through the Aguada torrent which races through a very narrow gap. One must enter the depression down one of the few inclines left as a ladder by the erosive forces. The outer slope of these hills is gradual in many places and one can actually reach the lip in an automobile.

The plants are characteristic of the Monte with some Chaco influence like the White Quebracho (*Aspidosperma quebracho-blanco*) here at the southern edge of its distribution so reaching no appreciable size, but sprinkling the landscape with its typical globular-shaped crown. Creosote is the climax vegetation with *Larrea cuneifolia* dominating. There are also specimens of *L. divaricata* and the Jarilla Macho (*Zuccagnia punctata*) among the thornless species. A defense against herbivores here are the thorns which so many species possess, such as the Algarrobos (*Prosopis* spp.) and lesser trees like the Garabato (*Acacia furcatispina*) and the Espinillo (*A. caven*), the cacti and even the ground bromeliads like *Dyckia* sp. and *Puya* sp. which even have spiny leaves.

To the west of Sierra de las Quijadas runs the Desaguadero river, and between the two there is a great flood-plain with copses of Chañar *(Geoffroea decorticans)* and salt-flat species. It is an ideal area to cruise in a vehicle to discover the abundant fauna which includes flocks of Rheas, Maras or Patagonian Cavies *(Dolichotis patagonum)* and Woodland Maras *(Pediolagus salinicola)* - typically Chacoan - Elegant-crested Tinamou and flocks of Burrowing Parrots. With due protection the herds of Guanaco will soon increase as too the Collared Peccaries *(Dicotyles tajacu)* which are also a Chaco species, tortoises and boas to mention but a few species. The Desaguadero river and its adjacent Guanacache lagoons are home to a number of water-birds while in the thorny woods perching-birds abound. With its 150,000 has this would be the second in size of the parks in the region, and fulfill an important role in the conservation of this semi- arid, exclusively Argentine ecosystem.

Ischigualasto-Talampaya. Covering a trans-border area shared between San Juan and La Rioja is an area of special scientific interest, of scenic beauty and important to conservation. It includes Valle de la Luna where extraordinary discoveries about the flora and fauna of the Triassic have been made, now found to occur in the whole Ischigualasto watershed.

In that period this area was tropical and swampy, with ferns, conifers and lianas, and in keeping with the sumptuous vegetation the fauna of herbivorous and carnivorous reptiles was extremely rich and varied, as much terrestrial as aquatic; many were of respectable sizes, some upto five metres long. It is considered one of the richest deposits in the world of Theraspid reptiles, that is those from which mammals would evolve; the importance of data to be gathered there is such that it has attracted world-reknown scientists from within the country and abroad. Among the most noteworthy discoveries one can mention the foot-prints of the huge reptile *Rigalites ischigualastianus,* or the most abundant species like *Exaeretodon frenguelli* and *Scaphonyx sanjuanensis,* all Rhynchosaurs which fed on the fruit of cycads and ginkos and which became extinct with the appearance of various predatory dinosaurs.

Today this spectacular flora and fauna have been replaced by an arid region with the typical "bad-lands" where erosive

The Puma

agents have sculpted strange forms in the brick-red sandstone, and where isolated greenish or yellowish blocks repose on the surface, with scarps and cliffs of upto 200 m height showing the different strata, as also columns and obeliscs. There are gorges and such where streams and rivers flow during the brief periods when there is water.

The province of San Juan has created a reserve of some 50,000 has at Ischigualasto which has adequate control to safeguard this valuable heritage, and there have been, at times, initiatives to hand it over to National Parks.

Across the border in the province of La Rioja is the projected Talampaya National Park which is at present just a provincial reserve and lacks any effective protection. It is separated from the Ischigualasto by a range of low hills but possesses characteristics which are very similar. Rock-painting have here been discovered with some hundred or so drawings based on the present fauna - rheas' tracks, guanacos, lizards, as well as snakes - and stylized human figures and geometric designs. There have also been found

This cat was once widespread over all the country but today only survives where the rough terrain or the vegetation protect it from man to a certain extent.

remains of indigenous dwellings, stone walls, bits of pottery and stone tools. When Talampaya comes to be, it will be part of the largest area of Monte under protection - 230,000 has - in an area of tremendously varied interest, including the wildlife and vegetation which are typical of this particular biogeographic region.

The Ñacuñán Forest Reserve is in the province of Mendoza. With its 12,282 has it is in the centre of the province of the Monte with all its typical fauna and flora. It is patrolled and administered by the Institute for Arid Zone Research (IADIZA), and is where scientists carry out their projects.

The Patagonian Steppes - A Semidesert

"In calling up images of the past, I find the plains of Patagonia frequently pass before my eyes; yet these plains are pronounced by all to be wretched and useless. They can be described only in negative characters; without habitation, without water, without trees, without mountains they support merely a few dwarf plants. Why, then, and the case is not particular to myself, have these arid wastes taken so firm a hold on my memory?"

The desolate landscape which would so impress Charles Darwin, as he wrote in "The Voyage of H.M.S. Beagle", is an arid steppe of grass and bush which covers vast fractured plains. The stony soil lies naked between the sparse grasses and in the spaces between the thorny, small-leaved bushes which are adapted to the desert conditions and the violence of the wind.

The extensive Patagonian region of about 750,000 sq kms, covers the southwest of the province of Rio Negro, most of Chubut, nearly all Santa Cruz and the northern half of Tierra del Fuego. It consists of a succession of plateaux or terraces, flat-topped or slightly rolling, having cliffs or steep banks dropping into the valleys which descend from the Andes in a series of steps to the Atlantic.

The Patagonian Massif, together with the Brazilian shield, constituted the nucleus of eastern South America in pre-Cambrian times, as part of Gondwana. Over this crystalline rock which still pokes through and is found in the plateau of Somuncurá and the Pampa de Gastre, were deposited mantles of basalt and marine and terrestrial sediments, with successive rises and falls of sea-level during the Secondary and Tertiary. Patagonia is covered with a pebble-paving, the most extensive on earth, whose small rounded pebbles probably originated in the Andes and were brought down by water, wind or ice, varying in size between that of a wall-nut and an orange.

Uplift, which has been occurring since Mesozoic times, plus erosion and water action have all moulded the landscape. Plateaux alternate with areas of subsidence like the depression which contains the huge lakes Musters and Colhué-Huapi; the cores of old volcanoes and volcanic plugs stand like turrets and castles; a system of low hills, very eroded - the Patagonides - with the San Bernardo range being the most notable, originated from a sedimentary fold during the Cretaceous; deep canyon-like valleys, once the beds of rivers, and the inordinately wide valleys of certain rivers which flow west to east. These were carved by the waters from the ice-melt at the end of the Quaternary glaciation when the Andes were under a vast ice-cap.

The climate of this region has changed tremendously. In former times here there were humid tropical forests. During the Cretaceous for example, Patagonia was covered by Araucaria forests where dinosaurs roamed, all evident today in the fossil record; and the abundant paleofauna from the Tertiary leads one to suppose that environmental conditions were then more benign.

Since the uplifting of the Andes however, the climate has become extremely inhospitable: cold, with mean annual temperatures below 10°C almost everywhere, absolute minimun temperatures at -15°C or even less, frosts possible in nearly every month, snow in winter over most of the land; but it is the winds which are the dominant negative aspect as far as life is concerned.

Blowing almost continually from the west they sweep across Patagonia with force and tenacity, often reaching speeds in excess of 60 m.p.h. (100 kms.p.h.), and are the main reason for the dryness. Since the humid winds off the Pacific are forced to rise over the Andes, there they drop their moisture and, now dry, descend the eastern slopes sucking up any humidity, causing tremendous evaporation and creating very arid conditions in an area which only receives 100 to 150 mm of precipitation per year.

The very infrequent Atlantic winds also are dry, against all expectations, as they dump their humidity over the Falkland (Malvinas) Current and turn to fog on chilling. Even so the coastal areas do enjoy slightly more benign conditions aided by the dampness of, and temperature moderation by the proximity of the ocean.

The character of the vegetation of this Patagonian Province as it is termed by botanical geographers, is determined by the hostility of the climate and the poor stony-sandy soils which lack organic material and have a low nitrogen content. As in the Monte, only here more so because of the more demanding conditions, plants must be very well adapted to stand the drought and cold, and defend themselves from herbivores to protect the foliage which costs them so much effort to produce.

Cushion plants are numerous as such a structure helps preserve moisture and warmth and is less chastised by the wind. Some of these are not tight, compact plants but open and hemispherical like the Umbelliferous Neneo *(Mulinum spinosum);* others are indeed compact and "solid" like another umbel Yaretta *(Azorella trifurcata)* or the composit *Brachyclados caespitosus.*

Bushes or shrubs in this habitat tend to have small curled leaves to minimize evaporation, and waxy or resinous thick cuticles; they are usually covered with thorns; grasses grow in low clumps with tightly curled leaves, high in silicone content, thick epidermis, all of which makes them tough and spiky. They are of the genera *Stipa, Festuca* and *Poa* and receive the local name of Coirón. Another adaptation displayed by many plants in this region is that in good years only do they put out long shoots (macroblasts) which ensure growth, while short stems (brachyblasts) which get covered with leaves, are put out every spring. This results in an appearance of bushes of apparently undivided branches covered with tight "scale" leaves as in the Colapiche *(Nassauvia glomerulosa),* Mata Negra *(Junellia tridens)* and the solanaceous *Fabiana peckii.*

As can be expected in such a huge region which covers such a wide range of latitudes, the lacklustre and apparently homogenous botanical panorama in fact presents a marked diversification, passing from scrub to grassland and noticeably changing the composition of the various communities.

In the northeastern section (Chubutense sub-district) the characteristic plant community is the Quilenbai, Colapiche and Bitter Coirón Steppe, a sparse and low cover which leaves at least 65% of the soil bare and unprotected. Here the rounded bushes of Quilenbai *(Chuquiraga avellanedae)* half to one metre tall, dominate, interspersed with clumps of Coirón and dwarf Colapiche which barely gets off the ground but spreads to become equally dominant. Quilenbai is a composite whose small, tough, ovate leaves end in a sharp point which deters any herbivore and which in flower are covered with yellow blossoms. The Bitter Coirons *(Stipa humilis, S. neai* and *S. speciosa)* grow in tight, low, spiny clumps alternating with more tender grasses. The Colapiche is a composit which may well be taken to be stones because of its colour has branched twigs with such a disposition of tight scale-leaves that it does in fact look like the armadillo tail after which it is named (cola = tail, piche = armadillo).

Where the Quilenbai grows densest, as near the coast, they seem to cloak the landscape with a continuous green blanket infinitely speckled with yellow buttons in spring.

The uniformity of this vegetation is often broken by clusters of taller bushes, thorny and small-leaved like the Molle *(Schinus polygamus)*, the Patagonian Algarrobo *(Prosopis denudans)*, Mata Laguna *(Lycium ameghinoi)*, Calafate *(Berberis cuneata)* or the thornless Verbena *(Junellia ligustrina)*.

Salt bottom-lands and sometimes real salt-flats are frequent in the region, a result of the many past invasions by the sea, where special varieties of plants which can tolerate such conditions grow: two Zampas *(Atriplex lampa* and *A. sagittifolium)*, *Frankenia patagonica*, and, mainly near the sea, Jume *(Suaeda divaricata)*. They are usually accompanied by the Mata Laguna and Patagonian Algarrobo growing widely spaced on the clay soil.

The bottoms of valleys which are periodically flooded form green Vegas where the lawn-like vegetation is made up of *Juncus leseurii* and the grasses *Distichlis spicata* and *D. scoparia*.

Further south, and occupying most of the province whose name it bears, is the Santacrucense sub-district, characteristically a low-bush open steppe known as the Mata Negra, Colapiche and Bitter Coirón steppe. Here the Quilenbai has been replaced with the Mata Negra *(Junellia tridens)*, a verbenaceous plant over half a metre tall growing in irregular shapes, straggly and branching, covered with scalelike three-lobed leaves; the very dark colour gives it its common name (mata = bush, negra = black).

Towards the mountains, with elevation, and at the southern extreme of the continent the bushy steppe gives way to seas of grass, the White Coirón Steppe *(Festuca pallescens)*. This grass grows to some 20 to 60 cms tall, has tightly rolled leaves and constitutes the main part of the vegetation. It grows here, where it is colder but also damper, with annual precipitation between 200 and 500 mm. The soil is richer in organic material. This is called the Subandean district.

In the northwest of this botanical province, relief near the mountains becomes increasingly rolling and hilly, which provides a variety of habitats where different plant communities prosper, various bush and grass species dominating the landscape alternately. This western district is characterized by the abundance of bushes, many hemispherical like the Neneo which as a dominant plant gives whole hillsides its own pale green colour, and rounded like the composits *Haplopappus pectinatus* and *Senecio filaginoides*. The tasty grasses like the White Coirón, so sought after by cattle, have been grazed to such an extent that here they have almost ceased to be the dominant species of any vegetation community.

Many plants have developed resins or essences which make them foul-tasting to herbivores - mostly insects - such as Orégano *(Acantholippia seriphioides)* and Neneo. Some plants lack leaves, like Solupe *(Ephedra ochreata* and *E. frustillata)*, and there are thorny shrubs which chiefly grow on rocky outcrops, like Malaspina *(Trevoa patagonica)*, Duraznillo *(Colliguaya integerrima)* and the Calafate, a Berberis.

In the same way as the flora, the patagonian fauna is composed of species which are adapted to life in such a hostile environment, in which aridity is perhaps the most limiting factor. Here there are repeats of many of the animals found in the Monte, another biome where animal adaptations to this determining factor are essential. The most characteristic species of these steppes, abundant where they have not been over-persecuted, and visible because of their size or their movements, are the Guanaco, the Lesser (or Darwin's) Rhea, the Mara or Patagonian Cavy and the Elegant-crested Tinamou.

The Guanaco *(Lama guanicoe)* is the largest animal here and is gregarious; it is the dominant animal feature on the landscape. Usually in herds of four to ten females with their respective calves, they graze and browse while the male stands sentry on some prominent elevation if the countryside is rolling. His alarm call, to which the females immediately respond by alertness and motion, is a whistled whinney, which, with the noise of the wind, is surely the most typical of patagonian sounds.

Competition for these harems leads to violent battles between males during the breeding season in spring, when males spit in each others' faces, rear up on their hind legs to punch each other with their fore-"knees", bite the opponent around the neck and hind legs.

Bachelor herds can be very large, especially in winter; before they were hunted so drastically several hundred at a time could be counted.

Easily seen at a distance because of the height of the vegetation, herds of guanacos alternate with flocks of rheas, the "grand archaic ostriches" of South America. The patagonian species is the Lesser Rhea *(Pterocnemia pennata pennata)*, smaller, rounder, browner-plumed and these with white tips; it here replaces the Greater Rhea of the Monte, Pampas and Chaco. Flocks of young or chicks, a dozen or more accompanied by the male, are a fairly common sight.

Pairs or small groups of Maras *(Dolichotis patagonum)* can be found in northern Patagonia sitting up to keep an eye out for danger. The social organization is based on the pair which mate for life and are always close to each other. Females give birth by a communal den which gives refuge to the young of several females. Adults do not use the den; they graze in the vicinity, and the female visits it periodically to feed her young, escorted by her mate. She suckles them in a sitting position, her twins one on each side.

In the warmer months the Elegant-crested Timanou *(Eudromia elegans)* forms small flocks of three or four females while the male does the nesting and takes care of the chicks. In winter all group up into large flocks of upto one hundred birds. This species is typical in the northeast Quilenbai and Bitter Coirón steppes, in the Western and Santacrucense districts there is another large tinamou, the Patagonian Tinamou *(Tinamotis ingoufi)*. It lives socially, keeping loose contact with mournful whistles as flocks wander through the Mata Negra or southern grasslands.

Looking like small doves, tiny partridges or even perhaps quail, there is another family of birds, also exclusive to Patagonia and the High Andes, the Seedsnipe. They feed on vegetation - sprouts and seed - and flock together in winter. These are a clear case of convergent evolution as they are this continent's equivalent in many aspects, of the Sandgrouse (fam. *Pteroclidae*) of Africa, of completely different derivation.

The smallest species, the Least Seedsnipe *(Thinocorus rumicivorus)* is the only one of the four which is distributed throughout the region; another, the Gray-breasted *(T. orbignyianus)* is limited to the southern portion and the high

The Red-backed Hawk

This raptor (Buteo polyosoma) *is the most abundant raptor in the region. This young bird is on the steep banks of one of the many drainage canyons that descend from the plateaux. Feeding mostly on rodents these hawks of the open spaces behave and look more like small eagles.*

The Elegant-crested Tinamou

The partridges and quail of other continents are replaced in South America in their role of ground-birds with a predominantly herbivorous diet, by the Tinamous. The abundant Elegant-crested Tinamou (Eudromia elegans) *is the most common in Patagonia, the Monte and Chaco regions.*

The Mara or Patagonian Cavy

An inseparable pair of Maras (Dolichotis patagonum) *visits the den where the young of several pairs live as in a communal nursery, in order to suckle their twins. In this case the den was shared by eighteen young Maras, probably the progeny of some nine pairs of adults.*

Petrified "Forest" Natural Monument

The Bosque Petrificado, to give it its spanish name, is in the Patagonian Steppes. From the eroding slopes appear, with time and gentle weathering exposure, the petrified trunks of Araucarias some 150 million years old.

Andes, the White-bellied *(Attagis malouinus)* descends to the steppes in winter flocks.

At the base of the bushes one can find the Least Cavy *(Microcavia australis)* the smallest of the three wild guinea-pigs. They live socially in small groups, using a common burrow with several entrances, spend several hours a day in the shade of the bushes, darting at full tilt across open spaces.

Much less visible and obvious are several other rodents like the Tuco-tucos (several species of the genus *Ctenomys*) and Coney Rats *(Reithrodon auritus)*, or the lovely Darwin's Leaf-eared Mouse *(Phyllotis darwini)*, large ears and long tail, proceeding in leaps and bounds, a behaviour similar to Gerbils and Kangaroo-rats from other arid parts of the world.

The rocky slopes of western Patagonia and especially in the tumbled rockfalls at the base of cliffs are home to the shy but charming Mountain Viscacha *(Lagidium viscacia)*. This is a rabbit-sized rodent with large ears, and a thickly hirsute upturned tail - a punky squirrel. It runs along and leaps from rock to cornice with fantastic agility, takes refuge between and under the rocks, in burrows, and spends most of the day sunning. It feeds on grasses in the vicinity.

Often seen are the armadillos which are represented in this region by two species with hair, the Patagonian Pichi *(Zaedyus pichyi)* and the Hairy Armadillo *(Chaetophractus villosus),* the former also found in the Monte, the latter also in the Pampas.

Predatory mammals here are the same as those of the Monte; they feed on furred and feathered herbivores, but are becoming ever more rare in Patagonia. The Puma and the Red Fox are today restricted to the Andean foothills where relief and the woods provide a refuge from man. Geoffroy's and the Pampas cats as well as the grisons are hard to find. The Gray Fox and the Patagonian Hog-nosed Skunk *(Conepatus humboldtii)* which were abundant not so long ago are victims of the great demand for their pelts.

The strangest little predator is a small patagonian marsupial *(Lestodelphis halli)* rather like the murine opossums from further north, barely 14 cms long (without the tail), which in spite of being endemic is only known from a few specimens which have been found. Obviously virtually nothing is known of its habits, though small birds may be its main staple.

The only bat in the Patagonian steppes is the Big-eared Brown Bat *(Histiotus montanus)* of wide distribution. Here it roosts in bushes.

The most abundant bird of prey is the Red-backed Hawk *(Buteo polyosoma)* which has the habit of perching on telephone poles alongside the roads as observation posts to spot its prey, making it an unfortunate target for the many drivers of vehicles who carry rifles. The raptors of this biome are virtually those of the Monte: the Black-chested Buzzard-Eagle, the Crested Caracara and the Chimango Caracara, the American Kestrel, Aplomado Falcon, Peregrine Falcon and Cinereus Harrier. Whereas they are fairly rare in the eastern part of Patagonia they are quite abundant in the foothills of the Andes in the west.

Some of the owls which have wide distributions in the rest of the country are here too - the Burrowing Owl, the Short-eared Owl *(Asio flammeus)* and the Barn Owl *(Tyto alba)* as well as a form of the Great Horned Owl *(Bubo virginianus magellanicus)* typical of the Andes and Patagonia, nesting often on the ground between the scrubby bushes, and frequenting canyons.

Particularly beautiful is the Tawny-throated Dotterel (*Oreopholus ruficollis*), singly or in family groups patrolling the upland grasslands in search of the invertebrates on which it feeds, running a distance to stop and hold itself very upright, courser-like, displaying the black horse-shoe on its belly. Its mournful whistled flight-call is one of the most haunting of sounds.

In tree-less areas such as this many birds must become runners, living on the ground. So there are several ground Furnariids like the Common Miner (*Geositta cunicularia*), the Band-tailed and Scaly-throated Earthcreepers (*Eremobius phoenicurus* and *Upucerthia dumetaria*), all dull-coloured brown-jobs like the rest of the family, though the Band-tailed Earthcreeper does hold its chestnut and black banner-tail aloft and brightly visible when running around. The tyrants (*Tyrannidae*) also have a genus of terrestrials, the Ground-Tyrants (gen. *Muscisaxicola*) with several species out on the flats, on rocky slopes or in the high Andes, the Puna and Patagonia. The Chocolate-vented Tyrant (*Neoxolmis rufiventris*) is one of the largest and most robust birds of the family and falcon-like in flight. These characteristics, coupled with its plumage, a striking combination of black, silvery-gray and cinnamon, make it stand out in the short vegetation which it frequents.

Grass seeds and the small fruits of the bushes are a resource exploited by the Fringillids, the other family of well-represented birds of the Patagonian steppes. The Common Diuca-Finch, (*Diuca diuca*), Gray-hooded and Mourning Sierra- Finches (*Phrygilus gayi* and *P. fruticeti*), the Black-throated and Yellow-bridled Finches (*Melanodera melanodera* and *M. xanthogramma*), the latter descending from alpine habitats in winter, the Patagonian Yellowfinch (*Sicalis lebruni*) and the ubiquitous Rufous-collared Sparrow (*Zonotrichia capensis*) are the commonest of this family.

Many of these birds have to migrate out of Patagonia during the inclement winter months, so flocks of Dotterels, Seedsnipe, Chocolate-vented Tyrants, Diucas and Mourning Sierra-Finches all move north into the Pampas grasslands or the Monte.

The most conspicuous migration is that of the Upland and other Sheld-geese (*Chloephaga picta* et al.). In summer they breed in the wetter valleys and around lakes and ponds, moving up to the verdant grasses of Buenos Aires during the winter. Here they feed on the tender winter wheat which mostly does it good in spite of what the farmers say.

The lesser, cold-blooded creatures of the region include some modest *Liolaemus* lizards sunning on a rock or clump of vegetation under which they generally have their burrows. About 20 cms long, on spindly legs ferocious Lion Lizards (*Leiosaurus belli* and *Diplolaemus bibronii*) resemble nothing so much as miniature tyrannosaurs, with their huge heads and small beady eyes. Their main food is beetles. A small nocturnal gekko *Homonota darwinii* feeds on arthropods, especially spiders; one single species of poisonous snake *Bothrops ammodytoides* is small, timid and not aggressive seldom being more than 40 or 50 cms long, with an up-turned snout.

It must be fairly obvious that this semi-desert offers little promise for agricultural practices, so the habitat has not been substantially transformed; sheep-farming however, which is the only possible exploitation on an extensive scale, through overgrazing by too many sheep, is causing a continuous and aggravating generalized desertification of the environment.

For a long time wildlife survived together with this type of activity - except the greater carnivores which were virtually eradicated, and a reduction of the numbers of Guanacos - but today there is such a demand for pelts and rhea feathers that most farm workers and itinerant labour gangs supplement their wages with a bit of trapping on the side.

The regression of wildlife in this biogeographical province is today accelerating and it is mandatory therefore to ensure its survival in nature reserves. The biome is insufficiently represented in the system of preserved areas of the country. A small area of the Andean district is covered in the buffer National Reserves surrounding the National Parks of Los Glaciares and Perito Moreno, but these reserves have human settlers and sheep-farming with the corresponding alterations. The Western district is represented in a tiny belt around Laguna Blanca National Park. The Santacrucense district has a fraction preserved in the Bosque Petrificado Natural Monument which is too small to be of any real value.

The most diverse vegetation and abundant wildlife of this Patagonian Province is to be found in the Chubutense district which has no portion protected nationally though the province of Chubut, realizing the value of its wildlife to attract tourism, has made a start by declaring the Valdés peninsula a provincial Reserve.

This landform is of particular interest because, thanks to the marine influence which moderates its climate and raises the ambient humidity a point or two, there is a great abundance and variety of vegetation communities and, consequently of fauna. Bordering as it does on the Monte, it is to a certain extent an ecotone where several plants of the Monte are present, such as the Piquillín (*Condalia microphylla*), the Alpataco (*Prosopis alpataco*), and Monte Negro (*Bougainvillea spinosa*). There are special attractions like great salt-pans some 50 m below sea level, a 20 kms long "Chesil" beach - all storm- and current-accumulated rounded stones and pebbles which has created some islands that have never been grazed.

The shape of the Valdés peninsula itself is conducive to good management as a nature reserve with a bottle-neck entrance for control of access and check-points. The whole system could be globally managed with the litoral where, as will be seen in a later chapter, there is much to be protected, as well as a marine park, which Golfo San José pretends to be.

Another valuable area to be studied for its possible inclusion as a National Park is the Somuncurá plateau, some 15,000 sq. kms of basalt uplands, partly in Rio Negro and partly in Chubut. The elevation is around 1,000 to 1,500 metres above sea level with peaks of upto 2,000. Communities on the plateau are typically Chubutense though the slopes at between 900 and 500 m are ecotonic between Patagonia and the Monte.

Barely exploited - it has a little sheep-farming on it - Somuncurá has healthy populations of many patagonian species, even of the Red Fox (*Dusicyon culpaeus*), but enriched with a number of endemics: in the numerous small temporary lakes from rainfall and snow-melt accumulating in clay-bottomed depressions, which last for a few weeks, can be found *Atelognathus reverberii*, a small speckled frog with round reddish spots on a gray background; most water filters away through the basaltic substrate and reappears in springs along the slopes, giving rise to streams which are the home to yet another endemic frog, *Somuncuria somuncurensis;* the spring which gives rise to the Valcheta stream is the only habitat of a curious little scaleless fish the Mojarra Bronceada

Darwin's Rhea

This smaller species also known as the Lesser Rhea (Pterocnemia pennata) is, after the guanaco, the most visible form of wildlife of the patagonian scene. It is still abundant in certain areas in spite of the hunting pressure it undergoes for its feathers for making dusters.

The White-tufted Grebe

This grebe (Podiceps rolland chilensis) *is one of the six species of grebes found in Argentina and its white "ear"-tufts are the distinguishing feature. Widespread and abundant it can be found on just about any body of water in the country.*

Chiloe Widgeon

Perhaps the smartest duck on the patagonian lakes Anas sibilatrix, *is also the most abundant. As the name* sibilatrix *implies its call is an attractive whistle. They are grazers and this bird is selecting parts of the aquatic weed washed up on the shores of Laguna de los Escarchados.*

Andean Ruddy Duck *This diving species* (Oxyura j. ferruginea) *is characteristic of many of the lakes in the foothills of the Andes including Laguna Blanca. During courtship the male (with the black head and bright blue bill) points its tail skywards and, inflating its neck, beats its breast and says "come 'ere".*

or Desnuda *(Gymnocharacinus bergi)* a relict species of the *Characidae*; lastly the rocky slopes of the meseta are the home of a race of the Mountain Viscacha *(Lagidium viscacia somuncurensis)*.

From this it can be seen that both Valdés peninsula and the Somuncurá plateau have geographic and biological characteristics which make them ideal areas to be part of the National Parks system.

Bosque Petrificado Natural Monument. The first land plants appeared in the Devonian period some 350 million years ago. Since then they have been evolving to become what we see today. During this long period there have existed all manner of woods and forests formed by many different shapes and sizes of trees belonging to different taxonomic groups, many of which are now extinct.

Some of these forests have been submitted to petrification, whereby the most perfect fossils are obtained. In order for this to occur a medium rich in mineral salts (such as siliceous and calcareous) must be present, as well as little or no destructive organic or chemical processes, so that the plant remains can eliminate all their water and replace it molecule by molecule with mineral solutions without changing the shape, nor losing the structure of the original sample. The parts of plants which best support this process are the woody trunks and branches.

In Argentina there are several petrified "forests", all of different composition; the best examples are in Patagonia thanks perhaps to the constant and intensive volcanic activity which provided the basic material for petrification, silica.

A wonderful example is the Madre e Hija (Mother and Daughter) petrified forest about 150 kms west of Puerto Deseado, declared a Natural Monument in 1954. The trunks are in the La Matilde geological formation of the upper-middle Jurassic (about 150 million years ago). At this time conditions here were very different from what they are today, with a temperate and uniform climate as the Andes mountains did not exist and the moisture-bearing winds off the Pacific supported an exuberant vegetation. In the east there was no Atlantic and South America was joined to Africa. The first invasion of Patagonian latitudes by the sea was at the end of the Cretaceous and beginning of the Tertiary.

Some volcanic cataclysm buried extensive forests like this one and many of the trees were fossilized. Most of the trees fell (or were knocked down) and lie in an east-west direction though some of the stumps were left standing in which position they were turned to stone, roots and all. Cones were preserved also, both the male and female ones in which the embryos - the most delicate part - can still be detected. The petrified remains belong to Araucarias, described as *A. mirabilis* on the basis of the cones found, but there are also other conifers and fungi of the fomitoid group which lived on the bark of the trees like horizontal, semi-circular shelves or platters. There are also sedimentary rocks with the imprints of leaves, especially fern fronds, Benettitales (now extinct) and Cycads, there being no record of angiosperms which only appeared in the Cretaceous, the next geological period. In the same formation there are fossil skeletons of primitive anurans.

The "forest" occupies an extensive area of the Monument of 10,000 has. Some of the trunks are 30 m long and 2 m in diameter.

Waterfowl Sanctuaries: Laguna Los Escarchados Wildlife Reserve and Laguna Blanca National Park. Patagonian bodies of water, both the countless salt lakes as well as the fresh water ones - these last more numerous at the foot of the Andes where they are fed by snow-melt - all have a rich avifauna which includes many of the species that occur in the Pampas: the Black-necked Swan *(Cygnus melancoryphus)*, the Coscoroba *(Coscoroba coscoroba)*, the Brown Pintail *(Anas georgica)*, Yellow-billed Teal *(Anas flavirostris)*, Red Shoveller *(Anas platalea)*, and the striking Chiloe Widgeon *(Anas sibilatrix)* which here forms large flocks, as well as endemic specialities like the Crested Duck *(Lophonetta specularioides)* and the Flying Steamer-Duck *(Tachyeres patachonicus)*. The Chilean Flamingo is also regular, often in small flocks and never anywhere for long; there are breeding lakes known in the region.

The shores of the lakes are habitat for a number of small shorebirds, some migrant visitors from the northern hemisphere like Baird's and the White-rumped Sandpipers, Wilson's Phalarope, others local, like the Two-banded Plover *(Charadrius falklandicus)*, the Rufous-chested Dotterel *(Zonibyx modestus)*, or the scarce and little-known Magellanic Plover *(Pluvianellus socialis)* which might be a family on its own for all its aberrant behaviour and physionomy. The Magellanic Oystercatcher *(Haematopus leucopodus)*, inland and breeding in summer, is found on upland short-grass meadows in the proximity of lakes where its pure whistled calls can be heard.

Many of the patagonian lakes have a dense growth of the submerged water-weed *Myriophyllum elatinoides* whose tips just break the surface and carpet the whole area with their red colour. This weed is the material and the anchor for the nests of such as the Red-gartered Coot *(Fulica armillata)* and colonies of the Silvery Grebe *(Podiceps occipitalis)* all of which build up the floating platform nests on which copulation takes place and where the eggs are incubated. In the strong winds of Patagonia coots' nests are sometimes torn from their moorings and drift downwind through other coots' territories before they fetch up on the shore, which gives rise to severe and splashy battles.

Grebes are excellent divers and propel themselves underwater with their feet with flattened toes in search of their prey, the fish and invertebrates on which they feed.

Their attractive plumages with ear-tufts as in the White-tufted Grebe *(Podiceps rolland)* or bright chestnut and black of the Great Grebe *(Podiceps major)*, as well as their behaviours and courtships with "dances" along the top of the water for example, and their habit of carrying the chicks on their back even, when small, while diving, all make grebes one of the favourite birds on any lake.

The attention of ornithologists was riveted on this family when in 1974 Maurice Rumboll described a new and very different species of grebe from a lake in southern Patagonia. This newly-discovered Hooded Grebe *(Podiceps gallardoi)* is white with dark gray back and a black head, topped with a frontal crest of chestnut. It is a sociable species which also nests on water-weed platforms. Activity around the breeding colony is permanent and accompanied by the birds' whistled and trilled calls.

There is an estimate, based on recent counts on many lakes in the area, of a total population of around three thousand birds. Nobody yet knows where they go in winter as they must leave the breeding lakes which all freeze over. The lake they were originally discovered on, Laguna de los Escarchados, has been declared a wildlife refuge by the province of Santa Cruz. As it depends on snow-melt from heavy winter snows which occur only periodically, the lake is often almost dry which led to the search in the whole region. The reserve which had a permanent ranger while the grebes nested there, also protects the other water-birds and shorebirds. The Argentine Wildlife Foundation was instrumental in getting the legislation through for its protection.

Some 2,000 Black-necked Swans nest on Laguna Blanca, and it was with a view to protecting these that the lake was declared a National Park in 1945. It is set on a plateau studded with gentle, conical hills, with volcanic soils and a great basalt sill spread beyond the north shore. The vegetation is typical of the Western district where thorny, dry bushes dominate the grasses. There are Neneo, Duraznillo, Molle, Charcao and other Senecios, *Haplopappus pectinatus* and a composite *Nasauvia axillaris,* which here replaces the congeneric Colapiche. There are also grassland patches of almost pure Bitter Coirón. On parts of the shores as well as along the few feeder-streams more luxuriant grasses grow, like the Pampas Grass *(Cortadeira pilosa),* and in damp bottom-land the boggy Mallín vegetation dominates with reeds and sedges. In the water itself there are two water-plants, one the already mentioned *Myriophyllum* and the other is *Potamogeton pectinatus.* They constitute luxuriant underwater grazing for the myriads of birds on the lake. Uprooted and washed up on the down-wind shore they are the beds of rotting vegetation which attract the insects on which so many shorebirds feed.

The abundant and varied avifauna is the chief attraction of the park, and the Black-necked Swan plays a leading role, with its pure white body, black neck and red caruncle at the base of the bill, larger in the cob than in the pen, in the adults than in young birds. In August and September when the breeding season begins they build their nests on the banks with the available vegetation, or preferably on the small islands. There they lay three to five eggs and incubate them for 35 days before they hatch. Cygnets follow their parents on the water, but when tired, wet or cold take to the comfort of the parents' backs, under the wings with the little head sticking up through the feathers. Another characteristic duck in the area is the Andean Ruddy Duck *(Oxyura j. ferruginea)* a "stifftail" with remarkable courtship and territorial displays. In September the males, with tails erect and necks inflated, drum on their breasts with their bright blue bills producing a peculiar rapid dum-dum-dum-dum followed by the vocalization which sounds like "come 'ere". Chases across the surface and other such energetic aggressive displays are all witnessed by the females in complete calm. Even the human intruder might be displayed at to remove himself.

The multitudes of birds there, including nesting flamingos, great numbers of ducks, shore-birds and so on are no match for the quantity of Silvery Grebes which inhabit the lake. Coots and swans take second place. Come spring these small grebes congregate to build true colonies of nests - several of over 200 nests each - which are the stage for frantic comings and goings, squabbles and fights, borrowing and stealing, chasing and being chased, courting and copulating - a wonderful oportunity for anybody to be initiated into the fascinating subject of bird behaviour.

Strangely enough the fauna of this lake did not include fish of any kind but one day in the recent past a van drew up

Laguna Blanca National Park

Foremost among sub-Andean lakes not only for its size but for the numbers of Black-necked Swans (Cygnus melancoryphus) which can be found there is Laguna Blanca. Other birds which are particularly abundant on the lake are the Silvery Grebe and the Red-gartered Coot.

and tipped a load of fingerling trout into the lake, without permission of any kind and completely against the law, then drove off, all while the ranger in charge was trying desperately to get clear instructions from head-office over the radio. The effect of this is yet to be evaluated, but it is expected to throw the whole ecosystem out of tilt. Already the endemic aquatic frog *Atelognathus patagonicus,* known only from this lake is feared lost.

Another aspect of interest in the Park are the cliffs in certain parts where birds of prey nest: the Red-backed Hawk and the Peregrine Falcon.

As in so many other cases, the borders of the 11,250 has park are not adequate as a sector of the lake is not included and the park cannot properly function till this error is made good.

Another problem is the intrusion of sheep from the neighbouring farms which overgraze and destroy some of the vegetation needed for nesting birds as well as starting erosion. This will have to be remedied as soon as possible with a fence all around the park. Settlers must be removed and the grasslands allowed to recuperate.

There is little infrastructure to cater to the tourists' needs but it is hoped to build a shelter, a visitors' centre and so on in the not-too-distant future, as well as tracks or roads for visitors to explore the park in their vehicles.

The Hooded Grebe

Podiceps gallardoi *was not discovered until 1974. Restricted in summer to the high plateaux around Lagos Argentino and Viedma, nobody yet knows where it spends the winter. Its reduced numbers and its vulnerability made it one of the target species of the conservation effort of the Fundación Vida Silvestre Argentina (Argentine Wildlife Foundation). The reddish hues of the surface of the water are from the weed (*Myriophyllum) *which grows there and which is the essential nesting material on which these two birds are resting.*

Laguna de Los Escarchados

*High on the patagonian plateaux where Coirón (*Stipa*) steppe vegetation dominates is this lake protected as a nature reserve where the emerging tips of the aquatic vegetation* Myriophillum elatinoides *give it a red hue during the summer months.*

Two-banded Plover

*Though most of the shorebirds found in the country in summer are migrants from North America, there are some local species also. Like this Two-banded Plover (*Charadrius falklandicus*) mostly nest in Patagonia and move northwards in winter. Plovers' nests are notoriously hard to find as the eggs are so well camouflaged for the habitat.*

The Atlantic Seaboard

From the southern tip of the River Plate estuary Argentina has an extensive sea coast - more than 4,500 kms long - off which lies an immense continental shelf under the Atlantic Ocean with several recognizable steps at 35, 80, 110 and 140 metres depth. The continental shelf has an area of about one million sq. kms, and is 869 kms across at its widest. From it, islands rise to the surface and emerge, such as the Falkland (Malvinas) archipelago. The waters over the shelf are essentially sub-antarctic being brought up by a cold current flowing north which is called the Falkland (Malvinas) Current. This meets the warm Brazil Current in the northern part of this Argentine Sea where certain subtropical species of fauna exist.

The sub-antarctic waters are rich in nutrients and so support an abundant plankton - mostly diatoms which are drifting microscopic algae, foraminifera and copepods. Specially interesting members of the zooplankton are the larval stages of "lobster krill" *(Munida gregaria)*, a decapod crustacean with its lobster-like claws which give it its common name, and because it congregates in huge masses comparable to the better-known krill *Euphausia* spp.; all these constitute an excellent source of food for fishes, birds and marine mammals.

Two further important links in the food-chains of these seas are the small fishes of the Clupeiform order, which feed on zooplankton, the Fuegan Sardine *(Sprattus fuegensis)* and the little Anchoíta *(Engraulis anchoita)*, 12 and 17 cms long respectively as adults. In vast migratory shoals they are the food of the larger predatory fish of commercial value: Hake *(Merluccius merluccius* and *M. australis)*, the Long-tailed Hake *(Macruronus magellanicus)*, Pollock *(Micromesisitius australis)* and Abadejo *(Genypterus blacodes)*. Mammals and birds also feed on them, as do pelagic predators from the north - the Mackerell *(Pneumatophorus japonicus)* and the Bonito *(Sarda sarda)*. Pelagic plankton-feeders include the Pampanito *(Stromateus brasiliensis)* and Palometa *(Perona signata)*.

If we were to represent all living beings in this sea as a pyramid, each placed according to its feeding relationship to others, the tip would be occupied by the marine mammals: the Southern Fur-seal *(Arctocephalus australis)*, the South American Sea-lion *(Otaria byronia)* and the Southern Elephant Seal *(Mirounga leonina)* of the pinnipeds; the most common cetaceans are the Southern Right Whale *(Eubalaena australis)*, the Bottlenose Dolphin *(Tursiops truncatus)*, Dusky Dolphin *(Lagenorhynchus obscurus)* and Peale's Dolphin *(L. australis)* - the former in the northern waters, the latter around the Falklands (Malvinas) and Tierra del Fuego - Commerson's Dolphin *(Cephalorhynchus commersoni)* or affectionately Puffing Pig to the Kelpers, in coastal waters from San Jorge Gulf southward, and the Orca *(Orcinus orca)* which is the super-predator which takes pinnipeds, dolphins and on occasions even whales.

The presence of a shoal of Anchoítas would attract Dusky Dolphins which herd the fish upward and keep them against the surface for easy capture, while oceanic sea-birds take advantage of this situation - the Black-browed Albatross *(Diomedea melanophrys)*, Giant Petrels *(Macronectes giganteus)*, White-chinned Petrels *(Procellaria aequinoctialis)* and various Shearwaters *(Puffinus* spp.). Coastal birds do likewise if within range - cormorants, gulls and terns.

Patagonian coasts present many different aspects: sandy beaches, mud-flats, gravel beaches, shelving rock-ledges, cliffs with bars or shoals at their base, rock platforms exposed only at low tide which drop metres into the sea, of fine sedimentary rock with inclusions of many marine fossils. In many places tide-pools are common with their own vegetation and faunas. In the cliffs the sedimentary strata are easily distinguished, sandstones alternating with tuffs and clays, layers of fossil oysters and sea-shells, all covered with today's sand-dunes.

The diversity of the substrate (hard or soft bottoms) and the great tidal range - 7 metres between high and low, and in some placed upto 14 - offer a gamut of environmental niches for organisms adapted to different time-periods of being covered by the sea or exposed to the air, or different surroundings and habitats. Thus on rock ledges the small limpet *Pachysiphonaria lessoni* lives, as grow the beds of the three mussels, the large *Aulacomya magellanica,* the middle-sized *Mytillus magellanicus* or the tiny *Brachyodontes purpuratus,* and barnacles of the species *Balanus psittacus,* though these shell-fish are much less important here than in other places, as on the Chilean coasts.

In nooks and crannies between the rocks the local octopus *(Benthoctopus tehuelchus)* find refuge as do Meros *(Acanthistius brasilianus),* serranid fishes which can reach half a metre in length and weigh 3 kgs.

While on the stretch of shore exposed by the tides there are abundant but small benthonic algae, below low-water mark the giant kelps grows into forests *(Macrocystis pyrifera, Lessonia* sp. and *Durvillea* sp.), but in the northern parts these are replaced with the Rhodoficeae *Codium fragile* and *C. decorticatum.*

The benthic fauna, that which lives on the sea-bed, includes, as well as those species mentioned, sea-anemones, star-fish like the Giant Magellanic *(Cosmaterias lurida)* some 30 cms. across, the orange *Ceramaster patagonica* and the Serpent Star *(Phioceramis januarii);* sea-urchins *(Arbacia dufresnei* and *Pseudechinus magellanicus);* bivalves other than mussels and their family, like clams *(Marcia exalbida, Chione antigua* and *Darina tenuis),* scallops *(Pecten tehuelchus)* and the oyster *(Ostrea puelchana);* gastropod molluscs like the conchs *Zidama angulata, Voluta* sp. and *Bullia* sp.; attractive nudibranchs, the armadillos of the sea *(Chaetopleura isabelli* and *C. tehuelche),* polichaets and a great variety of decapod crustaceans. Among the truly benthic crustaceans, the crawling bottom-dwelling crabs, the Centolla or Fuegan King- crab *(Lithodes antarcticus)* is the best known as it is a species of high commercial value, with long legs and all covered with spiny protruberances, the Centollón *(Paralomis granulosa)* which is starting to be exploited, the Patagonian Crab *(Platyxanthus patagonicus)* also harvested, *Peltarion spinulosum* and *Ovalipes punctatus* so abundant that they define a community of the sandy-bottomed infralittoral. There are also Hermit-crabs like *Pagurus comptus* and some others from the northern sector limited to below low-water mark *(Leucippa, Leurocyclus* and *Libinia),* and on sand as the previous species. Between the high and low tide-lines, the common crab is *Halicarcinus planatus* and above the high tide, in the supralittoral, *Cyrtograpsus angulatus.*

Some decapod crustaceans, though they are found on the sea-bed, are good swimmers and capture or find their food at intermediate depths. This is the case of the Bogavante, and in the north, of the two species of commercial value, prawns and shrimps *(Pleoticus muelleri* and *Artemesia longinaris),* typical of the sandy-mud off the coasts of Buenos Aires, which feed on diatoms and are therefore fundamentally herbivorous.

Some fish are almost exclusively bottom-dwellers like *Pseudorhombus isosceles* and *Xystreuris rasile*, plaice- or flounder-like flat-fish with oval bodies and both eyes on one side of the head, upto 80 cms long and weighing 8 kgs, lying on the bottom in perfect mimicry; or the rays - *Raja flavirostris, Bathyraja brachyurops, Psammobatis scobina* and the Torpedo or Electric Ray *(Discopyge tschudii)* among others - flopping over the bottom in search of crabs and worms though they do also take demersal fish, those of the water near the bottom.

A small shark which is typical of the Falkland Current is the spotted *Halaelurus bivius* whose body of upto 60 cms is pale chestnut with white and reddish spots.

A benthic-demersal fish of shallow water over a sandy bed is the curious-looking Cock-fish or Elephant Fish *(Callorhynus callorhynus)* which gets its name from its long snout in the shape of a trunk ending in a fleshy knob. Another fish of that same habitat is called the Turco *(Pinguipes fasciatus)* less well-known than the other fish of that genus the False Sea Salmon *(Pinguipes somnambula)* which can grow to one metre long and weigh 11 kgs, much sought by sport fishermen.

Scorpionfish *(Scorpaneidae)* are also typical of the sea-bed. They are grotesque, with heads and eyes disproportionately large and fins with bony radii, represented on this coast by the Cabrilla *(Sebastes aculatus)* and its close relative the Pig Fish *(Congiopodus peruvianus)* notable for its large and tall dorsal fin, also held rigid by bony radii, which looks like a sail.

The *Nototheniidae* is another family of bottom-dwelling fish, exclusive of these southern seas, here represented by 18 species in the genus *Nototheniia*, the Patagonian Robalo *(Eleginops maclovinus)*, Black Hake *(Dissostichus eleginoides)* and two species of the genus *Harpagifer*.

There are also demersal cephalopods - squid, mainly *Illex argentinus* and several species of *Loligo* - which being good swimmer move around in great concentrations providing a valuable food source for larger creatures.

The invertebrates and fish which live on the shore between the tide-marks attract a great variety of coastal sea-birds which seek them on these beaches, on the rock-ledges and especially in the microhabitat of the tide-pools.

These include the gulls: Kelp *(Larus dominicanus)*, Olrog's *(L. atlanticus)*, Brown-hooded *(L. maculipennis)*, and Dolphin Gull *(Leucophaeus scoresbii)*; the Snowy Sheathbill *(Chionis alba)*; the skuas *(Catharacta* spp.*)*; Common, Magellanic and Blackish Oystercatchers *(Haemantopus palliatus H. leucopodus* and *H. ater)*; plovers, chief among them the Two-banded *(Charadrius falklanicus)*, and several migrant sandpipers from the northern hemisphere where they breed: White-rumped and Baird's *(Calidris fuscicollis* and *C. bairdii)*, Sanderling and Red Knot *(C. alba* and *C. canutus)*, and the elegant Hudsonian Godwit *(Limosa haemastica)*.

Many other sea-birds find, on the patagonian shore, places to rest or even to nest. There are enormous penguin "rookeries" of the Magellanic Penguin *(Spheniscus magellanicus)*, tight Cormorant colonies *(Phalacrocorax atriceps, P. albiventer, P. magellanicus, P. olivaceus)*, and even small nuclei of *P. gaimardi* and *P. bougainvillii*, large concentrations of breeding Kelp Gulls with patches of the smaller Dolphin Gull, colonies of South American Tern *(Sterna hirundinacea)* which often include the Cayenne *(S. eurygnatha)* and the Royal *(S. maxima)*. Sometimes a pair or two of the Snowy-crowned Tern *(Sterna trudeaui)* join these colonies.

The pinnipeds already mentioned also haul out to breed on rock ledges, beaches, promontories. Species here are all gregarious breeders - sea-lions, fur-seals and elephant seals. These last start first in spring, the others in high summer. Activity is rampant in these colonies as bulls fend off challengers for dominance of the harems.

Sanctuaries for Sea-Lions. Rock ledges at the foot of tall coastal cliffs, or gently-sloping gravel beaches are the places where South American Sea-lions *(Otaria byronia)* haul out every year to establish their breeding colonies.

These pinnipeds, known collectively as "eared" seals have their four limbs turned into big flat plippers, the hind pair capable of being brought forward to serve as short legs on which the animals can walk, albeit with a heavy and clumsy gait, and even gallop for short distances. They show a marked sexual dimorphism, the males weighing twice as much as the females - some 500 kgs and 2.5 m long - and can be told from these by their "bull" necks covered with a thick coarse mane and their blunt, squarish snouts.

The general impression a colony or "rookery" gives at the beginning of summer is of a huge number of smaller females piled on a beach, attended by a much smaller number of males spread out among them and easily distinguishable from them. They are polygamous, each male keeping a harem of a variable number of females - ten or so being normal - each harem so close to the others that the borderlines are confused. Dominant "Sultan" males therefore spend a considerable proportion of their time in threatening postures and in short skirmishes which serve to reaffirm their claims.

Non-harem bulls spend their days on the outskirts of the colony in the hope of nabbing some female, and usually congregate in "bachelor clubs" a short distance away.

Territorial claims and battles are the most noticeable characteristics of the behaviour of male sea-lions. Having established their patch of beach during the spring they defend it actively against any challenger. Most of these battles consist of much roaring, snorting and teeth-rattling, a short lunge in the direction of the adversary, feints and sparring with a few snaps and bites exchanged, all followed by an immediate cease-fire. But occasionally there appears an earnest contender who is ready to dethrone the sultan and in these cases the contest takes a bloody turn.

Most of the pups are born during the first fifteen days of January; the new-born pup is about 80 cms long and covered in a soft black fur. For the first few days the pup is permanently cared for by its mother.

The females are not allowed to return to the sea until they have been mated, the male forestalling any attempt in this direction by catching the fugitive by the neck, lifting her and throwing her back into his territory.

Females come into oestrus some eight days after giving birth, when an intense love-match starts: sensual nibbling of mouth and neck regions incite the male, mutual petting with rhythmic movements of the necks, until the male, checking by scent that the female is ready, finally mounts. Holding her firmly with his fore-flippers he heaves his considerable bulk on top of her which visibly flattens her.

After copulation the female is allowed to return to the sea to feed. The pups then gather in nurseries where they spend a considerable time in play, waiting for the return of their mothers. Whenever a female emerges from the sea she emits a series of calls and, from the chorus of answers can identify

Cliffs and rock ledges

The coast at Punta Pirámides is an excellent example of cliffs resting on rock ledges at their feet, typical of many places along the patagonian seaboard.
Of sedimentary clays and sandstone they form stratified layers and often contain interesting fossils.

The South American Sea-lion

As is the case with all Otarids (eared pinnipeds) and other polygamous species, in the South American Sea-lion (Otaria byronia) there is a marked difference between males and females. The mane and "bull" neck of the males contrasts with the graceful figure of the cows which make up his harem. One of these uses a pup as a pillow.

Southern Fur-seal

Arctocephalus australis *are not seals at all but smaller sea-lions - their ears are visible and hind- flippers opposable. They breed on rocky promontories as here at Cabo Blanco, and on ledges at the base of cliffs. Virtually extirpated in the XIXth century for their pelts, they are making a slow come-back. This is one of the very few rookeries in continental Argentina.*

Orcas feed on Sea-Lions

Though many Orcas feed on marine animals, they take Sea-Lions from the beaches and adjacent waters near the breeding grounds. In the final lunge after these they often grind right up onto the gravel but are only briefly aground - on the next wave they wriggle back into the sea.

her own pup and goes to meet it. Before suckling is allowed however, she checks its identity by smell.

Gradually the pups learn to swim and can accompany their mothers to sea. They stay close till the next breeding season, suckling, though as from the sixth month they start taking solid food.

When all the females have been covered the sultan loses interest in the harem and the territory. Noticeably thin and weakened by these months of fasting and continuous battles he returns to sea to feed and recuperate. The breeding colony slowly breaks up.

These sea-lions do not migrate, they merely go out to sea for their food - squid, crustaceans and fish. They return to haul out on their beaches at any time of the year to rest. At these times there is no segregation whatever and males, females and young pile up in the sun in a single pack. It is only in August or September that the males' territorial instincts become manifest.

Several of these breeding areas are protected as provincial reserves where access for the public is limited to certain observation areas, under the supervision of wardens. These are at Punta Bermeja in Rio Negro province which has an interpretive centre, at Punta Norte and Punta Pirámides on the Valdés peninsula, and at Punta Loma, all these last in Chubut province.

The Seas Around Valdés Peninsula. San José and Nuevo gulfs are north and south of the Valdés peninsula; they are more like maritime lagoons because of their narrow mouths, and because of being fairly shallow, with marked tidal fluctuations. Their shores are a succession of gently- shelving beaches alternating with tall cliffs. These gulfs are the meeting places for the giants of the seas. After a summer and autumn at sea far from here, the Southern Right Whales *(Eubalaena australis)* arrive to breed.

These are true whales in the sense that they have the filtering balleen plates in lieu of teeth, which strain out the food - plankton and small fish - taken in with a huge mouthfull of water.

Right Whales were the first to be hunted by man. Their slow speed, their proximity to the coasts and their property of floating when dead because of their fat, made them the "right" whales to kill with the then less-sophisticated weapons at man's disposal - oared long-boats and hand-held harpoons. The fat and the balleen were highly valued, the latter for corsetry, watch-springs, umbrella ribs, furniture and so on.

For centuries the most active whaling industry was that of north-east North America, taking Right and Bow-head Whales along the shore, and later harvesting the Sperm Whale at sea. By the middle of the XIXth century the whaling fleet included some 700 vessels off the coasts of Argentina, harvesting the slower whales (Right, Sperm, Humpback). In this way the Southern Right was virtually exterminated last century, even before the advent of modern whaling.

About 1935, when an international agreement was signed with the aim of regulating whaling, Right Whales were so scarce that they were given total protection. Since then their recuperation has been slow, perhaps because not all the whaling fleets complied with the regulations established by the International Whaling Commission. Today it is estimated that the world population is under 5,000.

The Right Whale is upto 16 m long and weighs 50 tons.

It has no dorsal fin on its wide back, and the pectoral flippers are huge and trapezoidal. The most striking

The Southern Right Whale

Arriving in winter and staying for five months thereafter, these Southern Right Whales (Eubalaena australis) *which are among the most scarce, come to the shallow waters around Valdés peninsula to breed. Mating takes place here; in the lower photograph several males pursue a female. These whales often stand on their heads with tails out of the water and "sail", or vigorously slap the surface. They also breach, leaping high, almost out of the water on occasions, to fall back with a mighty splash.*

characteristic is the huge head, one quarter of the total length, with the lower lip high and arched to permit of an enormous gape, and the strange callosities, lacking in other species. These consist of thick patches of rough whitish skin on the head, where a few stiff hairs grow, and where certain crustaceans and various other ectoparasites make their home. These callosities are found on the rostrum between narines and snout, on the lower lips and above the eyes, and are of different shapes in every individual which makes each whale recognizable. There is an irregular patch of white on the belly which is a further way to identify the individuals, both useful charcteristics as scientists are able to recognize well over 400 which visit these areas.

What is perhaps the largest surviving population of these whales comes to the vicinity of Valdés regularly. They seem to prefer fairly shallow and sheltered waters to calve and raise their young for the first few months of their lives. So the San José and Nuevo gulfs, and to a lesser extent the waters east of the peninsula, are, from July to November, the sporting grounds of these leviathans.

The pups start life as little giants, six metres long and weighing two tons. During its first few months the calf will never leave the mother's side. With close on a one year gestation and a further year lactating, reproductive cycles take at least two years, usually three.

Another activity which takes place here is mating. Groups of several individuals can be seen in close contact pushing, twisting and turning, one swimming under another. These are all attempts at copulation where the female is attended by various males competing between themselves for her favour.

When November comes around these southern oceans enter their period of highest production. The long daylight hours allow for a blooming and fast reproduction of the diatoms and other microscopic algae adrift in the upper layers of the sea, and these marine pastures give rise to the great population explosion of zooplankton. This is when the southern whales leave their breeding-grounds and go to sea to feed on the abundant krill. Scientists still do not know just where these "feed-lots" are.

The north and east coasts of Valdés are also the breeding-ground of the Southern Elephant Seals. This truly maritime species spends most of its life at sea, far from the coasts, feeding on fish and squid which it catches even at great depths; but they must return to land to breed and to moult. Their distribution includes all southern seas and their breeding colonies are on the ring of islands around the Antarctic. Valdés is the only continental site where they breed, and the most northerly, so is the only place where they are easily visited for observation.

This Southern Elephant Seal *(Mirounga leonina)* is the largest of all true seals. They are the exponents of a very marked sexual dimorphism, males can be over six metres long and weigh more than two tons, while females are between three and four metres, and never weigh as much as 1,000 kgs. Only the males sport the short "trunk" which has given rise to their name. This appendage is in truth an inflatable proboscis which develops with maturity and increases during the breeding-season. It changes from a flacid protuberance some 30 cms long into a gross sounding-chamber for the powerful bellows of angry males.

Like all true seals, Elephant Seals lack any external ear and their hind flippers are not opposable so are of no use for terrestrial locomotion. They move around on land therefore rather like huge maggots, lunging forward on their bellies, sometimes leaning on their fore-flippers but more for balance than effect, dragging their flippers behind them with the short tail between the "feet". In the water, on the other hand, the thrust comes from the vigorous waving of this caudal and hind-limb appendage.

The vast difference in size indicates that here too there is a polygamous reproductive system. Each beach-bull or dominant male has a harem of some 15 to 20 cows, sometimes many more, with which he will mate while preventing other males from doing so.

Towards the end of July adult males arrive at the beaches and lay claim to sectors of the gravelly shoreline as their territory. Females follow, and their gregarious instinct leads them to form groups which in turn attract more females. Once a female has entered a territory, the male will do everything possible to keep her there.

About five days after her arrival the female gives birth to a 40 kg baby 1.2 metres long covered in a sleek black coat. The newly-born start their rapid development thanks to the mother almost literally passing on her thick layer of blubber through her incredibly rich milk (50% fat as compared with a cow's 3 to 6%). Lactation however is remarkably short, 28 days being the average, during which the female loses some 350 kgs. Upon weaning the female abandons her pup at this tender age, and returns to the sea to recuperate from her fast. The pup meanwhile remains on the beach till hunger drives it to the water to learn to fend for itself.

Because birthing is not synchronized the breeding season can last up to 4 months till the end of November, though the height of activity is between mid-September and mid-October.

Beach-bulls are no doubt the most noticeable beings here. Only at eight or nine years is a male large enough to challenge for and keep a harem. In spite of an apparent indolence he is always alert for the intrusion of other males, those sub-adults which try to sneak a little cuddle with his mates, which he must drive away. The approach of another adult male to dispute his title to the harem is far from sneaky and is preceded by open challenges: roars determine by their volume which bull is the stronger. Face to face they rear up on the hind-part of their bodies, trying to gain the upper hand with greater height, chest to chest. Rhythmically swinging back and forth, one and then the other lunges forward, mouth agape, to bite the neck region of the opponent with the large canine teeth. The battle ends when one of the two, exausted, abandons the arena. There will have been serious cuts and lacerations inflicted not only around the neck, back and shoulders, but also on the face and even the snout, scars which nearly all the dominant bulls display. For all this, battles are never fatal.

The male covers every female of his harem 18 days after she pups. Lying beside her much smaller body he holds her still with his flipper on her back - a form of embrace - and draws her to him. He does not mount but lies beside her, otherwise he would surely squash her.

When the breeding season is over they return to sea to feed up and repair the wear and tear, coming back to the land only for their annual moult. This takes place during the summer, lasts 30 to 40 days, and involves loss of the hair and epidermis which peel away in patches. The rest of the year they are at sea, their real home.

Valdés peninsula is considered by many to be the most valuable natural area in the country as far as fauna goes, so its exclusion from the Argentine National Parks system is a

Terns

major deficit; according to the pertinent law, parks must include the main natural treasures of the country.

Its curious shape of an axe being swung into the Atlantic gives it a long shore-line with various types of coast: yellow cliffs with rocky reefs at their foot, dark beaches of clayey sand, or gravel, all forming bays, gulfs, islands, points, and making Valdés one of the more spectacular coastal landscapes in the country. It possesses one of the most extraordinary concentrations of marine mammals in the world: add to the myriads of pinnipeds and the quantities of whales other interesting cetaceans like the splendid Dusky Dolphins first discovered by Darwin in San José Gulf, Bottle-nosed Dolphins sporting in this same gulf, Orcas regularly patrolling the seas just off the breeding colonies for their prey, or the ocasional visitors like Beaked Whales (*Mesoplodon* sp.) or the very rare Tasman Beaked Whale (*Tasmacetus shepherdi*).

An infinite number of sea-birds find their food off the shores of Valdés or a place to rest or nest on them; there are colonies of penguins, cormorants, gulls and terns, and even the

Terns nest in tight colonies. The most abundant is the South American Tern (Sterna hirundinacea) *with red bill and feet, and pearl-gray plumage. The Cayenne Tern* (Sterna eurygnatha) *with the yellow bill and black feet joins the colony to breed.*

giant oceanic nomads visit these gulfs - the albatrosses and petrels. As a testimony of the richness of the underwater world we have the fisheries, scallop harvesters, sport-fishermen and scuba-diving.

Aware of the importance of protecting this valuable natural heritage, the government of the province of Chubut has taken a number of measures to this end. Since 1967 a number of small reserves have been set up, policed by wardens and with the basic infrastructure for visitors at Punta Norte, Punta Delgada and Isla de los Pájaros (Bird Island) where Kelp Gulls and Olivaceous Cormorants nest amongst others, and which is connected to the mainland at low-tide. Access to the northern and eastern coasts was forbiden to visitors to the peninsula except where certain control can be exercised at Punta Norte, to minimize molestation of the colonies of breeding mammals there. In 1974 the San José gulf was declared a provincial marine park.

These laudable measures have had their ups and downs with the fluctuations of policy and economic priorities in the province. Often the personnel were untrained, support was lacking. A marine park of these characteristics requires a high degree of custody which demands a strong, well-equipped force of rangers and the development of sophisticated planning to allow visitors to enjoy the wildlife without interfering.

Whale-watching, for example must be permitted in such a way as to allow the visitor to get his experience without altering the peaceful surroundings the whales seem to require. It would be very regrettable - criminal perhaps - to drive the leviathans from their breeding areas, and accidents would be lamentable. There are more and more visitors each year and the pressure on some of the places or beings is becoming heavy. Private boats have been the cause of evident molestation of the whales as people try to approach to take their photographs and films.

So a careful study has led to licensees being given the concessions to operate under these circumstances, offering certain guarantees of safety and responsible behaviour. There is supervision by a warden in the area.

Cormorants

It is remarkable that six species of cormorants breed on the patagonian coast. The Rock Cormorant (Phalacrocorax magellanicus) of the lower left photo prefers cliff-faces and steep areas for its nest and can be identified by the red face and the black neck, sometimes speckled as in this specimen. The lovely Red-legged Cormorant (P. gaimardi - top) is restricted to the coasts of Santa Cruz province on the Atlantic side of the continent and is impossible to confuse - gray with red feet and face. The Guanay (P. bougainvillii - lower right) is the "billion dollar bird" of the Peruvian coast for its production of fertilizer. More slender, its eye is ringed green on a red face, and there is a white throat patch on the front of the black neck.

The changing policies, budgetary urgencies and so on that have to be faced at a provincial level are not good for the stability of management in a natural area such as this, so it is very important that the Nation take up its responsibility and declare it a National Park, freeing the province from the onerous costs and responsibilities of management, and guaranteeing a certain continuity in the protection of this special sanctuary.

Such a National Maritime Park would have to include all the coasts and the sea to a certain distance from the shores and thus cover a representative sample of the benthic habitat as well. It could complement a National Reserve on land, already covered in the previous chapter, to protect an area of extraordinary beauty and interest.

A Sea-Bird Sanctuary. Punta Tombo is a narrow point of land over three kilometres long and only half a kilometre wide or less which juts out into the sea. It is without doubt the home of the most formidable concentration of sea-birds on the whole of Patagonia's coasts. It consists of an outcrop of ancient crystalline base-rock (quartziferous porphyry and partly metamorphosed rock) covered in parts with sand, clay or shingle, with some wide beaches. Here sea-birds have established traditional colonies as well as certain other birdspecies which are not gregarious.

All around and at the base of the point is the largest colony of breeding penguins on the patagonian coast, estimated at some 600,000 nests. The Magellanic Penguin, a temperate climate species as are the others of its genus, nests in burrows to escape the heat as much as the scavenger and predatory birds. They also use the shade at the bases of bushes, so the colony is situated where the clay soil admits of burrowing, sometimes upto half a mile from the water. In the more concentrated areas the colony takes on the aspect of a city with upto 80 nests every 100 sq. metres. The noise and bustle here is urbanlike, especially as penguins at eventide stand beside their burrows and bray - not unlike donkeys - at the top of their voices. The periphery has nests at a density of 5 to 15 per 100 sq.m.

The well-chosen beaches by which all penguins pass going to or coming from feeding at sea, are crowded, rather like man's most popular bathing resorts, as many penguins pause there to preen, or even to rest.

These penguins return to their "rookeries" at the end of August or early September and usually head for their last-year's nest. After a period of courtship and reconditioning of the burrow, the two eggs are laid and the male and female take turns incubating. When, some 40 days later the chicks hatch, both adults tend to their needs. They share guard-duty while the chicks are small as the gulls and skuas, sheathbills and caracaras are continually patrolling in search of some unguarded nest. The consort at sea catches as many fish as it can and returns to the nest to feed the chicks a partly-digested meal regurgitated directly into their beaks while the parents' roles are reversed.

When the chicks, now covered in gray down, are over one and a half months old, their food requirements are such that both parents are engaged in procuring it, while they, unguarded, are too large to be easy prey for the birds mentioned above. The parents go to sea after breeding and hunger drives the chicks to seek their own food in the element they will be in for more than half their lives. Adults return to moult later in the season, a period when they loiter around the colonies and on the beaches, in filthy moods and irritable. When this is completed they set off on the yearly migration following the schools of Anchoítas, up the coasts to Brazilian waters off Rio and Cabo Frío. The colonies stand empty and the scavenging armadillos and foxes move in till next spring.

Further out along the point from the penguin colony is the breeding area of the Kelp Gulls *(Larus dominicanus)* with well-spaced nests a fair distance from each other, while the tight or compact colony of the Dolphin gulls *(Leucophaeus scoresbii)* sits in certain squabbling dignity on a rocky outcrop in their midst. Skuas *(Catharacta skua)* have their mere scrapes for nests well distributed along the whole point to keep their chicks away from the neighbours who sometimes display a cannibalistic tendency. To defend their nests and chicks they perform an effective aerial attack which is somewhat like dive-bombing on any threat within their territory, even on humans who are sometimes hit with the wing, powerful, though glancing blows. Dispersed along the shores are the nests of the oystercatchers, Blackish *(Haematopus ater)* on the rocks, Common *(H. palliatus)* on the sand. The most individualistic are surely the Chubut Flightless Steamer Ducks *(Tachyeres brachypterus leucocephalus)* who defend their territories ferociously.

A spectacle to rival the penguins is surely the colony of cormorants where upto a maximum of 5,000 nests, each looking like a miniature volcano, cluster together at "pecking" distance one from another - the length the stretched-out neck and beak reach, times two. Adult cormorants when walking around look somewhat like penguins. On the nest they sit to incubate the two or three white or chalky-blue eggs or stand to feed the chicks which poke their head and neck right down the parent's gullet to reach the food in the crop. The colony is on a small plateau at the tip of the point, covered with a layer of guano which has accumulated over the years. There three species of cormorant nest: the most abundant using the flat top of the rise is the King Cormorant *(Phalacrocorax albiventer)* while on the low rock-faces around the perimeter are the Rock Cormorants *(Phalacrocorax magellanicus),* and, at their peak there were 70 pairs of Guanay Cormorants *(P. bougainvillii)* discovered here by Francisco Erize in 1967. These are the "Billion dollar birds" of the guano industry of Peru.

This concentration of breeding birds makes Punta Tombo one of the most remarkable bird-sites in the world. It is easily accessible though some distance from the towns. With all the attendant species of gull, skua, cormorant, oystercatcher and so on, and the tameness of the birds which have become quite accustomed to visitors, all make observation of bird behaviour and the interactions of the several species easy and rewarding. The relationships between predator and prey, between scavenger and scavenged; gulls stealing eggs or chicks, sometimes under the very beaks of the parent birds, only to be pursued themselves by skuas to rob them of their loot; the competition between Kelp and Dolphin Gulls and the cheeky Snowy Sheathbill for the contents of some stolen egg spilt on the ground after the original thief had purposely dropped the egg from a height to break it.

Such a place becomes a great tourist attraction and as such very vulnerable to interference, and it is heartening to see the provincial department responsible for these affairs take action to contain the wanderings of visitors to a small but adequate and well-planned area to see enough of the birds to satisfy all but some demanding naturalist. It has not been man however who has affected the cormorants, but the periodic storm which

Cormorants at Tombo

Now off-limits because of diminishing numbers, the spectacle at the very tip of Punta Tombo of thousands of Cormorant nests, rivalled the penguins. These King Cormorants with the yellow caruncles, blue eye-ring and crest in breeding plumage are the most abundant species.

The Great Skua

A pair of Skuas (Catharacta skua) *performs a triumphal territorial display to confirm the ownership of the baby penguin they have just taken from a Kelp Gull.*

The Magellanic Penguin

This penguin (Spheniscus magellanicus) *has no very elaborate display but each mating is preceded by a series of characteristic rituals.*

The Dolphin Gull

Here seen copulating, this small gull (Leucophaeus scoresbii) *is endemic of the extreme southern South American coasts.*

washes over the very point; when they rarely occur in summer the whole colony is washed away - in winter it would not matter very much. Numbers of these and the species which most depends on them, the Dolphin Gull, are down, but may climb back up again as there are no visitors allowed nearly that far.

Punta Tombo has been a provincial reserve since 1979 when the warden's lodging, the visitors' centre and bathrooms were constructed at the instigation of the New York Zoological Society, prime mover for conservation measures in this wonderful province.

Again, as this is the most important area for birds on the coast, it should become a nationally protected area, probably a Natural Monument, so as to benefit from the guarantees for conservation that only the Nation can offer.

Santa Cruz's Coastal Sanctuaries. Of the several wildlife sanctuaries along the coasts of Santa Cruz there are two which merit our attention as being exceptional - the ria at Puerto Deseado, and Cabo Blanco.

The ria at Deseado is dotted with islands on which numerous sea-birds nest: Isla de los Pájaros has a respectable and attractive penguin colony while Isla Quiroga is the breeding area for skuas, gulls and a few pairs of penguins, and on the clumps of the salt-tolerant bushes *Atriplex vulgarissima* and *Chenopodium scabricaule* cormorants nest.

The most spectacular area is the Barranca de los Cormoranes on a peninsula from the south shore of the ria, where on the narrow ledges nest the Red-legged Cormorants *(Phalacrocorax gaimardi)* and Rock Cormorants *(Phalacrocorax magellanicus)*. The Red-legged Cormorant is typical of the Humboldt Current up the west coast of South America, in northern Chile and southern Peru, but curiously enough it is also present in small numbers in a very restricted area of the patagonian coast between Cabo Blanco and San Julián Bay, but the most spectacular site is here, on the quartziferous porphyric cliffs where several hundred of the elegant gray bird with white markings and brilliant red feet nest.

The very lovely small Commerson's Dolphin is a regular in this ria which the local authorities declared a Nature Reserve in 1977, but so far it has only been a declaration of intent as no action whatsoever has been taken to effectively protect the area.

The extreme southern headland of San Jorge gulf, the largest "bite" out of Argentina's Atlantic seaboard, is Cabo Blanco. It consists of two huge block-like hills which at 44 m above sea level tower above the generally flattened aspect of the land thereabouts. The greater of these is crowned with a lighthouse manned by naval personnel, and before it emerge small, steep, rocky islets on the tops of which gather one of the last surviving breeding populations of the Southern Fur-Seal *(Arctocephalus australis)*. This animal was implacably hunted to the very verge of extinction in the XVIIIth and XIXth centuries for its much-valued pelt, and though it has been protected now for decades its recuperation seems to be slow. The reason the pelt was so sought after is the under-coat of fine fur through which grow the longer, coarser guard-hairs.

This pinniped belongs to the family of "eared" seals as does the Sea-lion, is smaller, with a sharper snout, more noticeable ears and is not found on beaches but on rocky ledges and steps at the foot of cliffs.

To complete the spectrum of attractions of this place there is a small colony of South American Sea-lions, several nests of Red-legged Cormorants and especially the large colony of two species of very similar cormorants, the King Cormorant *(P. albiventer)* and the Blue-eyed *(P. atriceps)* which proffers a great oportunity to investigate whether they are or not one and the same species with two forms, as some scientists believe, or two distinct species divided by ecological or behavioural differences.

In principle Cabo Blanco is a widlife reserve created by the National Government in 1937, but this has never gone beyond that stage either.

These two Santa Cruz areas have exceptional characteristics which, together with the Valdés peninsula, would constitute a chain of nature reserves very representative of the mammalian and avian faunas of the Atlantic seaboard.

(Islas Malvinas). The insular characteristics and the priviledged position these islands occupy out on the continental shelf make them an ideal base for sea-birds and mammals which breed here in vast numbers, mostly on the smaller outlying islands. Several species of penguin, albatross and petrels which do not nest on the continent do so here in numbers. The land birds are also of special interest, because of the tameness of many species; the vegetation likewise will be of interest to the visitor in its adaptation to the dominant weather conditions of the archipelago.

Though there are a number of private nature reserves and sanctuaries declared by the occupying British authorities, the Argentine government never has planned any project for a National Park or equivalent reserve, considering it more opportune to settle the political question first.

The Southern Elephant Seals

Colonies on the northern and eastern shores of Valdés peninsula are the only continental breeding ground of this, the largest of seals. During the breeding season the large males defend territories to which the females come to pup. These harem bulls have a short trunk - larger in the older specimens - which acts as resonance chambers to amplify the bellows. A big bull spins on his stomach to face an intruder or challenger (above). In an attempted copulation the male holds the female down by pressing her with the weight of his head (lower right).

The roof of America - Puna and high Andes

The variety of habitats which make up the Neotropical Region can hardly be accounted for unless one bears in mind the uplifting of the Andes during the Tertiary; it is the longest chain of mountains in the world, and the second in height. Those animals which lived during the Miocene and Pliocene - 20 to 30 million years ago - suffered a habitat transformation so drastic that many became extinct while others adapted and evolved to become the species we find today on the South American continent.

At that time the ocean we know as the Pacific covered an area much larger than it does now, and its American shores consisted of a series of eroded rolling highlands. The splitting up of Gondwana into the Antarctic, Africa, Australia, part of India and South America during the Mesozoic, caused pressure at the interface of the plates. The marine plates were subducted - pushed under the continental plate where the continental sedimentary rocks buckled and rose, creating the Andes range.

Of all the biotopes which were generated by this modification, the phytogeographical provinces of the Puna and the High Andes are most closely connected as they not only share a number of characteristics, but are the very essence of all the area they occupy, and obviously, both possess transition communities in appreciable quantities.

PUNA PROVINCE

The tectonic forces which created the Andes produced parallel ranges of mountains between which high, wide, flat-bottomed valleys exist today. They were filled in and levelled during the Quaternary by the products of erosionary action, sand and sediments, forming the intermontane plains at 3,400 to 3,800 metres above sea-level, dominated by ranges and peaks all around them which reach elevations of 6,000 m. Scattered throughout this "altiplano" are outcrops of rock and small hills or ranges giving the whole area a semblance of roof-tops - the well-earned title of this chapter.

The high plains themselves have no drainage other than a system of endorrheic watersheds which commonly end in salt-flats or saline flood-plains.

The lack of organic material is one of the characters of the soils of the Puna - immature, skeletal, sandy or stony.

Near the salt flats they have a high proportion of soluble salts and clays with gypsum deposits in layers, while in marshy conditions peat replaces humus.

Extremely arid, with a dessicating dryness overall, the few torrential summer rains together with the great diurnal temperature range because of radiation, are the distinctive features of the cold, dry climate of the Puna, consequences of the elevation and of being "closed" in.

The daily temperature range averages more than 20°C and can reach 50°C, and the mean monthly maximums and minimums oscillate between 21°C and -3°C. This variation favours the mechanical disintegration of the rock and accumulation of bits and chips like scree.

From April to October there is no rain, so the remaining six months share the average annual precipitation of 100 to 350 mm though there are great local as well as yearly fluctuations - in the salt flats it very seldom rains at all, for example.

This habitat, where dryness and cold join the desert's stillness and quiet, occupies a wide area of Argentina from the Bolivian border to the north of Mendoza province, including western Jujuy and Salta, and northern Catamarca, and extends southward through the high western mountains of La Rioja and San Juan. The Puna Province is bordered alternately by the Pre-puna of deep valleys which lead up to the Puna, by the Yungas and by the Monte; it is dominated by the High Andean Province above.

As in any other rigorously hostile habitat the living organisms have had to adapt to be able to survive. In the Puna it is the shortage of water, of oxygen, the poor soils, the strong solar radiation during the day, low nocturnal temperatures, the uneven distribution of the undependable rains, the lack or ambient humidity in the atmosphere, and certain nutrient defficiencies which all together limit the possibilities for life. However it is wrong to believe that there is a poor flora or fauna; it is the very harshness of the environment that has led to a certain variety, and a number of endemic forms.

The Puna has many points in common with the Monte and with Patagonia, provinces which have already been covered. Especially with this last there is much affinity, and many of the species of plant which appear in the climax vegetation are found in both. There are more genera of Puna plants found in Patagonia than the other way round, but interrelationships are very great, greater even than with the neighbouring High Andean province.

The vegetation here is preponderately that of the steppes and seems uniform and homogenous even though the great number of species is surprising. For plants the scarcity of water, and seasons not propitious for reproduction are important obstacles and are common to other habitats, but in the Puna the solar radiation (which melts the ice on the lakes by mid-day, even in winter, though they freeze over again at night) has required, in combination with the other factors, that beings evolve in several ways.

To avoid transpiration - an imperative precaution - leaves are smaller as in Patagonian species. Stomata are reduced to minimize the interchange of gasses, and also, consequently, the photosynthetic process, which nevertheless is sufficient to keep the plant alive. Again we find curled leaves, either on themselves or around a twig or branch, and cell walls are thickened.

Plants also develop inordinately deep root systems which contrast with the proportionately small aerial parts. There are bushes like those of the genus *Fabiana* which are but 20 cms tall but have roots over 2 m deep. The same is the case of the Canjía *(Tetraglochin cristatum)*, a Rosaceae which is somewhat larger, has spine-shaped leaves and grows in rocky places; legumes like the Churqui *(Prosopis ferox)* as the name implies also very thorny, as are *Adesmia schickendantzii* and many others.

The storage of water in the tissue is another well-known strategy which is seen in the cacti - some 20 species on the Puna. Among the plants with thick ryzomes, bulbs or tubers on the roots for storage are Achicoria *(Hypochoeris meyeniana)*, the species of native Potato *(Solanum acaule, S. megistracolobum)*, Esporal *(Pennisetum chilense)* and the salt-grass *(Distichlis humilis)*.

There are species, mainly woody plants, which drop their leaves for winter, while others shed them shortly after they form; some short-stemmed plants, and some which never have leaves like the Brama or Pingo-Pingo *(Ephedra breana - Gnetalae)* of many branches which has medicinal properties.

In all these cases photosynthesis is carried out in the branches and stems.

There are also plants which die off above ground during the dry, and whose buds remain at ground level - the hemicryptophytes, like most of the grasses, and the geophytes with their buds below ground like *Hoffmansegia gracilis* and *Calandrina punae*. Other are pterophytes, dying off in the drought and reappearing from seed when conditions are favourable, latent perhaps for years at a time, like Brama *(Bouteloua simplex)* which completes its cycle - germination, sprout, flower, seed - in less than a month, the grasses *Eragrostis nigricans* and *Aristida humilis*, the amarillidiaceous Puya-Puya *(Eustephiopsis speciosa)* and the nettle *Urtica chamaedryoides*.

Cushion-plants as described in the chapter on Patagonia are another way of organizing against adverse conditions. A typical species is the Yareta *(Azorella compacta)* which thoughout the region is over-exploited for the combustible property of the resin which covers it and which burns well. When one considers that the growth-rate of this plant is less than one milimetre per year, the destruction of these centuries-old plants is a real affront to nature. There are small plants which prosper under the protection of bushes or dry grasses. The spacing between bushes and isolated clumps of lesser plants is such that much of the soil is not protected. As in every region where vegetation is sparse, herbivory has led to many species defending themselves against grazing or browsing with thorns, hard spiny leaves (like the Iros - *Festuca ortophylla*), or with irritating hairs as in nettles and plants of the genus *Cajophora*.

Accumulations of salt in the soil inhibit plant growth there as, too, the grazers who would eat these plants. The Chenopodiaceae like the Cachiyuyo *(Atriplex microphylla)* and the Cachial *(A. madariagae)* have solved this problem in an interesting way. Small organs like the stomata, called salt glands, eliminate sodium chloride through hiperosmotic secretions, thus avoiding the loss of water. Insects which feed on these highly salt plants excrete the excess of salt through Malpighian tubes to maintain their osmotic balance.

There are many Compositae in the Puna flora. The importance of this family lies in its high production of dry seeds which feed an important number of birds and granivorous mammals.

The structure of the vegetation is simple as there are only a maximum of two strata and often barely one. Except in the cases of obvious protection against wind or grazers, where tender plants grow beneath tough bushes or grasses, the relationships of dominant and secondary plants is never clear as also the succesion between communities. This is in keeping with the poverty of the soil where the only plant material is in the peat-bogs.

The landscape is devoid of trees except for the copses of Queñoa *(Polylepis tomentella)* a rosaceae with loose scolled bark rather like a badly rolled cigar, which grows to 5 m at the headwaters of streams, in sheltered canyons and certain slopes between 3,500 and 4,300 metres. This is another species in serious and accelerated regression as it is used by man for fire-wood and as a building material.

The most common climax community is a bushy steppe where species do not grow to more than one metre tall, where Tolilla *(Fabiana densa)* dominates. This is a solanaceous plant with yellowish tube flowers, growing with Añagua *(Adesmia horridiuscula)* with pinnate leaves and triple thorns, and Chijúa *(Baccharis boliviensis)*, a resinous composit symmetrically divided in two, with linear leaves and white flowers. This climax community sometimes grows with scattered specimens of Queñoa, or, lower down, with Churqui and Cardoon *(Trichocereus pasacana)*. Also present are bushes like Rosita *(Junellia seriphioides)*, Mocoraca *(Nardophyllum armatum)*, Pingo-Pingo and Anagüilla *(Adesmia spinossisima)*. Changes in the dryness of the soils dictate variations, with one or other of these species dominating.

Another frequent steppe community is that of the Chijúa on the plains at greater elevations and on rocky slopes which includes some of the previous species as well as others like Iluca *(Krameria iluca)*, a very branching bush with liliaceous leaves, Cola de Zorro *(Cassia hookeriana)*, the medicinal Muña-Muña *(Satureja parviflora)* and, in steep places the Poco Cardoon *(Trichocereus poco)*. A common weed of these communities, advancing as a consequence of over-grazing by sheep and goats, is the Garbancillo *(Astragallus garbancillo)* with its blue flowers and toxicity attributed to the selenium it contains.

In La Rioja there is a plant community with Tramontana *(Ephedra breana)*, Cola de León *(Verbena seriphioides)* and Lampaya *(Lampaya schickendantzii)* with carpets of *Stipa* lawn-grasses.

The vegetation of those places where water forms peat-bogs as it seeps down slopes is dominated by grasses, reeds and sedges in uneven patches (genera: *Scirpus, Juncus, Eleocharis*, etc.). On the edges of streams and salt-flats where salinity is slight, grow patches of salt-tolerant lawn-grasses while in drier places nearer the concentrations of salt, grow extensive patches of Chillahua *(Festuca scirpifolia)*, one of the communities which best cover the ground - some 60% average in this case - and where the only orchid in the Puna is to be found, *Aa paludosa*.

Over the whole area, near water and in the bottoms of valleys and depressions, communities known as "tolares" are frequent. They belong on deep sandy soils with good drainage and are composed of Tola or Tola Vaca *(Parastrephia lepidophylla)* and Tola del Río *(P. phylicaeformis)*, middling to tall, with leaves like tight scales, much sought for fire-wood and associated with Muña-Muña and the Pampas Grass *(Cortadeira speciosa)*.

Many other thorny steppes complete the inventory of the Puna's plant communities. It is worth noting that in spite of the limitations this paucity of plant cover imposes on the animals, as well as those already mentioned, numerous other species have overcome these difficulties and live there.

Because of their size and abundance the local members of the camel family are most noticeable and as a consequence man has led to their becoming ever-rarer. As to the taxonomy of this group, not all authorities agree on the species that exist. Two of these are reduced to domesticity: the Alpaca *(Lama pacos)* which is scarce in Argentina, and the Llama *(Lama glama)*, while the Guanaco *(Lama guanicoe)* and the Vicuña *(Vicugna vicugna)* are the wild species.

They are well-adapted to the habitat in many ways: they have large hearts, their blood is rich in red corpuscles (it is the only family of mammals in which these are elliptical as in amphibians and reptiles), their warm woolly coats are proof against the rigors of the altitude's temperatures. It is the Vicuña which lives highest in the mountains. Its elegant shape, useful for detecting danger at a distance, its colours blending in with the tones of these steppes, its agility and its stamina, its nostrils which can be closed against dust-storms;

The Andean Condor

This huge New-world vulture, (Vultur gryphus) is without doubt the symbol of the Andes Cordillera occuring throughout its length. The Condor spends hours at a time soaring at height on thermals or relief-induced up-currents to search for the carrion on which it feeds. In many areas of the northern part of its range it has been extirpated, but still exists in some numbers in Argentina, more commonly towards the south.

The Vicuña

Undoubtedly this is the most important mammal of the Puna; excellently adapted to the rigors of the environment it here fills the role of the most efficient of the large herbivores. The value of its very fine wool took it to the verge of extinction, but happily, conscious of this irrational situation, the authorities have set aside vast areas for its conservation, and it seems to be responding well by increasing.

all these are further adaptations to the medium in which it lives. As a result Vicuñas are perfectly in tune with the vulnerable habitat they depend on. They walk on two large cushions per foot which are gentler on the vegetation than hooves which trample and harm; the powerful lower incisors are strongly enamelled on the outer surface for optimal use of the vegetation, cutting neatly, not tearing or loosening roots as do the introduced domestic animals like goats, sheep, horses and donkeys. The Vicuña's incisors grow throughout nearly all its life so they can take advantage of the grasses with high silica content.

This species, the smallest of the group, has a fleece of the finest pale cinnamon wool which has been the cause of its coming within a fraction of extinction. Markedly territorial, family groups made up of one male with three to eight females and the calves of the year, wander from grazing areas to rest areas but avoid rocky places, preferring grassy habitat, (Tola or Chillahua). Flocks of bachelor males are not territorial but wander around and through family territories which are stable from year to year.

In the times of the Incas the abundance of Vicuñas was such, and the greed it ignited in the conquistadors so evident that Garcilaso de la Vega in his "Comentarios Reales" points out, regarding the great hunts (chacu) the Inca organized: "At a certain time of the year, after the breeding... some areas being better grounds for hunting than others, over twenty, thirty and forty thousand head could be counted, a wonderful thing to see and of much rejoicing. This is what there was then: now let witnesses tell the numbers that have escaped the slaughter and waste of the blunderbusses, for today one can hardly find Huanacus and Vicuñas except where they have not been able to reach them."

The alarming situation of the Vicuña led to a series of measures taken for its recuperation, which is only now beginning. Protection over vast areas has led to an estimated population of some ten thousand in the country today as well as protecting other species and a variety of habitats.

The only natural predator of the Vicuña is the Puma, so adaptable that it is found at great height, while young animals could conceivably fall prey to the Red Fox *(Dusicyon culpaeus)*. The largest of the Hog-nosed skunks *(Conepatus rex)* is a Puna endemic, with its broad white back; the local form of the Pampas Cat *(Felis colocolo budini)* and the Grison *(Galictis cuja)* complete the roll of puna carnivores.

Feeding on the available plants there is a great variety of rodents, including a number of *Cricetidae*. Local forms of this family include a puna mouse *(Eligmodontia hirtipes)*, some large-eared rats *(Phyllotes sublimis leucurus* and *P. osilae)*, the Chinchilla Rat *(Chinchillula sahamae)* with its short tail and living among rocks. Another rock rodent is the Chozchoris *(Octodontomys gliroides)* with its paint-brush tail, which, like the other rock-rats *Abrocoma cinerea,* are agile in climbing bushes as well.

The concentrations of the underground Tuco-tucos *(Ctenomys frater* and *C. opimus)* are spectacular in the areas where they turn over the earth, wiping out the vegetation and even hindering road construction. These "ploughed" areas, with animal tramplings become the places where seeds lodge to start new cycles.

The rest of the mammals includes several cavies, an armadillo *(Chaetophractus nationi);* a ubiquitous Free-tailed Bat *(Tadarida* spp.) and a Big-eared Brown Bat *(Histiotus)*. There are representatives of 23 families of birds which can be found in the Puna. The ground between the bushes and clumps of grass is the habitat for the Lesser [Puna] Rhea *(Pterocnemia pennata garleppi)* which may be a seperate and endemic species, the Ornate Tinamou *(Nothoprocta ornata)* and for various furnariids - the Puna Miner *(Geositta punensis)* and the Slender-billed Miner *(G. tenuirostris),* the White- winged Cinclodes *(Cinclodes atacamensis)* and the Rock Earthcreeper *(Upucerthia andaecola);* tyrants like the Black-billed Shrike-tyrant *(Agriornis montana),* the Puna Ground- tyrant *(Muscisaxicola juninensis);* and fringillids like the Puna Yellow-finch *(Sicalis lutea),* the Band-tailed, Black-hooded and Red-backed Sierra-Finches *(Phrygillus alaudinus, P. atriceps* and *P. dorsalis* respectively) all sharing the variety of vegetation communities covered above, with some species of doves.

It is hard to visualize a member of the woodpeckers on these arid areas but the Andean Flicker *(Colaptes rupicola)* gregarious and terrestrial, has adapted to these habitats, digging in the ground for the larvae of beetles. It is here forced to tunnel deeply into banks (and the walls of adobe houses); it sometimes uses these "nests" as colonial roosts. This habit saves energy and effort, and conserves bodily warmth, as do, in the extreme, hundreds of passerines which on occasions pile into crannies.

Certain humming-birds also tend to use burrows, like the Hillstars, Andean and White-sided *(Oreotrochilus estella* and *O. leucopleurus),* which save energy by reducing their body temperature, slowing down their metabolism to one twentieth of the normal. These birds counter the dearth of flowers by feeding mostly on insects and spiders.

Near water one could encounter the Puna Ibis *(Plegadis ridgwayi)* somewhat larger than his counterpart on the plains, with the Andean Lapwing *(Vanellus resplendens)* which nests on the ground. Another charadriiform, the Diademed Sandpiper-Plover *(Phegornis mitchelli)* is much rarer and prefers the peat-bogs and floodable shores, while the Puna Plover *(Charadrius alticola)* is somewhat smaller and frequents the beaches of the lakes. The Puna Snipe *(Gallinago andina)* is to be found in damp, open ground while the Andean Avocet *(Recurvirostra andina)* nests on the wide beaches. It has a long, up-turned bill with which it catches its tiny prey at the surface of the water by continually "sweeping" from side to side.

The Andean Goose *(Chloephaga melanoptera)* is the largest of the genus and the least inclined to take to the water, a member of yet another genus shared with Patagonia. The Puna Teal *(Anas puna)* nests in the grasslands and shares the lakes and marshes with other endemic forms of waterfowl. The same habitat is home to several coots - the Andean *(Fulica ardesiaca),* the Giant *(F. gigantea)* this last being enormous and laying eggs which are much valued by the locals who rob the nests; and the very rare Horned Coot *(F. cornuta)* with a peculiar protuberance from its shield above the bill, which nests in deep water by piling up a pyramid of stones on the bottom till, above the surface, the nest is high, dry and sure not to drift in the strong winds.

Saline lakes are the preferred habitat of the three species of flamingo found in Argentina: the Chilean *(Phoenicopterus chilensis),* the Andean *(Phoenicoparrus andinus)* and James' *(P. jamesi).* This last species which was believed to be extinct, was rediscovered on Laguna Colorada in Bolivia, near the Argentine frontier, in 1957. The three species, which normally feed in different areas, do not compete with each other for food even when they are together as the filtering apparatus in their bills is different in size in each species so they trap different-sized organisms.

The largest flying bird, if wing-surface area is the criterion, is the Andean Condor and it is on the rising thermals of air-currents displaced by the relief that we find it here soaring at great heights to search for the carrion on which it feeds, hours of flying for little expenditure of energy. The likely source would be the camelids, large rodents or rheas. It is unjustly accused of crimes against livestock for the killing of which it is inappropriately equipped - turkey-feet and no strong, sharp claws. The raptor species which are true predators are the accipitrids - the Puna Hawk *(Buteo poecilochrous),* the Black-chested Buzzard-Eagle *(Geranoaetus melanoleucus)* and the Cinereus Harrier *(Circus cinereus),* or the falconids - of which the most conspicuous is the Mountain Caracara *(Phalcobaenus megalopterus)* which is black and white, and which shares with the Aplomado Falcon *(Falco femoralis)* and the Burrowing Owl *(Athene cunicularia juninensis)* a taste for insects and smaller birds.

Using the sunny hillsides as places for gathering warmth, certain lizards prosper, especially those of the important genus *Liolaemus,* mostly *L. multiformis,* and a colubrid snake *Tachymenis peruviana.* The cold and dryness affect batracians to the extent that some have had to evolve to live exclusively in the water where temperature variations are moderated, or by developing strongly granular and thick skins as is the case of *Bufo spinulosus,* the tree-frog *Hyla pulchella andina* and several species of *Telmatobius.*

Fishes here are small and have hooks on their gill-cases to help them adhere to the stones and even to climb *(Pygidium spegazzini, P. alterum* and *P. boylei).* Insect-life also shows interesting adaptations like the dark colours and hairiness which allow them to take better advantage of solar radiation. As is to be expected, on the open ground there are many running species.

HIGH ANDEAN PROVINCE

This portion of the Andean-Patagonian domain extends all down the west of the country from the north to Tierra del Fuego, and takes in a belt which starts in Jujuy at 4,400 m above sea level, descending towards the south. Because of the many affinities with the Puna it is convenient and perhaps necessary to deal with them here, especially as all the protected areas in the region cover both provinces. Nevertheless the Southern High Andean district will be dealt with in the chapter on the Subantarctic Province.

The high andean peaks from the border with Bolivia to the province of La Rioja make up what is called the Quichua High-andean District where the height to which vegetation grows is determined by climatic circumstances, especially the presence of snow, which in certain areas has permitted the growth of flowering plants upto 5,600 m above sea level. Temperatures which are much colder than on the Puna, can change abruptly, and there are usually high winds. Snowfall, which is not regular, contributes to the humidity of the ground, and melt-puddles are the chief source of drinking water for the animals.

A steppe-type vegetation dominates, with all the adaptations already mentioned. Climax vegetation communities are mostly composed of grasses like Iros *(Festuca ortophylla* and *F. chrysophylla)* as well as the various

The Andean Hillstar

The only way any hummingbird can live in these environments is by developing certain adaptations: it nests in holes in banks, feeds mostly on arthropods, and goes into a torpid state when conditions are adverse thus reducing its metabolism to a bare twentieth of the normal rate. This male Hillstar (Oreotrochilus estella) *was at over 3,550 m on Aconquija.*

Queñoa

Queñoa (Polylepis tomentella) *of the rosaceae family grows to 5 m tall in copses in canyons and protected places where water from springs or streams is available. It is the only species of tree and is becoming drastically rare because of its use as fire-wood and building material.*

High-Andean Landscape *At Abra del Lizoite about 4,000 m high, the desert aspect of the high Andes is dominant, for all that the typical small lake from snow-melt seems to indicate otherwise.*

Snow on the High Steppe

A snowfall during the night has covered this Iros (Festuca sp.) grass of stiff and sharply pointed leaves. In the middle-distance is Laguna Grande on Huanca Huasi, Aconquija, at 4,250 m.

species of Poa *(P. gymnantha, P. lilloi, P. muñozensis)* and Stipas *(S. caespitosa, S. frigida* - Viscachera -) with many other species growing under their protection. In places the Iros are associated with the legume called Goat's Horn *(Adesmia nanolignea)* a ground creeper, while in others they do so with the Bitter Coirón which is also associated, on sandy or rocky slopes, with Viscachera. There are deserts with nothing but lichens or semi-deserts of ground-hugging annuals or biennials with their buds right close to the ground - associations of vegetation which grow at the highest elevations.

In eastern Catamarca and a section of northwest Tucumán which would include the Aconquija range and the Calchaquí heights, there is another patch of this district, but the information available on the associations of plants is as yet scanty. No doubt when the pertinent research is done it will be evident that this sector is particularly interesting - an island with marked endemism.

As for the Cuyano High Andean district from San Juan, Mendoza and northwestern Neuquén, the elevations are between 2,200 and 4,500 m, the climate is similar to the previous sectors, with frosts year-round and snow mainly in winter. These extreme conditions lead to the steppe vegetation being found only in the bottom of the valleys and a few areas of the slopes; on peaks and the rest of the slopes grow ground creepers, cushion plants or others adapted to the snow and screes.

Different species of grasses dominate the steppes, mostly another coirón *(Stipa speciosa)* together with *Poa holciformes, S.scirpea, S. tenuissima, S. vaginata, S. chrysophilla,* etc. Other manifestations of the same structures are the small bushes like Leña Amarilla *(Adesmia pinifolia),* a slender-leaved legume, *A. obovata, A. uspallatensis,* or the composit *Nassauvia axillaris.*

Scree-slopes are the habitat of the lovely ground- creeping Nasturtium *(Tropaeolum polyphyllum)* with surprisingly beautiful flowers.

Though often present in Puna ecosystems, certain animals belong to the High Andean habitats. The Puna Tinamou *(Tinamotis pentlandii),* large and gregarious, eats seeds, larvae and arthropods and nests on the ground as does the Bare-eyed Ground-dove *(Metriopelia morenoi);* cryptic colouration and the habit of "freezing" may help it to escape from the extremely rare Andean Cat *(Felis jacobita)* which also captures rodents and is somewhat like a lynx. Its habits are almost unknown.

On the rocky slopes the whistled call of the Mountain Viscacha *(Lagidium viscacia)* is often heard, but it also comes from the near-by patches of grass where the animals feed. Its relation, the Chinchilla *(Chinchilla brevicaudata),* only survives in the wild in tiny pockets well to the west and at great heights specializing in feeding only on grasses of the genus *Festuca.*

Continuous hunting pressure has reduced the populations of Taruca - the Northern Andean Deer - *(Hippocamelus antisensis)* to the danger-point. This species is smaller than the southern Huemul, and is grayer. It lives up by the snow-line, feeding on the mosses and lichens *Gyrophora, Acarosphora*, etc. and prefers valleys with richer vegetation in the more out-of-the-way areas.

Another very interesting inhabitant of these regions is a reptile, the little iguanid lizard *Phymaturus flagellifer* which has taken to a vegetarian diet. Its tail is covered with conical scales which gives it a prehistoric look.

PREPUNA PROVINCE

The Prepuna corresponds phytogeographically to the Chaco Domain though its general aspect, in the northwest of the country and at elevations of between 1,000 and 3,400 m, is obviously influenced by the lay-out and orientation of the dry valleys and their slopes, all of which it covers. The climate is dry and warm with rainfall in summer.

On the permeable stony or sandy soils of the region grow bush steppes in which dominate species of the *Compositae* family like *Gochnatia glutinosa* or *Aphyllocladus spartioides,* or legumes like *Cassia crassiramea*. But it is the abundant cacti which give the Prepuna its special character: whole slopes and terraces where giant specimens of *Trichocereus pasacana* or *T. terscheckii* grow, looking like candelabra standing over the surrounding shrubs.

Among the animal species peculiar to this habitat is *Octomys mimax* which lives in the ubiquitous dry stone walls, with its tail ending in a cluster of longish hairs which look like a little brush. The Brown-backed Mocking-bird *(Mimus dorsalis)* with its chestnut back and white flashes in the extended wings prefers the open bush where it captures insects and their larvae, or finds small fruit.

None of these areas are very populated, there being isolated communities of more or less pure indigenous people belonging to the Quechua-aymará group and known as "coyas". Their main activities are pastoral, barely above subsistence level. Llamas do not interfere with Nature and can even be considered an added attraction in certain of the protected areas. However, the tendency to replace this camelid - as also the sheep - with goats is to be lamented because of their well-known tendency to create deserts where not properly managed. There is some mining as an alternative source of income for the regional economy.

Tourism has not reached these areas yet because of the lack of adequate hotels and roads, though there is much pressure on the more accessible parts - the spectacular coloured-rock canyons of the Prepuna as at Humahuaca and Río Toro.

As has been implied above, the relict populations of the Vicuña have stimulated actions to protect this species, and some 4 million hectares have been declared provincial reserves to that end. Though there is no National Park yet, some are planned, and they will conserve important areas of representative habitats.

Carahuasi National Park Project. In the north of the province of Jujuy, just east of Rinconada, is the area being studied for this possible national park named after the village in the southeast of the area, some 60 kms west of Abra Pampa, 90 from La Quiaca and 280 from San Salvador de Jujuy, the provincial capital. It covers about 163,000 has. and its borders are along natural features between 22°15' and 22°50' S latitude, and between 65°48' and 66°12' longitude West.

It includes the Pozuelos Lake which is already a Natural Monument and Ramsar Site, formed by the endorrheic drainage system of the Cincel River and others, some 15,000 has in area where water birds congregate in geat numbers. It is a place of great beauty and importance where the three species of flamingo nest as well as the rare Horned Coot.

The land areas contain good habitat for Vicuñas whose local population does not exceed a few hundred individuals which have not recently been hunted because of the enlightened attitude of some of the local land-owners, and so have become relatively tame. The Tolilla and Chijúa bushy steppe is the most abundant, followed by the Tolars growing among dunes. The transition zones between the provinces of the Puna and the High-andean are also well represented. Among the threatened vegetation it is good to report the presence of copses of Queñoa so very much cleared by man for fire-wood, and extensive areas covered by cushions of Yareta.

The biological diversity of the chosen area is assured by the inclusion of the Cochinoca and Quilchagua ranges where the characteristics of the province are well represented, as well as a string of fresh-water lakes behind Lagunillas. This way there will be sufficient area to conserve such variety, overcoming the danger of extinctions produced by insularity.

There is a greater human population than the general aspect of the landscape would lead one to believe could be supported there, thanks to their resistance to emigrate. Their pastoral activities would not interfere with the ecosystem, nor would the mining which is calculated to give-out within a certain time-lapse. The whole watershed of Pozuelos has been declared a Biosphere Reserve to guarantee a more rational management and preservation of the resources. The Administrative Centre could be at Carahuasi though at present there is a ranger station at the south end of Pozuelos. A Visitors' Centre is planned.

Aconquija National Park Project. In western Tucumán and spilling over into eastern Catamarca, are the Aconquija mountains, the heart of this project. It would include all land above the 3,000 m contour, descending in the south to include a section of pedemont forest. In the north it includes the Calchaquí heights and to the south ends at the Catamarca mountains of Yutuyacu and Medanito. If eventually it should be contiguous to the planned provincial park of Cochuna its biological interest would be enormously increased, with a surface of some 200,000 has between the two provinces.

Among the more important and relevant functions of this park would be the conservation of the headwaters of all the major river in the region, whose volumes are regulated by the singular vegetation there. Not least among the important aspects are the endemic species: the local Taruca Deer or the frequent new discoveries which appear as the area is explored (e.g. the lizard *Liolaemus huacahuasicus* which is endemic). Another point of interest are the many archaeological sites which would be included.

In must be stated that this would be an easy project as there are no settlers at those elevations, and there is a paucity of mineral wealth. Initiatives have recently been re-started and have the approval of a number of influential locals and communities.

The main roads of access to the national park would be along route #307 in the north, #65 in the south, while from Catamarca it would be along national route #63.

San Guillermo Provincial Reserve. A real step forward was taken by the province of San Juan when, by decree # 2,164 dated 22nd. June 1972, what was to be one of the largest natural area conserved in Argentina was created. In 1975 law #4,164 was passed for the expropriation of 860,000 has belonging to the San Guillermo "farm", tucked in the northwest corner, surrounded on one side by Chile and on the north and east by the province of La Rioja. Its elevation varies between 2,100 and 5,300 metres and it had been severely overgrazed by cattle and sheep.

Puna Lakes

These are the habitat of an avifauna which is particularly interesting for its many endemisms. Saline lakes are preferred by the three species of flamingos, the most abundant Chilean Flamingo (Phoenicopterus chilensis) *is seen here.*

The sector where the reserve is located belongs to the geological unit known as Frontal Cordillera where there are enormous flat-bottomed valleys - the flats of San Guillermo or the Hoyos (small holes), the last alluding to the activities of the burrowing Tuco-tucos which have dug up vast areas. Other flats are de la Paila, de los Leones, Batidero and others. The National Department of Wildlife has contributed actively in this whole project.

Cardoon

Flowers give the cardoon a gentle touch, quite opposed to the general aspect of its thorny impregnability, a defence against herbivores.

Stands of the Cardoons

Such stands of Trichocereus *are typical of the Prepuna. They look like giant candelabra and have a tough skin which limits transpiration, converting them into reservoirs of water in a habitat where this is generally lacking.*

The real importance of San Guillermo is due to its great population of camelids, estimated at 10,000 of which 4,200 are Vicuñas - some 60% of the total Argentine population of the species - which is enough to create interest in the well-being and proper functioning and administration of the reserve. Here too there are flocks of the Andean form of Darwin's or the Lesser Rhea, often associate with the camelid - a parallel behaviour to the African Ostrich which keeps company with the ungulates of that continent.

Information regarding the area is still incomplete as the biology is still being studied but there seem to be tendencies toward endemism, for example with the Paint-nosed Mouse *(Neotomys ebriosus)* and the hunting ants *Aurancomyrmex tewer.*

New Provincial Reserves. As in the case of San Juan, the other provinces with populations of Vicuña have been working towards their conservation. Thanks to inter-provincial agreements abetted by the National Government, growing concern for the survival of these valuable native animals has led to the protection of other areas.

Continuing to the north from San Guillermo the reserve of Laguna Brava has been created by La Rioja, with some 405,000 has, and the housing for properly-trained rangers for its protection. It has been offered by the province to the National Parks Administration and seems to be headed that way.

In Catamarca which was the traditional centre of the illegal poaching activity and commerce in these valuable fine wools, in a laudable about-face, poachers are today punished with all the power of the legislation, and a reserve of some 600,000 has - Laguna Blanca - has been managed with such success that there is an overspill into neighbouring areas of the rapidly-expanding population of Vicuñas. More recently still it has been the province of Salta which has set aside an area in the vicinity of San Antonio de los Cobres of some 1.4 million has known as "Los Andes" to the same ends.

In this way the very symbol of the most valuable, appreciated and vulnerable species serves as an umbrella to protect the other beings in the area, some of which may also be in danger or very rare. Great areas of the roof of America today protect these resources, giving good reason to hope for their recuperation.

Subantarctic Forests

Eloquent words were those of Francisco P. Moreno: "Each time that I have visited this region I told myself that should it be made public property for ever, it would soon become a centre for social and cultural activities, and thus an excellent instrument for human progress. The natural phenomena here encountered are starting to attract scientists who, happy with their fruitful research and the magnificent scenery of lakes and torrents, of giant forests, of steep mountains and permanent ice, all in an unbelievable geographic setting... ..consitute a unique set of circumstances favourable to my present purpose in this beautiful piece of the Andes." This paragraph is taken from a letter in which he renounces title to a portion of three square leagues (7,500 has) of land he was awarded in recognition for services rendered to the nation, stating that it should be for a national park. It expounds upon the relevance of this area and of the rest of the region where it is located, to the destiny and use which he proposes for it. The area he donated "on the borders of the territories of Neuquén and Rio Negro, at the western end of the main fjord of Nahuel Huapi lake... contains the most interesting aggregate of natural beauty" which the Perito Moreno had encountered on all his patagonian expeditions, states the document which is dated 6th. November, 1903. The fascinating landscape was, above all else, the factor which led to the preservation, for the first time in Argentina, of a significant sample of its natural heritage.

A narrow strip all along the southern Andes on their east-facing slopes, occupies the phytogeographic region known as the Subantarctic Province. This starts in the north of Neuquén at about 37°S and runs for 2,200 kms to the tip of the continent to reappear on Tierra del Fuego and Staten Island. Its widest part in Argentina is but 75 kms across and in places it is interrupted by fingers of the Patagonian Steppe (south of 45°S) cutting right over into Chile.

The characteristic vegetation is cool-temperate woods of evergreen and deciduous trees which grow thickly over the slopes and certain valleys, alternating with crystalline lakes, tumbling rivers and imposing mountains, some with extensive glaciers descending from their peaks.

It is remarkable to note the relationships of this flora with that of Oceania and the fossils found in the Antarctic, which speak volumes for the former contiguity of the continents. According to the theory proposed by Alfred Wegener in 1912, these southern continents were all one. It has been stated in the chapter on the Puna that South America, Antarctica, Africa, Australia and India's sub-continent were part of the super-continent Gondwana. Starting some 65 million years ago Africa and India drifted north at a speed of a few centimetres per year, then the rest "scattered", Argentina leaving Africa in a westerly direction at the speed that fingernails grow - averaging about 4 cms per year. When united, the continents shared species, plants possessing the most noticeable likenesses. For example the beeches of the Southern Hemisphere, the *Nothofagus* genus, is today distributed in the Patagonian Andes, Tasmania, southeast Australia, New Zealand, New Caledonia and New Guinea while their fossil remains are to be found in Antarctica. Many other genera of plants are shared, like *Lomatia, Laurelia, Aristotelia,* or *Hebe*; also in the fauna are parallels to be found - the family of frogs *Leptodactylidae*, the genus *Galaxia* of fish, many invertebrates such as wasps of the sub-family *Thyniae*, some spiders *(Migidae),* and the genus of opilionids *Nuncia*.

For this interaction to occur there must have then reigned meteorological conditions very different from today's. It has been pointed out in the chapter on Patagonia, that during the Eocene the climate was warm and subtropical and the region which was totally covered with forests, started to cool off. Also, between the Cretaceous and the Eocene the first step in an event of substantial importance was taken - the uplifting of the Andes. The second great moulding of the land occurred in relatively recent geological times under conditions which were much colder than at present, during the Pliocene. Three glaciations successively advanced and retreated, altering each time the distribution of the forests.

The mountain range deflects upwards the humid winds off the Pacific, which on cooling as the air rises, dump precipitation on the west-facing slopes, and a smaller portion just over the crest. In the rain-shadow are the Patagonian Steppes and there is an abrupt cut-off between these two biomes.

Geologically there are two identifiable sectors in the southern Andes. The northern sector goes from latitude 39°40'S to 42°30'S which is characterized by Tertiary and Quaternary volcanic rocks and blocks of crystalline base-rock, the southern sector goes from there to Tierra del Fuego and here Mesozoic marine sedimentary rocks predominate, dating from the time when the area was under the sea, with insertions of volcanic rocks and intrusive grano-diorites.

Physiographically there is not any marked difference between the two sectors - both show glacial relief and reach no very pronounced heights. There are bare, steep pinnacles, sharp arretes, U-shaped valleys sometimes with enormous lakes dammed up by terminal moraines and others still occupied by the glaciers as if to demonstrate the tremendous forces which sculpted the landscape. Innumerable streams tumble down the mountainsides, with waterfalls and rapids, in deep canyons or through the woods, generally flowing into one of the numerous lakes. From these lakes flow the big rivers which, like the Santa Cruz, cross all arid Patagonia and flow into the Atlantic, though some take the shorter route to the Pacific.

The climate is now cool-temperate, influenced by latitude and by the height above sea-level, but at the same time being moderated by the proximity of the two oceans, or even to a certain extent by the huge lakes. The mean temperatures go from 13°C (max.) and 3.4°C (min.) in the north to 9.9°C and 2.2°C respectively in the south.

It has already been said that the westerly winds dominate, and that the mountains form a rain-shadow so precipitation drops off markedly and abruptly towards the east. The vegetation reflects this, but is also affected by the orientation of slopes, and by the many microclimates. Thus in Puerto Blest the Mountains are lower permitting the penetration of damp air so 4,000 or more milimetres of rain- or snow-fall is measured yearly - one of the wettest areas in the country. Here the vegetation known as the Valdivian Rainforest grows. Further east, on Isla Victoria in Nahuel Huapi lake, the average is between 1,700 and 2,000 mm while in the city of Bariloche on the edge of the woods, only 1,000 mm fall - all this in a 50 kms transect, while out on the steppe a few kms further, the rainfall is in the region of 300 mm per year.

In the northern sector of the Patagonian Andes the wet season is winter with about 70% of the total precipitation, while the summers show a marked negative water-balance where evaporation exceeds rainfall. Towards the south the systems change to where, in Tierra del Fuego, the 600 mm of annual precipitation are distributed about evenly throughout the year. In spite of the lesser rainfall, the weaker solar radiation allows humidity to remain about constant all year.

Snowfall, which increases with latitude and elevation, occurs throughout the region and - like frosts - in any season.

Though South America is part of the great Neotropical region, phytogeographers refer to these subantarctic regions as part of another, the Antarctic Region which includes New Zealand, Antarctica and the Sub-antarctic Islands. This classification is amply justified if one considers the enormous number of endemics, families *(Desfontainiaceae, Gomortegaceae,* etc.) as well as genera *(Nothofagus, Fitzroya, Austrocedrus, Coriaria, Grilinia, Weinmannia* and many more).

There are plants which are without doubt of southern origin and have spread northwards, even as far as southern Brazil *(Drymis, Gunnera, Azorella, Araucaria, Ourisia, Fuchsia, Pernettya)*. By the same token Amazonian vegetation has invaded southwards as can be demonstrated by the presence of the cane *Chusquea*, genera of *Myrtaceae* and others.

Almost throughout, the patagonian Andes' forests of Argentina are made up of a limited number of species which, even though they vary with latitude, are ecologically similar as far as animals are concerned. Hence though the flora varies, the fauna is fairly constant in the whole extent of the region. However, the diversity of habitat, the various strata which make up the different ecological niches, are provided by the different stages the species grow through.

A fully developed tree is fundamental for the structure of the forest, providing the shade, and as a result the humidity necessary, and above all, protecting the soil. In its branches different birds will build their nests. As it grows old it will provide food for different wood-boring insects which in turn will feed the woodpeckers. Cracks, hollows and gnarls will increase and provide refuge for more species of wildlife. Then woodpeckers, owls and other hole-nesting species will find places to nest. Hibernating species like the Monito de Monte *(Dromiciops australis)* will find a safe haven to sleep its lethargic sleep all winter. When branches are blown down or the trunk falls, the tree on the ground will not have completed its usefulness since it accumulates humidity for fungi to grow as well as a refuge for frogs, arthropods and other invertebrates. In drier places lizard will seek refuge under it, or other insects or perhaps even a snake or small rodent. Little by little the wood will decompose and nutrients will be released to return to the soil and repeat yet again the eternal cycle of life.

It is only right to point out that in places like these the processes are slow; long periods elapse between one state and the next. From the dead tree to the fallen tree years might pass, as also between its fall and reabsorbtion by the soil. These are the signs of a natural balance which differs greatly from man's eternal hurry.

In the Sub-antarctic Province there are four recognizable districts, each with its own peculiar characteristics and species, though transition is gradual from one to the next.

In the north the Pehuen District is defined by the tree of that name, known in English as the Araucaria or Monkey Puzzle *(Araucaria araucana)*. This magnificent conifer raises its columnar trunk upto some 45 m in height and from the top grows the umbrella crown with its slightly drooping branches. The greatest and densest concentrations grow at between 900 and 1,800 m above sea-level in cold damp areas. Some trunks reach 2 m in diameter and the largest are over 1,000 years old. In certain areas they grow spaced out among Lenga *(Nothofagus pumilio)* forests with an understorey of the Colihue cane *(Chusquea culeou)*, or isolated specimens are

Sheld-Geese

These attractive geese of the genus Chloephaga *which is endemic to southern South America, are abundant in the patagonian Andes and frequently form huge winter flocks. In the photograph one can see Ashy-headed Geese (C. poliocephala) - gray heads and chestnut neck and breast, the four in focus - and Upland Geese (C. picta) - male white, female orange-footed, both out of focus right.*

The Magellanic Oystercatcher

The Magellanic Oystercatcher (Haemantopus leucopodus) *is a coastal species which comes inland to breed in the warmer months. It is often found on grassy meadows unassociated with water, in the Andes valleys. The intense black of its upper-parts and the yellow ring around the eye distinguish it from the Common Oystercatcher* (H. palliatus).

A Common Snipe

The enormously long bill of this Snipe (Gallinago paraguaiae) *is characteristic and serves to prod for invertebrates in the mud of grassy bottomlands where it also nests. Snipe are widely distributed throughout the world.*

Thick Forests *Under gigantic Coihues which can grow to 40 m there are often thickets of Colihue cane which make progress difficult off the trails.*

dotted through the countryside. The large seeds are edible and are the food of several birds and animals as well as for the native peoples who were even referred to as "Piñoneros". Entering a stand of these archaic and ancient trees has been likened to visiting a cathedral, such is the aura emanated by them.

The district of the Deciduous Woods extends the whole length of the Patagonian Andes and has its eastern border in the ecotone or transition zone with the steppes, and the western at the upper limit at the tree-line. Characteristic species are the southern beeches of greatest distribution found all through the southen Andes, the Ñire or Low Deciduous Beech *(Nothofagus antarctica)* of the edges of marshy bottomland and peat-bogs and at greater elevations. It is the smallest species of the group, and a pioneer, as it is the first to grow after forest fires and as glaciers retreat. The other species is the Lenga, the Tall Deciduous Beech *(Nothofagus pumilio)* which grows to be a big tree in the forest, but at a certain elevation which diminishes from north to south, grows in stunted form, dictates of the wind and snow-load in winter. It seems to be the species most tolerant of the cold. The foliage of these two species, come autumn, turns yellow, orange, golden and wine-red before it falls, at the upper limit of tree-growth and around the lower valleys, as if bracketing the evergreens with unusual and spectacular beauty.

Even though the Tall and Low Deciduous Beeches intrude into the steppes along river-valleys, the Andean Cypress *(Austrocedrus chilensis)* is typical of the ecotone though only found in the north of the region. This conifer forms dense cypress groves, growing upto 20 m tall which in the higher-rainfall areas prefers the drier north-facing slopes.

Polenization is by the wind and on breezy days there is a veritable golden fog in the woods, such is the density of pollen grains in the air. This species has been severely over-exploited for its wood, and now merits protection. The Maitén *(Maytenus boaria)* tends to grow with the cypress in the transition zone, in mono-specific groves in some damper places. This tree with its globular crown of intense green and its tiny yellow flowers, is much appreciated by cattle which graze it, cutting off its drooping skirt at the height they reach. It occurs also in the hills of San Luis and Córdoba.

Together with the Maitén, Chacay *(Chacaya trinervis)* also invades the steppes along the river valleys, while on the more open land Espino Negro *(Colletia spinossisima)* with white and very perfumed flowers, forms dense and impenetrable clumps of bushes. There are two tree Proteceae: the Radal *(Lomatia hirsuta)* of large coriaceous leaves which in the ecotone grows as a large bush, but where the rainfall is greater is a tree much sought for its gray, speckled wood, and the Notro or Chilean Fire-Bush *(Embothrium coccineum)*, a small tree or large bush, which flowers in late spring when it gets covered with brilliant scarlet-orange flowers. Other colourful touches in the landscape are given by the flowers of the climbing Mutisias, *M. retusa* is lilac-pink, while *M. decurrens* bright orange; by the Amancay *(Alstroemeria aurantiaca)* at the lower edge of the woods, forming a colourful golden border between these and the ecotone, where the dominant species are Coirón and Neneo mentioned in the earlier chapter on Patagonia.

In the damper areas, where rainfall exceeds 1,500 mm per annum begins the district known as the Valdivian, which is most developed on the Chilean side of the cordillera. It extends as far south as Lake Buenos Aires at about 47°S - which is as far as the dominant species grows, the Coihue Southern Beech *(Nothofagus dombeyi)*, reaching over 40 m tall and 2 m in diameter. It has small evergreen, coriaceous leaves, and is adaptable enough to become the dominant pioneer plant in deforested areas - fire or landslide - as is also the Ñire. Coihues grow from 500 to 900 m above sea-level where the Lengas begin. Within the Coihue forests also grows the occasional patch of Ñire in swampy conditions, and on rocky outcrops, Andean Cypresses, or Radals.

Because of the dense understorey of the Colihue Cane which grows several metres tall and in a continuous thicket, it is difficult to walk through these forests unless on a trail; it is said that the cane - a solid bamboo with lanceolated leaves - flowers every 40 years or so, all together, then dies off. Together with the cane there are to be found bushes with attractive flowers - Berberis such as Michay *(Berberis darwinii), Pernettya mucronata, Azara lanceolata,* or the Espino Negro *(Colletia spinossisima).* In clearings other flowering plants grow such as *Oxalis valdiviensis* with its five-petalled yellow flower, the Yellow Violet *(Viola maculata),* or the "pouched" flowers of the Lady's Slipper, local *Calceolarias,* long-stemmed and tall here, but in the steppes growing to barely the height of the flower because of the strong winds.

Ferns, mosses, lichens, liverworts and fungi are profuse on the ground, among fallen boughs and trunks or at the foot of the trees, flourishing during the wetter months of autumn and winter, when the forests are dripping, creating perfect conditions for these plants. Amongst the fungi Darwin's Fungus or Llao-llao *(Cyttaria darwinii)* is the most noticeable, growing on twigs, branches and trunks of the trees where it causes the tree to produce a "burl" deformation as a reaction to the fungus. The sporing head is about the size of a golf-ball, spherical, creamy-yellow later turning orange. It was eaten by the natives though virtually flavourless. When ripe the globular surface becomes honey-combed with holes.

An indication of the long history of these forests is the fact that here has evolved a genus of hemi-parasitic plant which is exclusive to them - *Misodendron*, with various species which grow almost spherical in shape, pale green, on the branches of these southern beeches. There is also the Quintral *(Tristerix tetrandrus),* frequently but not exclusively on the Maitens, with its lovely red tube flowers virtually year-round, an important source of food for hummingbirds. The lichen commonly known as Old Man's Beard *(Usnea* spp.) is simply epiphytic and uses trees for support. Being pale whitish-green it looks ghostly in the quiet woods where it grows in some profusion.

But though the Coihue forests are the most extensive in the district in Argentina, it is the Valdivian Rainforest which is the maximum exponent of this southern district. This formation crosses the cordillera from the west in the lower passes that also give rise to the higher rainfall which in certain places exceeds 4,000 mm annually. This humidity moderates the temperatures necessary for growth. Here the vegetation is exuberant, the number of species increasing markedly. Ferns and epiphytes grow through the thick wads of mosses on the trunks while climbers and lianas give the whole a rainforest-look as they dangle from the heights.

This habitat developed as a result of the benign climates of the Tertiary and even today possesses certain tropical characteristics within the cool-temperate southern regions.

Within Argentina the places where these cool rainforests exist are few, and they are all up against the Chilean border. One of its species, possibly the most imposing species of these southern areas is the local equivalent of the red-woods, *Fitzroya cupressoides,* which grows to 60 m in height, 3 m

The White-throated Treerunner

In behaviour like a nuthatch or a small woodpecker, covering trunks and branches in its search for the invertebrates on which it feeds, this Treerunner (Pygarrhichas albogularis) *flips off bits of bark or dislodges lichens to discover its tiny prey. It is a member of the typically South American Furnariid family and nests in holes in rotting wood which it digs itself.*

The Magellanic Woodpecker

This Woodpecker (Campephilus magellanicus) *is the largest of this family in South America and is exclusively of the Patagonian Andes; it feeds on the larvae of wood-boring insects it finds in the abundant dead wood. The male has the red head; the female's is black with a forward-curling wobbly crest.*

The Black-chested Buzzard-Eagle

The largest bird of prey (excluding the Condor) in southern South America is the Black-chested Buzzard-Eagle (Geranoaetus melanoleucus). *Here we see it perched on a Ñire* (Nothofagus antarctica). *In flight the wide wings and the short tail of adult birds are diagnostic, as well as the dark chest.*

diameter. Certain specimens are upto 3,000 years old. When these grow half-way up slopes and are covered with *Usnea* they are more strikingly visible than in the boggy bottomlands where they normally are found.

Other conifers of the area are the Ciprés de las Guaitecas (*Pilgerodendron uviferum*) found growing on flooded land in the northern sector and reappearing in a patch near Lago Argentino, the Maniú Macho (*Podocarpus nubigenus*) and Maniú Hembra (*Saxegothaea conspicua*) both Podocarps which share the generic common name merely on the basis of their external appearances. There also grow the Fuinque (*Lomatia ferruginea*) with its fern-like leaves on rusty-coloured stems and twigs which give it the specific name, and the Tique (*Aextoxicon punctatum*) with leaves and fruit like the olive, giving rise to its other vernacular name Olivillo; the Laurel (*Laurelia philippiana*), a large tree with perennial and fragrant leaves which grows to 35 m, and the Lingue (*Persea lingue*) which grows extremely slowly.

The lianas and climbers have in the Pahueldín (*Hydrangea intergerrima*) their largest representative, sometimes as thick as a man's arm, with tiny white flowers in umbel-like heads. Pil Pil Voqui (*Campsidium valdivianum*) has long, red tubular flowers some 5 cms long. Though the understorey is dominated by the cane, at the forest edge, in clearing, or along the banks of streams the vegetation of this bush stratum becomes lavish with a multitude of species like *Fuchsia magellanica*, Fire-bush, *Azara lanceolata, Pernettya mucronata, Ovidia pillopillo* and many others.

The herbaceous stratum includes many ferns like *Dicranopteryx cuadripartita* looking like a miniature palm upto 40 cms tall, the Adder's tongue fern *Ophioglossum vulgatum, Blechnum chilensis* and *Lophosoria quadripinnata* which grows in damp places with fronds upto 3 metres long.

In size it rivals the Pangue (*Gunnera chilensis*), of the Halorrhogaceae family with its gigantic round leaves and edible stalks like rhubarb which also grows in wet places along the stream-banks. There are not wanting plants with beautiful flowers like the *Calceolaria* and *Mimulus luteus* whose yellow flowers are speckled with red, *Lobelia tupa* as well as several species of ground orchids like *Codonorchis lessonii* with white flowers, or *Gavilea lutea*. The strange white spider-flower of *Aracnites uniflora* of the *Burmaniaceae* family, a saprophyte lacking chlorophyl grows on the rich litter material of these forest floors. A varied epiphytic flora covers the trunks with mosses, lichens, liverworts and ferns like *Serpyllopis caespitosa* or various species of the delicate *Hymenophyllum* genus and even an iris, *Luzuriaga radicans*.

In the northern section of these beech-forests, growing among the Araucarias, the Coihues or the Lengas, occasionally in pure stands, there are two further species of *Nothofagus*, Roble Pellín (*N. obliqua*) and Raulí (*N. alpina*), both notable for their size and nobility. The west-facing slopes have the most important concentrations, but in Argentina there are good stands between lakes Quillén and Lacar. Roble Pellín is the species of *Nothofagus* adapted to stand the warmest climates and folds its leaves to avoid excessive evaporation, while young leaves are hirsute to the same end. It grows on south-facing slopes on the edge of the steppe. The Raulí is the beech with the largest leaves and more precise requirements: higher humidity and lower temperatures. It does not grow as far north as the previous species, but the southern limit of distribution of both is about 40°S.

Both are deciduous and give these woods an outstanding beauty in autumn. This, coupled with their adaptation to different levels of precipitation, makes us wonder whether to place them in the Valdivian or the Deciduous Districts.

Pure stands of the myrtle Arrayán *(Luma apiculata)* are a well-defined type of woodland and occur in very few areas. They usually like the wetter sites and grow on the edge of the water. A close relative, Patagua *(Myrceugenia exsucca)* actually grows in the water and is the habitat for the nests of certain aquatic birds.

The southernmost District is the Magellanic which starts at 47°S where the Coihue gives way to the very similar Guindo *(Nothofagus betuloides)* growing upto 35 m tall and often accompanied by Lenga (mostly the dominant), Ñire and Winter's Bark *(Drymis winteri)* which proved to be an effective antiscorbutic for the early seamen. Surprisingly this genus reappears in the Araucaria forests of southern Brazil, an element of a subantarctic invasion.

These southern forests are easy to walk through as here the cane is missing though there are several bushes like *Tepualia stipularis, Pseudopanax laetevirens, Berberis ilicifolia* and *B. microphylla*. Even the herbaceous stratum is more impoverished than further north, with a little fern *(Blechnum penna-marina)* carpeting the soil of certain sectors, while the *Calceolaria* is ever-present as are the orchids, mosses and lichens. *Usnea* is the most noteworthy of these.

Peat bogs are characteristic of the District, and in them we can identify several communities according to the associated plants, like those with Balsam Bog *(Bolax gummifera)* or where reeds predominate. The dominant plants however are different species of moss of the genus *Sphagnum* which give rise to an extremely spongy damp soil of ochre and red tones. Often peat-bogs are surrounded by silvery-gray dead trees which were unable to stand the permanent wetness. In these bogs grows a small creeping conifer, *Dacrydium fonckii* as, too, a small "carnivorous" sundew *Drosera uniflora*.

The fauna of the Subantarctic Province is less rich than the other regions. In its uniformity, it allows us to cover it as a whole, as variations from one district to the next are generally unimportant.

The most noteworthy mammal is surely the Huemul or Andean Deer *(Hippocamelus bisulcus)*, an adept swimmer of robust shape, huge long and wide ears, a rich earthy brown in colour. The stags which have small simply-forked antlers and measure about a metre at the withers, seem to be in charge of the small family groups in which the does are accompanied by their yearling fawn, though there are historic records of large herds. In winter the Huemuls descend from the "alpine" slopes and meadows where they have spent the warmer months, and seek refuge in the woods to graze in the valleys. It is just here that man now runs his domestic herds, so the Huemul has been displaced and much of its habitat reduced, as also its distribution and numbers by the diseases of domestic animals and by the hunting of which it has been prey.

Another deer which, alarmingly, is becoming ever rarer is the tiny Pudú *(Pudu pudu)*, the smallest deer in the world, males of which barely reach 40 cms and weigh a maximum of 9 kgs. It is dumpy in shape and its antlers are merely small spikes which just emerge from the long chestnut hairs on the crown. It has evolved to live in the deepest undergrowth and canes, especially in the Valdivian Rainforests. Its timidity before the advance of man is one of the reasons for its rarity, but perhaps more important is the competition and persecution by introduced animals - it is an easy prey for dogs - both domesticated and feral.

There was a breeding station for the endangered Pudu on Victoria Island in Nahuel Huapi lake where the National Parks administration and the Argentine Wildlife Foundation together, with guidance from the New York Zoological Society, bred up numbers of this species for release into areas of their former range raising hopes that the deer will continue to be part of the Argentine fauna. Research later showed that numbers in the wild did not justify the effort so the programme was closed and the deer released into the wild.

There is a third mammal in the area which is also seriously threatened with extinction, the local otter, *Lutra provocax*, victim of trapping pressure for its pelt, but perhaps more affected by the introduced trout which compete with it for the crayfish which were the most important item of its diet. Other members of the mustelid family, the Grison *(Galictis cuja)* and the Patagonian Hog-nosed Skunk *(Conepatus humboldtii)* are better off.

The depths of the forest, where it is active at night and rests during the day in holes in the trees, is the habitat of the little-known marsupial *Dromiciops gliroides*, tiny, insectivorous. It may be more abundant than it seems because of its secretive nocturnal habits. It has a short velvety coat, hibernates during the winter and produces a maximum of four young in spring. It is called "kongoy" by the local Mapuche indians, and held in certain superstitious awe by them.

Three widely-distributed bats can be found in this area: the small Mouse-eared Bat *(Myotis chiloensis)*, the Big-eared Brown Bat *(Histiotus montanus)* and the Red Bat *(Lasiurus varius)*.

On their part, rodents are abundant, especially the *Cricetidae*. According to recent research theirs would be the greatest vertebrate biomass in these forests though it probably varies tremendously with seasonal availability of nourishment. Several genera are present like *Abrothrix, Reithrodon* (in the ecotone), and *Oligorizomys*, this last with but one species, the long-tailed *O. longicaudatus,* a species which can climb, with a tail over twice the length of the body and head together. Genera of burrowing mice are common to this area and the Pampas, well-adapted in their strong claws and short tails. Nocturnal and more like a Hamster to look at than a true mouse, they have a velvety fur and are omnivorous. Two species occur, *Chelemys macronyx* and *Geoxus valdivianus*. Interesting endemics are *Irenomys tarsalis,* a mouse with a long, tufted tail and an agile climber which lives in the denser parts, and *Euneomys chinchilliodes* barely known from a few specimens from Argentina.

As elsewhere, the Puma is the largest predator, but it is absent from the islands of Tierra del Fuego and is everywhere actively persecuted by livestock ranchers. The Red Fox *(Dusicyon culpaeus)* is more of a forest species than the smaller Gray *(D. grisseus)* which is basically of the open spaces of Patagonia and elsewhere. For these species, as also for *Felis guigna* the endemic spotted cat, the National Parks offer a haven as well as constituting a real genetic bank before the pressures from the unmitigated slaughter, as much for their pelts as based on unfounded accusations. In the ecotone and also in certain valleys, populations of Guanacos survive, though the same applies to them as has been stated immediately above.

As in other biotopes the birds constitute the most visible animal life. The birds of the area here covered are almost the same throughout the region, with some exceptions. This even distribution of birds is particularly evident in species of

Colihue Cane

Cane dominates the forest floor in the wetter areas of the northern Patagonian Andes. It is solid and used in furniture-making, and provides the habitat for a species of interesting though hard-to-see bird, the Chucao Tapaculo.

Chilean Firebush

This small tree (Embothrium coccineum) grows in clearings and on the edge of the forests by streams and lakeshores. In late spring it becomes a show when covered with scarlet flowers.

The Torrent Duck *Rapids, waterfalls and white-water throughout the Andes are the habitat for the wonderfully adapted Torrent Duck (Merganetta armata a.); this race is endemic to the southern regions. It dives to capture the invertebrate larvae on which it feeds, finding them under rocks and stones. The drake (black and white) differs markedly from the chestnut and gray female.*

Early show highlights the tints appropriate to the season - rich yellow, orange and red of the leaves of the deciduous Antarctic Southern Beeches: Raulí, Roble Pellín, Lenga and Ñire.

water-birds. Thus the Great Grebe *(Podiceps major)* which is the largest of the grebes, can be found on virtually any open water of these 18° of latitude and prefers quiet bays to nest and breed, differing from other populations only in the nest which is not floating, but by the shore and built up from the bottom. Other grebes in the area are the White-tufted *(P. rolland)* and the Pied-billed *(Podilymbus podiceps)* which appear to depend on the emergent Patagua in which they nest.

Wildfowl (ducks, geese and swans) are amply represented and include the Black-necked Swan *(Cygnus melancoryphus)*. Some forms are restricted to the salt water of marine situations as is the Kelp Goose *(Chloephaga hybrida)* which feeds on sea-weeds and whose male being pure white, is more easy to spot than the dark female which blends into rocky shoreline backgrounds, or the Flightless Steamer-Duck *(Tachyeres pteneres)*, a gigantic version of the similar, daintier species of wide distribution which can fly, the Flying Steamer-Duck *(T. patachonicus)*. There are other geese of the same genus as the Kelp Goose distributed throughout the subantarctic forests, the Upland Goose *(C. picta)* which prefers to breed in the ecotone, and the Ashy-headed Goose *(C. poliocephala)* which is found on the shores of the rivers and lakes and which nests upto several feet off the ground in the crutches of trees, in hollow logs, or in the thick base of a bush. Both species are considered to compete with sheep for the tender grasses and for grazing new wheat on their winter migrations. Whether true or not, studies seem to indicate quite the contrary. But their numbers are down, perhaps because there is no legal bag-limit or season for hunting them. Many sportmen come from abroad to shoot them. Also illegal pesticides are being used against them.

The most attractive and interesting member of this family *(Anatidae)* is surely the Torrent Duck *(Merganetta armata)* which exists in various races all along the Andes. It is wonderful to see it in its habitat which is the tumbling white-water of mountain streams and rivers to which it is so well-adapted. Its huge webbed feet, long, stiff tail and slender shape all help it to live in this turbulent habitat. It feeds by

Autumn's first snow in Lanín

Once found only on the coast, the Kelp Gull *(Larus dominicanus)* has now extended its range inland by feeding on materials which man unwittingly but carelessly provides, like rubbish or the carrion of dead domestic livestock. For some years it has been breeding inland in the forested area and, being a strange new predator, has had an adverse effect on certain bird populations. At the same time the Olivaceous Cormorant *(Phalacrocorax olivaceus)* has been benefitted by the introduction of the exotic trout which have become its main source of food, so it is persecuted by fishermen. It would be worth their consideration that the actions of the birds is probably an important factor in the development of the game-fish population.

The general ornithological panorama of the forests varies but little as the species are nearly all found throughout the patagonian Andes. It is strange though, to encounter in these cool forests species of families of birds one automatically associates with warmer or even tropical habitats, like parrots and hummingbirds. The Austral Parakeet *(Enicognathus ferruginea)*, medium-sized and olive-green with reddish belly and golden undertail, flies by in noisy flocks, and a hummingbird, the Green-backed Firecrown *(Sephanoides sephanoides)* is active on all but the most bitter days, feeding in winter on the nectar from the Quintral's flowers which they seem to be permanently fighting over.

On the forest floor midst the thickets of cane, lives a family of birds which carry their tails erect - the Rhinocryptids. This South American family of birds feeds on insect eggs and larvae, and includes the Chucao Tapaculo *(Scelorchilus rubecola)* which emits a noisy many-syllabled call, while the Andean Tapaculo *(Scytalopus magellanicus)* announces its presence with a continuous "Ket-lék, ket-lék, ket-lék...", permitting one to identify a bird one is not very likely to see. You can be forgiven for believing that the little gray shadow which darted along the log was a mouse. The largest member of the family is the onomatopaeically named Huet-Huet *(Pteroctocos tarnii)* of a dark general colouration with deep chestnut on crown, breast and belly. The first and last of these birds do not get south as far as Tierra del Fuego. Except in Australia where they do not exist, woodpeckers are present in all woodlands and forests, and here there are three species. The largest is the Magellanic *(Campephilus magellanicus)*, all black but with white on parts of the flight-feathers; the male sports a brilliant red head and short crest, while the female an exaggeratedly long black, floppy, forward-curling appendage more appropriate for some cartoon character. They move around the forests in pairs or family groups and their drumming can be heard a long way off - a sonorous double "budum". The Chilean Flicker *(Colaptes pitius)* is a bird of the more open woodland and somewhat terrestrial, medium-sized and is not found on Tierra del Fuego; nor is the smallest species of the family in these forests, the Striped Woodpecker *(Picoides lignarius)*, all speckled black and white in an orderly striped pattern.

Another exclusively Neotropical family is the Furnariids which here has a wide variety; a species which behaves like tits, the Thorn-tailed Rayadito *(Aphrastura spinicauda)*; a nuthatch equivalent is the White-throated Treerunner *(Pygarrhichas albogularis)*; others behave like dippers at the edge of the water and on the rocks in mid-stream - the Dark-bellied Cinclodes *(Cinclodes patagonicus)*; there is even a species exclusive of the dense cane-brakes, Des Murs' Wiretail *(Sylviorthorhynchus desmursii)* with an amazing long two-shaft tail. Of the great American family of insectivorous birds the most abundant is the White-crested Elaenia *(Elaenia*

diving for the aquatic larvae of insects which it finds under rocks and in turning over stones. It nests in hollow trees or banks directly over the water and the ducklings have their own adaptations - huge egg, long incubation, long stiff tail unfurls from the cramped shape of inside the egg to serve as a rudder, long days exercising before eventually jumping into the raging world beneath. The species seems to be becoming rarer though reasons for this are not clear; perhaps just general disturbance is enough.

The Lake Duck *(Oxyura vittata)* uses ponds and lakes with a certain amount of emergent vegetation as its preferred breeding habitat as do many species of the genus *Anas*, one of which, the Spectacled Duck *(Anas specularis)* is typical of the region, nesting on high mountain tarns and corrie lakes or by big rivers, and descending to the big lakes for the winter. There is however only one species of resident heron, the Black-crowned Night-Heron *(Nycticorax nycticorax)* of widespread distribution throughout the Americas, which is active at dusk and in the dark.

albiceps) whose mournful whistled call is so typical of the summer months, while the rarest is the Patagonian Tyrant *(Ochthoeca parvirostris)*. Fringillids include two very abundant songsters: the Patagonian Sierra-Finch *(Phrygillus patagonicus)* and the Black-chinned Siskin *(Carduelis barbatus)*. The edges of the forest and its clearings are frequented by the Austral Thrush *(Turdus falcklandii)* and the Chilean Swallow *(Tachycineta leucopyga)* while clearings and the ecotone have another spectrum of passeriform birds: two tyrants - the Tufted Tit-Tyrant *(Anairetes parulus)* and the Fire-eyed Diucon *(Pyrope pyrope)*; a Furnariid, the Plain-mantled Tit-Spinetail *(Leptasthenura aegithaloides);* and two Icterids, the Austral Blackbird *(Curaeus curaeus)* and the Long-tailed Meadowlark *(Sturnella loyca)* the last being also a bird of the steppes. A peculiar oddity sometimes seen as far south as Tierra del Fuego is the Rufous-tailed Plantcutter *(Phytotoma rara)*.

The commonest raptors are the Bicoloured Hawk *(Accipiter bicolor)* and the White-throated Caracara *(Phalcobaenus albogularis)*, this last basically a scavenger, and the American Kestrel *(Falco sparverius)* and Black-chested Buzzard-Eagle *(Geranoaetus melanoleucus)* in the more open areas. All these are diurnal. At night the small Austral Pygmy-Owl *(Glaucidium nanum)*, the widely-distributed Rufous-legged Owl *(Strix rufipes)* and the ubiquitous Barn Owl *(Tyto alba)* are all active though the first is frequently seen in the daytime, often being mobbed by smaller birds. Tha Andean Condor *(Vultur gryphus)* is still fairly common in the whole area, the Black Vulture *(Coragyps atratus)* abundant, but both tend to operate out towards the steppes where there is more livestock, and therefore carrion.

The damp grassy valleys where the geese are found are also the habitat of the Southern Lapwing *(Vanellus chilensis)* as well as the Black-faced Ibis *(Theristicus melanopis)* which feed on invertebrates and a little plant material all day, returning in the evening to communal roosts in trees in the thick of the woods, or on cliffs. Its call is one of the sounds which one will remember - a metallic "tank" in flight and a chattering of the same quality when in groups on the ground or in trees.

Most of these species migrate during the Winter to escape the conditions of that season, some merely descending to the lower elevations in an altitudinal migration, while others move northwards to more benign climes though it is suspected that a number of species can "hibernate" or go into a state of torpor for a few days during the severest weather.

There are some relatively abundant batracians in the region like the smallish, gray-green woodland toad *Bufo spinulosus* and two leptodactylids of the genus *Pleurodema* both called Four-eyed Frog because of their prominent black lumbar glands which look like eyes at the wrong end. *P. bufonina* is agile in spite of its looks, does not grow more than 5 cms long, and is the southernmost batracian in the world. Its eggs are encased in irregular gellatinous strings and usually float at the surface of the ponds, while the more slender, thinner-legged *P. bibroni's* 400 to 500 are covered by a foamy jelly.

Furthermore, in the wetter areas live other species which are usually less known and rarely found in our country. Characteristic of the Valdivian Rainforest we have *Batrachyla leptopus,* a climbing frog with "suckers" on its toes, pale green with dark markings which camouflages it perfectly for this habitat; *Hylorina selvatica,* another climber but of brilliant colours; a little dark-green toad with orange and yellow stripes

Coihue Forests

and markings *(Bufo variegatus)* is often found in the upper elevations near the snows. There are also some species of *Eusophus* and the extremely interesting Darwin's Frog *(Rhinoderma darwinii)* a "marsupial" species with a little tubular projection on its nose. Its colouration varies from bright green to sandy upper parts and boldly but irregularly marked cream and black on the belly, the first to tone in with fallen leaves of the floor of the forests where it lives, the second to startle any possible predator as it jumps into a stream and floats away belly up, the pattern speckled like shade and sun on the ground. Its most curious aspect however is the male's bucal sack. He "eats" the fertilized eggs that the female has laid as soon as the embryos move, and they develop in this pouch into miniature frogs, at which point, having escaped the danger stage as tadpoles, they emerge through the male's mouth.

Several *Liolaemus* lizards are the most visible reptiles as they move among fallen logs and rocks in the clearings and more open spaces, while other species of the same genus live on the bare slopes of the high mountains.

These trees dominate the Valdivian District from the height of the lakes (500 to 750 m) upto 900 m. The evergreen Antarctic Beeches grow to over 40 metres in height and 2 m in diameter.

There is but one snake found in the region: *Tachymenis chilensis,* uncommon, smallish, olive-gray with longitudinal black stripes, and slightly poisonous.

The fish fauna of the patagonian lakes - and of the watersheds they belong to - is poor in species diversity, but some species were, and even are, abundant. Either way it is becoming necessary to evaluate the situation as to the real impact that the introduction of the several trout and salmon have had on the native species.

Galaxia has two main species in this region, *G. platei* which occasionally reach 30 cms in length though usually smaller, and *G. variegatus,* olivaceous in colour, mottled with irregular black markings. *G. attenuatus* reported from Tierra del Fuego, Staten Island and the southern reaches of the continent is interesting in that it also exists in Australia and New Zealand.

In the same family of the *Galaxidae* is the genus *Haplochiton* known locally as Peladillas. *H. taeniatus* was supposedly a main part of the diet of said introduced salmonids. The native Perch (*Percychthys* spp.) is a valued species for its size, excelent flavour and as a game species, so much so that it is raised in fish-farms. Solid-looking fishes with the fins, especially the dorsal, with spines, a gentle gold or pinkish colour, they are voracious predators - cannibalism has been known. They reach a length of 40 cms which varies with the species, three of which are known from the rivers and lakes of the region, *P. trucha, P. vinciguerrai* and *P. colhuapiensis.*

Another species avidly sought by fishermen is the Patagonian Pejerrey *(Odontesthes microlepidotus),* a silvery colour and also bred in captivity. The Velvety Catfish *(Diplomystes viedmensis)* which has rarely been found of late, is the only representative of its family in the country, the other species being in Chile. Finally, mention must be made of the tiny catfish of the streams *(Hatcheria)* about which virtually nothing is known. As early as 1904 the first salmonids were introduced and today they are well-adapted in many places in Argentina. However their value as game fish and for the table must not let us forget that they are exotic species in the habitats they occupy here. As such it has been inevitable that they produce alterations in the ecosystems of the rivers and lakes they live in, which may not be understood in their entirety but for all that, are important. Mention has already been made of the probable fate of the Andean Otter thanks to the competition for food by the trout, as well as what is surely happening with our native species of fish. Sense will have to prevail, abstaining from sowing the salmonids just everywhere, in order to get research done in a few well-chosen places, on the real effect and the interference with ecosystems they cause. Moreover, in those bodies of water where these do not yet exist, a refuge must be maintained for the native species which are so very affected by the trout.

The introduced species include the Land-locked Salmon (*Salmo salar sebago*), originally from the U.S.A., which feeds on other fish, crustacea, insects and larvae as do nearly all the salmonids. Rainbow Trout (*S. gardneri*) are also from North America and can be identified with ease by its greenish and blue back which turns irridescent pinky-red at the lateral stripe and silver below. The Brown Trout (*S. fario*), originally from the Old World in 1921, is very well adapted. In Nahuel Huapi a Brown was caught which weighed 16 kgs, one of the four in the world which have reached that weight. Lastly the Brook Trout (*Salvelinus fontinalis*) from the U.S.A. is restricted by its intolerance of temperatures above 16°C to 18°C and prefers fast well-oxygenated water.

Though the variety of insects in the region is large, they are not abundantly visible in the forests, except the numerous microlepidopterae. In spite of this there are many forms which are wood-borers, using wood as the food for their larvae, in very different ways.

Coleoptera are prolific; there is a species of long-horned cerambicid (*Oxypeltus quadrispinosus*) which burrows within the twigs of the Coihue, and later cuts them off to fall to the ground where the larva completes its development; weevils (*Curculionidae*) with their ugly-charming appearance such as *Rhyephens maillei* or *Lophotus vitulus* whose larvae live between the bark and roots of several species of tree. *Anthaxia concinna* is a Buprestid beetle and feeds on the dead wood of the cypress while *Calydon submetallicum,* a black, brown and yellow cerambicid with long legs and antennae has a larva which burrows just between the bark and the outer layers of wood. Perhaps the most typical genus of coleoptera is *Chiacognatus* a Stag-beetle with large mandibles which is endemic to these southern Subantarctic Beech Forests. Predatory species are not absent, feeding on other species of insect. The carabid genus *Ceroglossus* is very characteristic, *C. buqueti* being but one. Also *Plagiotelum irinum* and *Mimodromius irinum* of the same family, are of predatory habits.

There are a number of noteworthy hymenoptera such as the bumble-bee *Bombus dahlbomi* one of the most conspicuous species, or various thynnids, *Elaphroptera* the main genus, ichneumonids, tentredinids and pompilids. In the *Nothofagus* woods parasitic hymenoptera are abundant and some genera are endemic, like *Ecphysis, Classis, Caenopelta* and others.

Butterflies are here represented by the genera *Vanessa* and *Yramea* which are common, and a saturnid (*Polythysana rubrescens*) with two red "owl's-eyes" surrounded by black. There are numbers of ants and flies as well as a rich fauna of aquatic insects, primarily in the streams which tumble off the mountains.

The whole area under discussion, and especially the National Parks within it, are a text-book example of the negative effect of the introduction of exotic species on natural native ecosystems. Here is found a veritable gamut of such species both for their population sizes as because of their distribution - a clear lesson to be learned. In the case of exotic plants, apparently less damaging than the exotic animals, some 140 species, mostly from Europe, have been identified, brought by the European settlers in nostalgic mood, without thinking of the imbalance they would some day cause. It was an attempt on their part to recreate the environments of their home-lands with plants and animals which were familiar to them.

For the tourist a view dominated by the Briar Rose (*Rosa moschata*) covered with blossom, or a field of daisies (*Chrysanthemum leucanthemum*) waving in the breeze could be aesthetically pleasing, but not for the ecologist who understands the detrimental effects of these on the native flora. The dreadful effect of the Water-hyacinth in Africa or the U.S.A. - a plant originally from South America - should be more than enough to stimulate reflection on the subject and a determination that it happen not again.

The effect of the modifications of the sub-antarctic biomes by the 12 or so exotic animal species which exist there, is very much greater. Any insertion of a "foreign" biological form into an ecosystem implies either the possibility of its success or its failure - in the latter case it will disappear. In the former situation lies the problem.

The list begins with the European Wild Boar (*Sus scrofa*); its rooting in search of food destroys the sod. Its tremendous aptitude for survival, high reproductive success, and omnivorous diet are some of the factors which contribute to its numbers. This animal is today widely distributed in the country and the obvious use of diseases like swine-fever is ruled out for the effect it might have on domestic pig-farms. Perhaps biological methods of control might be effective against the two populations of feral Rabbits (*Oryctolagus cuniculus*), one at each end of the region, in Tierra del Fuego and Neuquén (on lat. 37°S and in alarming expansion), for now. The effect of browsing by this species on the vegetation is especially noticeable in Tierra del Fuego National Park where several species of plant are forced to grow in dense, dwarfed mats, and whose reproduction is effectively stopped by these rodents. The possible use of Myxomatosis is being considered after its fair success in Australia.

Also in Tierra del Fuego the Beaver (*Castor canadiensis*) released in 1945 has spread through the whole forested area on both sides of the frontier which makes its control particularly difficult. It directly affects the woods by felling trees as much to feed on as to be used in the contruction of dams, and indirectly kills off whole areas by flooding. In one particular case which was studied, this species, which as all exotics has no natural predators in its new home, felled every single Lenga tree which housed a "rookery" of the Olivaceous Cormorant, a species which has now vanished from the area. Another foreign species in the Fuegan environment is the Muskrat (*Ondatra zibethica*), but its effect on the environment, for all its having been introduced in the same year as the beaver, is barely noticeable. There have been initiatives in recent years to introduce Reindeer (*Rangifer tarandus*) and even the Brown Bear (*Ursus arctus*) into the same delicate habitats, but fortunately they did not prosper because of public opinion.

Much more to be regretted is what has happened with the Red Deer (*Cervus elaphus*) from Europe and the Fallow Deer (*Dama dama*) in the Nahuel Huapi and Lanín National Parks, and soon will also happen in Los Alerces as the deer reach those latitudes. These two species, together with the Axis or

The Black-crowned Night-Heron

This race of Night-Heron (Nycticorax nycticorax obscurus) *is the only resident heron in the Biome; the species is cosmopolitan. The form here is dark.*

The Red-Gartered Coot

Found almost throughout the country, this species, like all coots, has to run along the surface to take off, displaying its lobed toes.

The Black-faced Ibis *This elegant ibis is to be found in short grass of the damper valleys where it feeds on invertebrates and some plant material, though it roosts and nests in trees or on cliff-faces.*

Spotted Deer *(Axis axis)* were released for hunting, to enrich the, in the opinions of some, "impoverished" fauna of the areas. In woods and forests which have never been exposed evolutionarily to the intense browsing of ungulates, these can only be detrimental to the cycles and renewal of the vegetation. Extermination in the vast areas where these deer exist today is probably out of the question for the lateral damage which would be produced as well as for the magnitude of the task, so for years the species have been managed as a recreational resource, for hunting in such a way that least possible damage be done to the National Parks.

The European Hare *(Lepus c. europaeus)* spread from a few nuclei of released specimens towards the end of the XIXth century, and has reached virtually all corners of the country, even the nunataks poking out of the continental ice-caps, some 20 kms across the bare ice. In many places it is so abundant that its grazing is a real problem. One count at dusk in the Los Glaciares National Park produced the figure of 522 hares in 19 kms of road.

It is not yet clear what effect the Californian Quail *(Lophortyx californica)* has on the environment. It was released as a game species and has spread throughout the ecotone in parts of Rio Negro and most of Neuquén. At the moment there seems to be no evidence of direct effects nor of competition, but the risks are clear and the effects may be detectable only years later.

The most complicated of the invaders is the recent aparition of the Mink *(Mustela vison)* which escaped of was released from mink-farms near Esquel in the 1970's. This small predator is semi-aquatic, and by 1980 it was found as far north as Bariloche, and down the Chubut river had reached the Atlantic; southwards there is no data. In places where the mink multiplies already many species of birds and mammals related to water have been driven out by direct predation or by its feeding on nests or young, including an important population of Coypus which lived in Los Alerces National Park. On the steppe it is just as harmful, feeding off the native wildlife; it has caused alarm among sheep-farmers whose lambs it has killed as well as chickens.

One of the main reasons for protecting these forests, as well as their aesthetic values and as genetic reserves, is as protectors of the soil and the watersheds where they play an essential role. They have two functions in this: first as contributors to the formation of soils by providing a continual shower of plant material, and secondly as a defender of the soil by protecting it from the elements.

The tree canopy saves the soil from the direct impact of the drops of rain which is a source of erosion, and it also moderates the extremes of temperature which would dessicate or destroy the organic substrate. The soil also acts as a sponge, slowly absorbing water which mostly passes through to the underground aquifers and to rivers and lakes. The organic material increases the capacity for absorbtion in the upper levels while roots do the same at lower levels. Roots also hold the soil together physically, avoiding thus its being washed away. This way too, floods and droughts are avoided in their extremes. It is not only the useful life of the vast dams built downstream that depend on the conservation of the forests, but also important areas under irrigation, like the upper Rio Negro valley.

The regulation of lumber extraction in those areas which allow for it must be effective and strict, especially as there are ample examples of excessive extraction of timber, and in some areas fire has completed the process of destruction. In the biotopes we are here covering, fire is a serious factor of

The Pudu

destruction and its results are irreparable in the time scale which man needs to function on - just think how long a stand of *Fitzroya* would take to be replaced. Contrary to the effect of fires in prairies or savannas where periodic burning is part of the natural cycle of the ecosystem, in forests where trees are of slow growth the consequences of fire are practically irreversible. Fires are known to advance underground through root-systems, tree-trunks as well as all other material is turned to ash and the gradient of the relief is such that it all washes away with the first shower of rain. In our National Parks such conflagrations occur more often than desirable and require prodigious efforts to extinguish. Even in countries which are pioneers in the field and which can count on resources which include the most up-to-date equipment, success is often only partial. Education and conservation are still the two actions which best can ensure that these catastrophes do not happen, and it is necessary to remember that it is often the landowner who sets the fire in an attempt to enlarge his grazing or simply "better" it.

This is the smallest deer in the world, barely 40 cms high at the withers. It is endemic and very rare. It lives in the dense undergrowth and browses on several bushes of these biotopes.

As was indicated in the chapter which dealt with the High-Andean Province, part of it is in the southern region above the tree-line. This limit upto which trees can grow gets lower the further south one goes, so in Neuquén it is at about 1,900 m while in Tierra del Fuego it is at 500 m. There is very little data on the climate of the region which is obviously cold all year-round. Temperature descends 6°C with each thousand metres ascended. This, together with the higher latitudes, makes snowfalls frequent in any season of the year, though only in winter is it all white. On cloudless days solar radiation is here very intense, so the vegetation must be adapted to these rigorous extremes.

Plants up here grow in sort of islands above the woods, till one reaches bare rock where but a few lichens can survive. As one proceeds southwards in the region the elements of northern influence drop out and only the species typical of the Patagonian Andes' forests survive to the point where the sub-antarctic forms exist in pure stands.

In the north, steppes of two species of grasses dominate - *Poa* and *Festuca*, accompanied by cushion-bushes or ground-creepers. One can mention Neneo, or *Berberis empetrifolia* with its small dangling yellow flowers, *Baccharis magellanica* in thick clumps against the ground, or *Pernettya pumila*, a ground-creeping heath. Among herbacious plants there are plants as diverse as *Viola* and *Senecio* like *S. julietti* with fleshy leaves and tubulous corolla, *S. portalesianus* or *S. poepiggi,* all with yellow flowers, and *Oreopolus glacialis* which has golden-yellow. Among all these stand out the compact cushions of *Lucilia araucana,* a silvery-leafed composit, or umbels of the genus *Azorella*.

Where the waters from snow-melt run or where they stand in bogs, there is a peculiar flora with pale green Maillico *(Caltha sagittata)*, Primavera *(Primula farinosa)* here with lilac flowers and a cluster of leaves at the base, *Pinguicula antartica* with very asimetrical violet flowers and one of the prettiest and most conspicuous flowers it is possible to find at these heights, the Waterfall Flower *(Ourisia alpina)* which grows right beside the tumbling torrents or streams and colours the banks bright red, offsetting the dominant green. Towards the south, starting in Santa Cruz, clumps of the Diddle-dee *(Empetrum rubrum)* grow, a low ground-hugging bush with tiny leaves, red, fleshy berries, adapted to very low temperatures.

At these heights the fauna is not abundant; visitors arrive in the warmer months. The Huemul or Andean Deer comes up here at this season, as too the White-bellied Seedsnipe *(Attagis malouinus)* which nests here, amazingly cryptic in its brown speckled plumage. A hummingbird, the White-sided Hillstar *(Oreotrochilus leucopleurus)* exists here as far south as the province of Santa Cruz, while the Yellow-bridled Finch *(Melanodera xanthogramma)* also nests, throughout the length of this habitat, lower further south, descending temporarily to sea level when weather so dictates.

In these Subantarctic Forests are the most important of Argentina's National Parks. Many of these are subject to economic activities. They include vast mountainous areas, ice-caps and glaciers and certain transition zones giving onto the Patagonian Steppes.

The beautiful forests alternating with lakes of crystalline water from the snow-melt on the magnificent mountains all make this last stretch of the Andes one of the most frequented tourist attractions. These parks then are the better-known of the system by the general public, to such an extent that these are for many the only parks.

But the scenic beauty and recreational oportunities must not detract from the fundamental reasons these areas were set aside as a chain of protected areas. In spite of what appears at first glance to be a general uniformity, here is a region with many different biological forms, a quantity of endemics, but above all, the forests play an imperative role in the preservation and regulation of the water-balance on which depends a large proportion of the hydro-electric output of the country.

Lanín National Park. The northernmost of the Patagonian Andes protected areas, Lanín takes its name from the volcano, now extinct, of 3,777 metres height. It is a typical cone-shaped mountain which is covered in great part by perpetual snows as it towers some 1,500 metres above all the surrounding peaks. From any angle it dominates the mountain panorama at these latitudes. The park was created in 1937 to encompass an area of 378,000 has of cordillera, by law #13,895; it is 170 kms long north to south and 40 kms across at its widest. Continuing southward it borders onto Nahuel Huapi, and both back onto the Chilean border. It includes three areas as National Reserve responding to political criteria rather than ecological reasons, which has led to many difficulties in its management for conservation.

The administrative centre is in San Martín de los Andes at the head of lake Lácar, a charming town which will satisfy the lodging and supply needs of any visitor. Most of the park has a pleasant climate because of its low general elevation; the annual rainfall is around 1,800 mm.

As well as its scenic attraction, Lanín has a number of forest communities which are unique in the country for their specific composition, reason enough for its existence. A succession of lakes, each with its own catchment, are protected, and each has peculiar vegetation communities which have been described in the introductory section of this chapter. They are off-limits to lumber industries and protect the waters destined to a series of dams downstream, like Alicura, Piedra del Aguila and El Chocón. The northern part of the park is the domain of the Araucaria with the mature

Lanín Volcano — *Though now is extinct, its classical white-cowled shape dominates all the mountain views of the northern part of Lanín National Park. It is 3,777 m high and towers above the other heights around it by some 1,500 m.*

trees having the peculiar parasol shape as they lose the lower limbs. Between lakes Ñorquinco, the northern limit of the park, and Huechulafquen there are dense, pure forests of this species, especially on the west-facing slopes and valleys, although there are patches of these woods in the ecotone to the east. Today the Araucaria is still an important part of the diet of the Mapuche tribes which live on reservations like Rucachoroi in the northeast corner of the park. Their traditional rights to harvest these seeds is respected as they contain a high percentage of carbohydrates and proteins supplemented nowadays with domestic animals.

In the north and southwest of the park, but notably absent in the central part, the Roble Pellín *(Nothofagus obliqua)* dominates large areas, sometimes growing with Coihues and at others with the Raulí *(N. alpina),* this last of high commercial value for its fine timber, which suffered from over-exploitation until recently. The new forestry regulations which are restricted to private properties within the reserve, tend towards conservation through rational management, especially of the Raulí which is restricted to this area in its distribution, there being no others in Argentina.

Around lake Tromen on the northern slopes of Lanín there are groups of Araucarias, while on the eastern side is the transition zone towards the steppes. These forests by the Chilean border flank an international pass in frequent use, while at the southern foot of the mountain lakes Paimún and Huechulafquen join where the steppes penetrate. There are few more memorable views than of the mountain in all its glorious cape of snow on a moonlit night. A little further west lake Epulafquen with the Currhué Grande on the south, has the famous thermal baths of Lahuen-Co which attract tourists, even from abroad. Thermal springs with high mineral content are often associated with vulcanism, which is confirmed when, at the western end of Currhué one encounters a river of lava, now cool and solid, which descends from the peaks through the woods. That volcanic detritus which comes from the now extinct Huanquihue volcano gives one a taste of what the area must have been like in the period that Lanín was active. A ferry-crossing allows one to complete a circuit of great interest, passing also by Currhué Chico lake, leaves the park and re-enters further south on the edge of Lake Lolog where there are good camping areas and where the fishing is rewarding. The whole Currhué circuit which is pristine still, offers remarkable landscapes and views.

Cruising on lake Lácar or driving along its southern shore leads one first into another indigenous reserve on the Quila Quina peninsula accompanied by an area of summer homes. On the opposite shore one can go as far as the western end of the lake at Hua Hum, the pass over to Chile which is kept open all year. Of special botanical interest is the lake Queñi area some 12 kms south, where the great variety of plant species and their exuberance make the trip through the forests and crossing beautiful rivers well worth while. The excursion is complete with a visit to Nonthué and Escondido, difficult to reach. To the east of Lácar there are pure stands of cypresses.

One can reach San Martín de los Andes, some 1,650 kms from Buenos Aires, along paved roads. The town which is growing, ought to be kept in harmony with the surrounding ountryside. It has an airport some 24 kms away at the foot of Chapelco with regular flights to Neuquén and beyond.

To the south the roads link up with those in Nahuel Huapi National Park. Other than the magnificent panoramas all the way, there is a point of special interest. Everyone knows that mountain streams flow together, but this is not the case of the Arroyo Partido. The stream bed is such that the stream divides just below the road-bridge that crosses it. The branch to the north is called Pil-Pil and flows into Lácar lake and thence to the Pacific ocean, while the southern branch becomes known as Arroyo Culebra and joins the Rio Hermoso, Lake and later Rio Meliquina, and down the rivers Caleufu, Collón Cura, Limay and Negro to the Atlantic. From the point where it splits therefore, the continental divide runs up the middle of the stream, probably the only place in the world that such a curiosity occurs.

As is the case of all the National Parks in the region, the fauna is as that described in the general section of this chapter. Historical records of numbers of Pudú and Andean Deer a few decades ago give a glimmer of hope that somewhere in the farthest reaches of the park there may be remnant populations of these endangered deer. The Red Deer was released here many years ago, and it spread rapidly with all the problems that that entails as mentioned before. For the autumn rut, hunting rights are auctioned some time before in Buenos Aires. This practice provides some few heads of international standard, but the system is under review as it is intended also to exercise some measure of control of numbers as well as offer sport.

This park, with its relatively warm climate, offers quality fishing to the visitor who can fish in most of the rivers and lakes, but this varies from year to year and the regulations are distributed with the purchase of the corresponding fishing permit, there being an arrangement with the Province of Neuquén that the permits are valid in both jurisdictions.

As has been said at the beginning, the areas where forestry or livestock exploitation take place were delineated without ecological consideration coming into play. Many valleys of value to the fauna - think only of the winter descent of Andean Deer for example - are part of the Reserve which covers about half of the protected area (some 184,000 has) with the consequent detriment to the natural heritage which should be protected.

Nahuel Huapi National Park. The piece of land of three square leagues (7,500 has) donated by Dr. Francisco Moreno to the Nation in 1903 and which gave rise to the present National Parks Administration, is today within the borders of this national park created in 1934 by law # 12,103. The park covers some 750,000 has of which 330,000 belong to the National Reserve. It is some 155 kms long north to south, against the Chilean border, and some 75 kms across at its widest. Bordering as it does onto Lanín National Park to the north and with several others over the mountains in Chile, the aggregate conservation value of this whole unit is immense.

The highest point in the park is Mount Tronador, an ancient volcano, at 3,554 metres above sea-level, called Tronador (Thunderer) because of the noises produced by avalanching snow and ice from its three ice-covered summits. Other peaks which can be mentioned are Catedral, Millaqueo, Cuyín Manzano, Crespo, López and others, none of which go above 2,400 m. Several passes exist through the cordillera to Chile, like the Vuriloche pass used in ancient times by the indians, or Puyehue with its permanent vehicular traffic.

There is an abundance of rivers and lakes, the foremost of these last being Nahuel Huapi with its 560 square kms and a maximum depth at well over 450 m. It is a typical glacial lake, the valley being dammed by a huge terminal moraine. Originally there were faults and subsidence, uplift and movements of the earth's crust, all subsequently carved and shaped and moulded by several glaciers which flowed together from the arms now called Tristeza, Blest, Machete, Huemul,

The Fire-eyed Diucon

Clearings in the woods and the ecotone have a wide variety of passerine birds which includes various tyrants like this Diucon (Pyrope pyrope) with bright-red irises.

The Dark-faced Ground-Tyrant

This Muscisaxicola macloviana *is yet another tyrant but of open spaces where it runs down its insect prey. To this end it has long legs. It is the most common of its genus.*

Deciduous Beeches *Lenga* (Nothofagus pumilio) *and Ñire* (N. antarctica) *line the eastern reaches of the patagonian Andes, and compose the Deciduous Woodland District. These species turn colour in the autumn and assume warm tones of yellow, gold and red.*

Rincón and Ultima Esperanza. Isla Victoria of about 3,100 has was left behind by the glacial erosion, having been uplifted.

This Nahuel Huapi lake drains along the Limay and Negro rivers to the Atlantic. The first stretch of the Limay, just as it flows into the headwaters of the Alicurá Dam, is through the magnificent Amfiteatro and the Valle Encantado, this last with strange sculpted rocks shaped by erosion. Some of these have received names alluding to their shapes like Dedo de Dios (God's Finger), el Castillo, el Penitente. Though the dam spared these formations it flooded a very rich area as is usually the case of ecotones. Dams don't only destroy habitat but are eye-sores for ever and upset the natural balance irreversibly.

Towards the north is the Traful watershed which drains into the artificial lake via the Traful river at Confluencia. To the south is the Rio Manso system gathering the waters from Tronador's east face. Several lakes form this watershed: Mascardi, Guillelmo, Hess, Steffen, Fonck, Felipe and others. Mascardi and Gutierrez, though they occupy the same valley are separated by a narrow bog whose waters can choose to drain either to the Atlantic or the Pacific according to which of these two lakes they settle for.

This is the park which receives the greatest affluence of tourism, virtually all operating from Bariloche where the park's headquarters are. The town, well-connected by road and air-services to the rest of the country, is some 1,700 kms from Buenos Aires. The municipal jurisdiction occupies a large sector on the southern shores of Nahuel Huapi lake and includes the peninsula of Llao-Llao and most of the loop-road called the Circuito Chico. Though there is an agreement between National Parks and the town about the management of the area, severe tourist pressure does seriously affect certain plant communities which it would be desirable to protect. Most visitors arrive attracted by the magnificent landscapes with mountains, lakes and forests as much as for the capacity and efficiency of the services to cope with land and water excursions. For these as well as for those who have more specific objectives in mind, Nahuel Huapi offers a variety of places to visit and of recreational activities.

There is a good network of internal roads which allows the visitor to benefit from his automobile. Also several trails or footpaths cater for those who prefer exercise and a closer contact with Nature, giving access to areas which are otherwise not reachable. A series of moutain refuge huts with the upland trails connecting them cater for those inclined to mountaineering at various levels of competence or difficulty.

Fishermen, during the summer months and keeping strictly to the rules, have an enormous choice of waters to try their skills, with good probabilities of success. Bariloche has a winter ski-resort on Cerro Catedral, some 20 kms distant, with modern facilities and pistes of varying demands. The development of new areas and pistes has unfortunately been done with little regard for the erosion factor.

One of the most important and interesting excursions is on the lake, visiting Isla Victoria and the Arrayanes woods on Quetrihué peninsula. For legal reasons and for the better consevation of these interesting woods, the area is park in its own right, Los Arrayanes National Park of 1,000 has. It is totally surrounded and administered by Nahuel Huapi National Park. These woods are almost purely of the Myrtle Arrayan of smoothe, peeling bark, giving the tree a spotted appearance, which, together with the light and shade speckling the ground and filtering through the canopy make a fascinating display which accentuates the beauty of the place. There is a boardwalk with containing rails to avoid

Nahuel Huapi Lake

Some 560 square kilometres in area this lake is well over 450 m deep in places. In the middle distance is Isla Victoria and the backdrop is provided by Cuyín Manzano, while in the foreground evident signs of civilization show that this is part of the mainland, within the municipal jurisdiction of Bariloche.

The Myrtle Woods of Arrayán

Quetrihué is one of the most visited areas in the Nahuel Huapi National Park where almost pure stands of this lovely tree (Luma apiculata) *show off its wonderfully coloured bark.*

Ferns and Mosses

Together with lichens and liverworts these are all denizens of the damp woodlands of the southern Andes. Old Man's Beard (Usnea spp.) *is a lichen which hangs from the trees in whispy grayish tufts, like the facial hirsute appendage of old Father Time himself.*

The Upland Goose

Chloephaga picta *is markedly sexually dimorphic, males being the whiter specimens. The Kelp Goose,* (C. hybrida) *on page 206 is even more so, though in the rest of the genus the sexes are virtually identical, as in the Ashy-headed on page 170.*

The Patagonian Hog-nosed Skunk

Conepatus humboldtii *is one of the few mammals in the biome and one of the most visible.*

The Southern Lapwing   *This shorebird* (Vanellus chilensis) *is the best-known bird in Argentina. It is everywhere that there is short grass in plains or valleys, north to south. Tero is its local name, onomatopaeic, and an alarm for all other beings to be warned of some danger or intruder.*

uncontrolled trampling of saplings and compaction of the soil or uncovering of the roots by erosion.

The other place visited on this excursion is the large Victoria island where visits are restricted to the central part around Puerto Anchorena. The area was once an arboretum and nursery garden which may with time be reactivated to supply only the native species for the needs of parks' recuperation schemes. What is left from bygone years are acres and acres of exotics and some fine specimens of the native flora. There are two restaurants to replace energies lost, a sheltered beach for the hardy to bathe, Playa del Toro, and picnic areas. Trails in this central part of the island lead to wonderful look-outs to admire magnificent views - or one can take the chair-lift. The island should return to the category of "park" rather than reserve as it now is, except for the central and service areas.

On the island there is a huge population of the three species of exotic deer which has caused serious damage to the habitat by eliminating the undergrowth and saplings of the larger trees. This has effectively destroyed the habitat of birds, small mammals and reptiles and batracians of the lower stratum, and as a consequence diurnal and nocturnal predators too have disappeared. There is a plan for the elimination of exotic species from the island. Nesting on certain cliffs which drop directly into the water are birds which many would call "penguins". They are in fact the Blue-eyed Cormorants *(Phalacrocorax atriceps),* usually a sea-coast bird. They nearly vanished in 1966, when for months after much volcanic activity, the waters were turned cloudy by the volcanic ash and they could not find their food - small fish and crustaceans. There is but one other inland rookery of these birds, in Tierra del Fuego.

Another excursion on the lake takes one along the arm going in a westerly direction almost to the border with Chile and the highest rainfall area in the country, Puerto Blest. It is here that the best example of Valdivian Rainforest is to be found, with gigantic Coihues. One of these is named "El Abuelo" (Grandfather), standing by the road between Blest and the close-by Frías lake, and is worth a visit. There is also a peat-bog by the road with *Fitzroya* and cypresses growing in an area which is apparently too wet for them as they are scrawny specimens. It was the construction of the road which has stopped the natural drainage and raised the level of the water which previously had run away over the surface. In the puddles and the road-side ditch one can find the little Darwin's frog as well as another batracian *Eusophus taeniatus*.

A pleasant walk around the bay at Blest leads one to the Cántaros river and to the attractive waterfall of the same name. On the way one can appreciate the magnificence of this special rainforest. Or one can take a boat to cross the bay, where, from the jetty on the other side there is a board-walk right up to Cántaros lake.

Towards the southern end of the park there is a paved road which goes from Bariloche, past Guillelmo lake to Villa Mascardi, from where a gravel road leads off to Lake Roca and the Alerces waterfall where the fortunate visitor might find the elusive Torrent Duck. The other branch of this road which leads northwest, after a spectacular run ends at the very foot of Tronador to continue onto the lower slopes as footpaths. There is a trail called Paso de las Nubes which leads from here to Lake Frías which we were at a paragraph or two back, in the area of Blest. There are magnificent views of glaciers and all the forces which moulded and sculpted the landscape of the southern Andes.

Lake Steffen is the place if it is peace one is seeking, along a road to the west from just beyond Guillelmo lake. In a bare seven or eight kilometres one descends over 400 m. By the road at a small lake called Hualahué, one should stop and spend a while watching the abundant waterfowl found there, especially the Great Grebe after which the lake is named (Huala = Great Grebe, hué = place). Further west, along difficult trails, Lake Martín waits for those who can do without superfluities, all the clutter of civilization.

North of the Nahuel Huapi lake there are also tourist circuits of varying length, two of which enter the neighbouring park Lanín and take in the Enchanted Valley near Traful. This is an area to keep watching for a Condor as there are nesting and roosting areas nearby, and abundant livestock to provide the carrion on which they feed. The little Californian Quail is also to be seen thereabouts with its peculiar forward-curling top-knot, like a question mark on the crown of the male in which sex the appendage is most developped.

Up the Traful Valley one reaches the Mirador which overlooks the lake and a vast panorama beyond, past Villa Traful, a hamlet of summer homes with a small population of permanent residents to cater for the needs of visitors. It is in these particular waters that fishing might reward the sportsman with a rare landlocked Salmon.

Across the valley the other road leads over Paso Córdoba, switch-back and precipitous, to enter Lanín park a few kms further on. Yet another variant is the road along the north shore of Nahuel Huapi which in its westward run to the Chilean border traverses all the plant communities in the park from the steppes to the Lenga forests.

The Seven Lakes (Siete Lagos) alternative which also enters the park to the north passes many different habitats from open grasslands to thick forests and dense cane-brakes. It passes scenic places like Pichi Traful and lakes Correntoso, Espejo (Grande and Chico) near which there is another patch of Valdivian Rainforest, Escondido, Villarino and Falkner. In certain inaccessible heights there are reputed to be vestigial populations of Huemuls (Andean Deer), and Pudu, as at the Portezuelo del Cajón Negro, for example. Along the north shore of Nahuel Huapi excursions pass by the small town of Villa Angostura near which are important ski-slopes, and excellent trophy fishing.

For Moreno's exhortation to be lasting, in this the largest of Argentina's National Parks the reasons for development and tourism will be gone if the natural wonders are misused. Urban development, settlers and private properties in a national park are subject to the principal objective which is the conservation of Nature; this is not always borne in mind, and with disastrous consequences.

Lago Puelo National Park. Originally this park was created as an annex to Los Alerces National Park in 1937, but since 1971, under law # 19,292 it is a park in its own right. With its 23,700 has of which 9,600 are National Reserve, it is the smallest of the South-Andean National Parks, but like them all it is mountainous and its scenery has glacial origin. Its waters drain through the lake and the Río Puelo to the Pacific. Along this valley there is a connection with the trans-andean forests and it is precisely in the botanical aspects that the interest lies. Only 200 m. above sea-level the temperatures are more clement and it is this characteristic as well as the connection just mentioned, which create the conditions for a plant community which is unique in Argentina. In the park are many species which are very rare or not found at all in any other part of the country.

Where the river Puelo flows out of the lake, nearly at the Chilean border in the extreme northwest of the park, the following species are found: Guevín *(Guevina avellana)* also known as Native Hazel (Avellano Silvestre); Tique *(Aextoxicon punctatum)* another tree recognized by the rust-coloured scales, especially on the under-side of the leaf which is oblong and 4 to 8 cms long; Ulmo or Urmo *(Eucryphia cordifolia)* with scattered white flowers; and the liana or creeper Voqui Blanco *(Boquila trifoliolata)*. Deu *(Coriaria ruscifolia)* is common in the park; it is fatally toxic to humans who eat its fruit and grows near the water in sandy or stony soils. The fruit hangs in bunches and is dark violet in colour.

Just about all over its surface this park has suffered forest fires; whole mountain slopes are today recuperating very slowly from the effect of these disasters, and there are places where the lack of vegetation cover has led to serious erosion.

Entrance to the park is from the north, the gates being but 19 kms from El Bolsón. At the head of the lake the visitor will find a splendid camping area, facing the towers of mount Tres Picos. The area does tend to flood in late Winter or early Spring at the time the snows melt. The occasional storm in summer might have the same result, because of the prompt run-off from the areas outside the park which have been clear-cut. Unlike the other parks, here there are never frosts in summer, simply because of the low elevation. Brair Rose is the most dangerous invader and has extended its occupation of grazed land enormously. There are no lake excursions but water-sports are allowed for those who bring their own craft. There once was a launch service to the Chilean border which ought to operate again.

The area of the park should include the Epuyén arm of the lake as imposing stands of Patagua grow there. The borders, determined without a thought to ecological units, must be redrawn to include the heights around. There ought to be a camping ground with certain comodities in the Río Turbio area.

Los Alerces National Park. In 1937 Los Alerces National Park was created for the main reason of protecting stands of the "Alerce" or Lahuán which contain the largest trees of this species in the country as well as being the most extensive groves. Many of these trees are thousands of years old, the giants of Andean Patagonia's forests.

The law which created the park probably saved these trees from extermination as their wood is much sought. Their slow growth-rate makes forestry with this species quite impractical. Many of the trees which were felled were over 3,000 years old.

Several mountain ranges are included in the park: Rivadavia towards the north, Pirámides in the centre and Situación, which runs north-south in the south-west area of the park. The outstanding peak is Torrecillas which at 2.253 metres above sea-level has been climbed but once. A number of rivers, lakes and streams are fed by rainfall but mainly by snow-melt in spring. Most of these drain into the Pacific Ocean along the Grande or Futaleufú river. A complicated system of lakes occupies the whole area. In the north-west Lake Cisne is connected with branching Lake Menéndez which in turn, over a series of rapids bearing the same name drains into the small but beautiful Lago Verde. Verde also receives waters from Lake Rivadavia in the north-east and all these drain down the wilderness Arrayanes river into huge Futalaufquen, the focal point of this drainage system. From the west Lakes Stange, Chico and Kruger form another chain which also connects with Futalaufquen in the extreme south, through the Estrecho de los Monstruos. From Kruger the Rio Frey recieves

Lago Puelo National Park

all this volume of water and flows into the recently-created Amutui Quimei. In the local Mapuche language Amutui Quimei means "lost - or vanished - beauty" for this was originally the site of a chain of lakes called Situación, Uno, Dos and Tres, and wonderful rapids below Situación. Unfortunately the big hydroelectric dam at Futaleufú flooded the whole area - 8,200 has - and all this pristine beauty was lost, hence the name. All the old landing-beaches have been submerged, and this with the rough water on the big lake make boat or launch trips unfeasible, which, together with the lack of access by land have rendered the affected area unusable for recreational purposes. Previously the value of the area for such ends was immense. At the same time the whole concept of a National Park was violated and an unfortunate precedent set.

Travelling from Esquel, the nearest town some 45 kms away, one first crosses an area of Patagonian Steppe after which the first trees appear which are mostly Maitens. This same species grows in copses within the park as far as Villa Futalaufquen where the park's headquarters are located. There too are the

Although small in area this park is particularly interesting for the plant species found there, the only place that many of them grow in Argentina. The mountain in the background is Tres Picos.

living-quarters for the park's employees, an Intepretive Centre and a camping area with certain commodities.

Though relief makes any consideration of a network of internal roads very unlikely, the park is well suited for exploration on foot along trails, and waterways give access by boat and launch to back-country areas which may be visited. The northern sector of the park is considered by many to be the most lovely area of all the Patagonian Andes' parks.

From Puerto Limonao some 4 kms north of the headquarters, launch excursions depart which are almost "de rigueur" for anyone visiting the park. First the whole length of Futalaufquen is cruised before reaching the all-too-short Arrayanes river. Along its shores and even standing in the water grow the Arrayán trees which give it its name, the launch almost touching them as it seeks the deeper channels for navigation. Species of bird and animal like the Coypus, Sheld-geese and King-fishers are usually seen on this part of the cruise.

At Lago Verde a short walk takes one across a spit of land to another launch for a further cruise on lake Menéndez. Rounding Isla Grande the wonderful glaciers on Mount Torrecillas come into view. Finally the end of the north arm of the lake is reached where, along trails, steps and walkways one can explore the Alerce forests. In places one walks under a canopy of Colihue canes, at another along the side of the waterfalls on the Cisne river, or the shores of Cisne lake, and through and between the giant Alerces. This is the best and only way to get any notion of their true size and grandeur.

In this very damp sector one finds vegetation typical of the Valdivian Rainforests and may encounter some of the fauna associated with it. The Chucao Tapaculo is here very tame and approaches the visitor to a minimal distance. Here too with luck one could find the Magellanic Woodpecker. Extremely difficult to see is the Guigna Cat which, though abundant, here has many melanistic (black) specimens and is extremely shy.

Repeated observations of the Andean Deer (Huemul) within the park give rise to hopes that with continued protection a decent population of this rare and threatened species might build up.

Stands of *Fitzroya* *These are the main reason for the creation of Los Alerces National Park. This local equivalent of the Redwoods of California grows upto 60 m tall and 3 in diameter, and thousands of years old.*

Araucarias

Araucarias grow in the northern half of Lanín National Park, in the Domain of the Pehuén, a spectacular tree which is unlikely to be confused with any other because of its typical umbrella shape, achieved when it loses the lower branches as they die off.

The Huemul

From the "alpine" zone above the trees this robust Andean Deer (Hippocamelus bisulcus) *descends to the valleys and forests in winter. It only survives far from human interference and habitation in the more distant reaches of this mountain chain.*

The Futaleufu dam is not the only human impact on the park since the presence of exotic species, inadvertently introduced plants and animals, have altered the natural harmony of the place. Of all these perhaps that species which threatens to have the most dire consequences is the Mink. It has been seen to hunt in packs and among its victims can be counted the many forms of water-birds and land vertebrates, and even the peaceful Coypu. The European Hare is found in the open areas even high on the mountains, while the European Boar prefers the denser vegetation. The concentration of settlers on the east shores of lakes Rivadavia and Futalaufquen has severely damaged the vegetation, mostly through overgrazing, and in some areas erosion is a matter for grave concern. The Briar Rose, spread by wildlife (especially the exotic boar and cattle) occupies large areas, growing in impenetrable thickets.

The road from the park to Cholila was drawn without considering the damage to the scenic values and the ecosystem, and it is another "trigger" for erosion, damaging the vegetation and producing a scar on the landscape.

It will be necessary in the future to re-draw the borders of the park to include an area to the north where populations of the Pudu and the Huemul are found as well as the "Alerce" forests by the Tigre river. The eastern border allows for grazing on slopes above the park so it will have to be extended to the ridge.

Francisco P. Moreno National Park. This park was also created early by law # 13.895 but had been forgotten for years because of its remoteness and inhospitable climate; the intense cold and strong winds are unavoidable at all seasons, and summer temperatures rarely reach 15°C.

The Park covers some 115,000 has of which 30,500 has on the eastern border are the Reserve. It is in the north-west of the province of Santa Cruz some 240 kms from the nearest town - Gobernador Gregores. Only in 1980 were the first parks buildings erected for ranger stations; previously this duty was performed on periodic visits operating from a caravan. The administrative duties are still performed by the superintendent of Los Glaciares in Calafate.

Wonderful forests where Lenga dominates, an interesting series of lakes draining both to the Atlantic and Pacific Oceans, glaciers, waterfalls, fossil-beds, one of the most spectacular mountains, interesting wildlife, and rock-paintings, all constitute reasons enough to value this park as an important part of the system.

The eastern reaches through which the access road enters the park are a high plateau steppe some 900m above sea-level and partly surrounded by hills, just as far as the edge of the first lakes.

The main physical feature of the park is the complicated drainage system of lakes Belgrano and Nansen where numerous arms and other lakes occupy narrow valleys between mountain ranges averaging 2,000 m in height; dominating all is the magnificent peak of mount San Lorenzo at 3,700 m, which, just north of the area, can sometimes be seen when the clouds clear. Three lakes in series - Mogotes, Península and Volcán - drain into Lago Belgrano whose eastern arms almost completely surround the Belgrano peninsula.

This is the main attraction for most visitors, as after crossing the very narrow isthmus one can wander over rolling, partly wooded terrain to reach wonderful panoramic points which afford views of magnificent scenic backdrops to the complex lake of bright turquoise waters. The copses and small

Guanacos on Península Belgrano

In the Francisco P. Moreno National Park Guanacos are an attraction as they are unafraid and numerous in the transition zone of this magnificent and unspoilt area.

The Flying Steamer-Duck

This species (Tachyeres patachonicus) is a member of a genus endemic to southern Argentina and Chile. Large birds and somewhat thick-set, this species actually can fly though the other species, found on the sea, do not, only "paddle"-steaming to escape danger across the surface.

Rock paintings

On the walls of a cave on the south side of the Río Robles in Moreno National Park, they usually represent hands, but some of the figures are of the local animals like this Guanaco. Obviously this beast played an important role in the lives of these indigenous peoples.

woods of Ñire trees provide shelter for picnics and camping. The Guanacos here have become so accustomed to visitors that they can be approached with ease to study their behaviour. The farm-workers and shepherds who have traditionally hunted guanacos culling the young, called "chulengos", in the first weeks of life for their hides, here are respectful of regulations. In winter the lake freezes over and the guanacos are reported to walk across the ice.

In its south-west corner lake Belgrano drops over a waterfall into Lago Azara and thence down a rushing river and over some fine rapids into lake Nansen. Out of the southern end of large lake Nansen, near the border with Chile, the waters flow to the Pacific down the Cabrera river.

One of the important biological aspects of the park is in this system of lakes as the entire fish fauna is composed of the native and original species, the exotic salmonids not having been introduced here. As almost the whole watershed is in the park and the waters reach the sea in an unpopulated area of Chile it may be fairly easy to keep things thus for some time; there is however in these distant chilean waters, an industry of salmon-farming in the fjords from which obviously many fish escape and pose a potential threat of uncalculable proportions.

To date only the eastern shores of Lago Belgrano are of easy access. Boating on the lake is extremely difficult as the waters are open to the dominant west winds. In the north-east of the park one can only go as far as the Rio Lácteo in a vehicle. This river forms a delta on flowing into Lago Belgrano.

There is difficult and limited access to the western reaches of the park along faint and unmarked trails. There, near lakes Nansen and Azara the Lenga forests grow. Being at considerably less elevation than the eastern parts here the climate is much more moderate which allows for the growth of the tall evergreen beech, the Guindo. The Tres Hermanos glaciers at the juntion of lakes Los Mogotes and Península is a spectacular feature to reward all those who have invested so much effort to get that far.

In the south-east of the park lake Burmeister receives water from many melt-water streams and drains eastwards along the Rio Robles, thence to the Chico and right across the Patagonian Steppes to flow into the Santa Cruz estuary and the Atlantic. A track for the tougher vehicles leads to the eastern end of the lake which is contained in a steep-sided valley where Lenga grows in a narrow belt, the trees being twisted and gnarled by the blasting winds along the lake. At times when it is possible to cross the Robles river one can visit caves and rock-shelters on the south side to see a variety of rock-paintings, and in a sedimentary layer in a nearby slope, the fossil bones of large vertebrates and the petrified trunks of huge trees which the visitor is reminded not to touch. Some pairs of Great-horned Owls nest in the area.

There are other valuable and interesting fossil-beds in the arid hills of the north-east corner of the park. There, on the Cerro Colorado there is a spectacular condor cliff where, low below the crest, ledges and shallow caves are the roosting places for these gigantic birds of which one can often see 15 or 20 individuals. Another interesting species which belongs in the area of tumbled rocks at the foot of the cliff is the Mountain Viscacha (*Lagidium wolffsohni*), this species being endemic to the extreme southern Andes.

The greatest concentration of wildlife in the park, as much in variety as in numbers, is to be found in the steppes where the dominant vegetation is the Coirón grass (*Stipa* spp.). It is an important breeding area for two species of Seed-snipe (Thinocoridae), the Least (*Thinocorus rumicivorus*) on the flats, and the Gray-breasted *(T. orbignyianus)* in rolling country. The White-bellied Seed-Snipe nests on the steep scree-slopes above the tree-line.

Darwin's Rhea *(Pterocnemia pennata)* is common on the road into the park and occasionally one might encounter a fox.

Very many small ponds which are filled by snow-melt in spring are scattered on the eastern steppes, providing much breeding habitat for wildfowl. Each small round pot-hole lake seems to hold its pair of Flying Steamer-Ducks *(Tachyeres patachonicus)* territorially keeping other individuals or pairs off the limited area. But what most draws ones attention is the summer population of Upland Geese which here breeds much later than in other places as a result of the elevation and subsequent late thaw. Like nearly all the birds here it must depart on migration to avoid the cruel winter.

Any visitor to the park will have to plan the trip with much thought to the necessities - equipment, food, and fuel for the vehicle - and will be rewarded by a truly wilderness experience, almost cut-off from civilization, an experience very few others will share. Roads within the park are few and rough and lead only to estancias and some of the areas already mentioned.

Because there are so few contacts with the outside world one will find the people there very helpful, the estancias willing to provide meat and help in times of distress or need. These farms close down for winter and leave a single care-taker. This winter withdrawal has not stopped all poaching, but nowadays, with permanent ranger presence this is changing. The whole Reserve is under review as there are farmers who are leaving the area as untenable under fluctuating economic pressures. It is hoped that areas like the Belgrano peninsula will soon be included in the park proper as exploitation ceases.

Los Glaciares National Park. Tucked close into the extreme south-west corner of Argentina's continental territory, an area blessed with natural features of great interest contains what is without doubt one of the natural spectacles of the world. Together with the falls at Iguazú in the opposite and subtropical north-east corner's rainforests, the Moreno Glacier is perhaps what any visitor to the southern cone of South America will remember most.

Basically Los Glaciares National Park consists of the Lago Argentino drainage complex and surrounding area, and the thirteen glaciers which descend from the adjacent ice-cap into lakes Argentino and Viedma.

Los Glaciares covers some 600,000 has of which 154,000 are the Reserve, divided into three separate sections. This park was also created by Law # 13895 - 1937 was a good year as many fine parks were added to the original nucleus.

On H.M.S. Beagle's epic voyage, the naturalist Charles Darwin and Captain Robert Fitzroy ascended the Santa Cruz river in 1833; they turned back only at a place from which they could see the vast mountain chain of the Andes as a backdrop to the west, so they did not actually discover the lake which was later to be called Argentino. It was only in 1873 that Lt. Valentín Feilberg arrived at the shores of what he believed to be lake Viedma, his mistake only being discovered four years later when Francisco Moreno named the lake for the Argentine flag because the blue waters were separated from the celestial blue by the white band of the ice-cap.

The spectacle of the glaciers must excite the wonder of any witness and stimulate much curiosity as to just where all this ice

The High Plains

The eastern sector of Moreno National Park, at 900 m above sea-level, is covered with a typical steppe vegetation, and surrounded by mountains and lakes. Rivers flow across the steppes.

A Span of Ice

At the snout of the Moreno glacier are the remains of the recent grounding. In its advances this glacier cuts off the drainage of an important arm of Lake Argentino only to be pushed away by the tremendous water-pressure from the upto 20 m difference on each side of the grounded snout, a spectacle that happens irregularly every so many years.

The Moreno Glacier

Being one of the few glaciers which are surging, it pushes up into the beech forests like a bull- dozer, tumbling trees and wiping the base-rock clean of any covering soil.

Lake Onelli

Into this lake three glaciers calve - Onelli, Agassis and Bertachi. Its surface is covered with small bergs which break off the glaciers' snouts during the summer months.

comes from. On the top of the Andes mountains, in an area some 350 kms in length, the year-round precipitation in the form of snow is much greater than the rate of melt and the resulting accumulation has given rise to the ice-cap, a mere vestige today of what it was some 12,000 years ago during the most recent ice-age. This accumulation of ice now sits in a wide, high valley between parallel ranges of mountains and is formed by season upon season of snowfall, compacted by the weight of subsequent falls into ice and forming the gigantic southern continental ice-cap. Up there the panorama and climate are Antarctic in character, with strong winds and snow-fields from horizon to horizon, interrupted occasionally by some emerging nunatak or peak which pokes through, too steep or slender to accumulate and hold ice. The ice-cap buts up against mountain ranges and spills over, flowing down valleys, drawn by gravity acting on its mass, and pushed by the ice from behind or above. The valleys' walls and bottoms are scraped, and this eroded material is carried along by the ice till, way below the snow-line and in far more temperate climes the rate of melt equals the rate of flow. The speed that glaciers flow is usually measured in metres per year. The accumulation of rock being dragged along the valley floor at the bottom of the glacier acts as abrasive material to steepen the sides and deepen the valley. Where two glaciers join or where the ice is forced over irregularities in the bed, cracking, piling up and general mayhem occurs on the glacier's surface - serracs.

Most glaciers in the park flow as low as the elevation of the lakes' surfaces at some 200 m above sea level (185 in Lago Argentino), and it is here that the forces of nature reach equilibrium, not so much by the rate of melt as that of attrition by large blocks breaking off and floating away. This tremendous spectacle can be watched from safe places close by the glaciers' snouts. One of these can be reached by car - the Moreno glacier - on the southern arm of Argentino lake. The snout of Moreno is some five or six kilometres across and about 185 high feet above water-level. It is 85 kms west of the town of Calafate.

In 1947 for the first time, and with increasing periodicity since then, the surging advance of the snout of this glacier grounded on Península Magallanes and cut the natural drainage of the lake. It pushed on up the slope bull- dozing the forest into so many piles of jumbled trees. The waters of the Rico (southern) arm of the lake rose some 50 feet or more - over 17 m -, valleys were flooded, whole areas of peat floated away with fences, sheep and even cottages on them. No amount of dynamite and blasting could blow away the grounded snout and in the end it was just the pressure of the backed-up water which seeped through weaknesses, melted away a tunnel and eventually demolished the "plug". This has recently been occurring every three years or so. The highest level attained by the waters of the Rico arm was 19 m above Lago Argentino. The destruction of the "stopper" happens in one short day and is a spectacle hard to imagine; for those lucky enough to have witnessed it, surely impossible to forget. Even when this does not happen the ice-face is permanently calving off great bergs and chunks of ice which crack off with the sound of artillery and hit the water with explosions as of a major barrage.

The advances and retreats of glaciers are obviously reflections of climatic variations, which in this case possibly occurred centuries ago because of the slowness of flow. The furthest point of advance is where much of the material carried along was eventually dumped in the terminal moraine, transverse to the valley and in many cases in Patagonia, holding back the waters of the large lakes. Argentino is but one such example.

It is erroneously claimed that Moreno is the only glacier which is advancing (that is more precisely "growing", - all glaciers advance or flow). This is not the case. Neighbouring Mayo glacier, somewhat off the visitors' circuit, as well as several in Alaska are also surging (temporary "growth"), though it is the general rule that there is a marked and rapid recession in most glaciers as is evident by the bare slopes along their sides. The visitor at the face of Moreno is in a wonderful position to observe all this from comfortable walk-ways, but before their construction several fatal accidents had occurred, the victims usually being washed off some perch too low by the water where the displacement waves from some big calving dragged them into the frigid lake. It is therefore unwise to leave the trails - and of course forbidden.

There are launch excursions which leave from Puerto Bandera along the south shore of the main body of Argentino lake. They sometimes visit the Moreno and Mayo sound. Most of these excursions however cruise up the north arm of the lake to Upsala glacier which is by far the largest glacier in the region. It is fed by several smaller glaciers in its 50 km run, and is nine or ten kilometres across the snout. Huge bergs break off here and are the source of wonderfully shaped sculptures in white or blue ice which sail down the lake on the prevailing westerly winds. Landing at Onelli to the west a short walk across a wooded moraine takes one to a small lake into which three glaciers flow: Agassis, Betracchi and Onelli. This last provides the jumble of small bergs to feast the eyes.

Another fjord which is part of the northern arm is Spegazzini which also deserves a visit to see the Spegazzini glacier.

To the north Lago Viedma - some 106,000 has in area - has its headwaters in what is a Mecca for mountain climbers, a complex range dominated by the stark and bare peaks of mounts Fitzroy and Torre and surrounded by other notable spires. These have been conquered on very few occasions and have exacted many a toll in human life. To the technical difficulties of sheer faces and overhangs in granite must be added the hostility of the climate, ice-walls, storms, avalanches and the hurricane-force winds. Fitzroy, part of the core of an ancient batholyth rising 3.775 m above sea-level, was venerated by the local tribes who called it Chaltén believing it to be a volcano. It is surely one of the most spectacular mountains. There are certain commodities available in the area - a visitors' centre, and camping areas which allow potential challengers or simply interested parties to approach more closely. The great Viedma glacier, second only to the Upsala, and the Moyano, both drop into lake Viedma.

The vegetation is of the Magellanic type with Lenga and Guindo predominating; there is a small patch here of the Ciprés de las Guaytecas some 1.000 kms or 8° of latitude from its main distribution, a curious "island".

The wildlife here is similar but there are some remnant populations of the Huemul to be found on Moyano fjord. Important troops of Guanaco survive in the eastern portion of the park's central area, and condors are abundant enough to be frequently seen. One must draw atention here to the great numbers of birds of prey, especially of the Black-chested Buzzard-Eagle, perhaps in answer to the population explosion of European Hares.

Close by Punta Bandera there is an interesting group of small lakes with a great show of waterfowl which includes a breeding population of Black-necked Swans, rafts of coots, grebes, ducks, especially the Ferruginous or Andean Ruddy Duck - interesting and unaffraid. There is another small lake near the village - a municipal reserve - equally interesting and with abundant bird-life.

Flocks of Upland Geese with small numbers of the Ashy-headed Geese graze on the short-grass meadows and around the lakes where Black-faced Ibis probe for invertebrates with their long bills.

Even in the proximity of the glaciers the woodland wildlife is typical of the whole area with flocks of Austral Parakeets, the occasional Magellanic Woodpecker or the Green-backed Firecrown (a hummingbird) all giving colourful but paradoxical bent to our ingrained expectations that parrots and hummingbirds belong in the tropics.

The park is reached along Route 40 from the north or from Rio Gallegos some 350 kms away, or along provincial route 290 directly from the east and Atlantic coast following the course of the Santa Cruz river valley. The Park's headquarters are in Calafate, a small town on the south shore of Lago Argentino which has lots of lodging and feeding facilities for the visitors and which is, via its airport, connected with Rio Gallegos, Tierra del Fuego and the rest of the country. A road leads due west from town towards the park and forks to Bahía Tranquila (the location of Punta Bandera) or to the Moreno Glacier, with side-roads also leading into the south-west corner via Lago Roca where there is a camp-ground, or beyond. In this vicinity there is a breeding colony of the ibis, and open woodland to search for such as the Austral Pygmy Owl, the Chilean Flicker, Austral Blackbird or Rayaditos.

One of the urgent needs of Los Glaciares National Park is an interpretive centre at the Moreno Glacier to explain to the visitor with clarity the internal workings of the glacier which they have before their very eyes.

The park was declared a World Heritage site by UNESCO and the MAB programme.

The main problem of the park has to do with settlers. Unfortunately farms tend to employ the type of hands who make poaching a way of life and alternative supplementary income so there is a serious management problem. Overgrazing adds to the deterioration produced by the Estancias' mere presence, and it is difficult to control. The redesignation as "park" of many of the areas where control is virtually impossible might be a solution as with the expulsion or simply the non-renewal of grazing permits, poaching and erosion as a result of overgrazing will both be stopped, or reduced. Only the fagile steppes would then be subject to such incongruities.

Tierra del Fuego National Park. Just a few kilometres west of Ushuaia, the capital of the island province of Tierra del Fuego, is Tierra del Fuego National Park. It was created in 1960 by law # 15554 and covers about 63,000 has. As in all the Patagonian Andes' parks its western border is the international frontier with Chile.

It can generally be stated that the fauna and flora of the park, as well as the physiography are similar to the rest of this system of southern parks; lakes, mountains and woods. The mountains here however run from west to east and are separated by deep valleys. The Beauvoir range is the northern border of the park, its peaks barely 1.000 m high. It is succeeded to the south by lake Kami (or Fagnano) occupying most of the length of a long glacial valley. Only a sector of the lake is in the park. Kami is only 90 m above sea-level and drains towards Admiralty Sound and the Pacific. South of the lake first lie the Valdivieso mountains and then the Martial

Peat Bogs *These are abundant in Tierra del Fuego. They are swamps where low temperatures and the acidity of the water virtually stop decomposition of plant material which accumulates, and on which grow the typical plant communities of mosses and lichens (Sphagnum included) as well as other plants.*

The Kelp Goose

On the sea-coasts of Tierra del Fuego National Park such as in Lapataia Bay, this, the most surprising of South America's geese is found. The Kelp Goose (Chloephaga hybrida) *feeds on sea-weeds exposed as the tide ebbs. The male is pure white while the female is dark in a delicate pattern of barring.*

Fallen logs

The slow rate of decomposition of the dead wood in these cold latitudes causes fallen logs to lie a long time before rotting away, so they get covered with mosses and ferns.

Llao-Llao

This is a fungus (Cittaria darwinii) *which causes the host-tree (Genus nothofagus) to produce burls on the branches, twigs or trunks as a reaction. It was eaten by the local indians.*

Range - an attraction for mountaineers - which drop down into Lago Roca and Lapataia bay on the Beagle Channel, dominated by the Pirámide range.

This is the only national park with sea-shores and coasts, the Beagle being a connection between the Atlantic and the Pacific. Here the "channel" is narrow, not exceeding six kilometres and within the park presents shores of varying characteristics.

Tierra del Fuego has but six species of trees: Guindo or the tall evergreen beech, Lenga and Ñire both deciduous, Fire-bush, Winter's Bark and Leña Dura *(Maitenus magellanica)*, but the general aspect of the woods differs hardly at all from that of those on the continent. The steep relief makes it difficult to build an extensive network of roads, but those that there are allow the visitor to reach the maritime coast biome at two points: the first on a road branching off Route 3 - the only access to the park - which after barely one kilometre reaches the shores of Bahía Ensenada, the Beagle Channel in the background and Isla Redonda sort of across the mouth of the bay. There are cruises from Ushuaia which sometimes visit this area and the island. The other access to salt water is at Lapataia Bay at the very end of Route 3 where the cruise ends. At both places the shores are as varied as the wildlife.

The Magellanic and the Blackish Oystercatchers *(Haemantopus leucopodus* and *H. ater)*, both with long, brilliant red beaks call to each other with strident but clear, fluted, whistled notes which can be heard from afar; the Kelp Goose *(Chloephaga hybrida)* white male and lovely dark-patterned female wait for the falling tide to expose the sea-weeds on which they feed; pairs of the Flightless Steamer-Duck flap-charge across the surface raising alarming splashes and spray as they beat the water with their wings - like the old paddle-steamers which gave them their name - in some intimidating territorial dispute with their neighbours over an invisible but very real frontier; further out the six-foot-across Black-browed Albatross glides by in swinging arcs; or the "bumble-bee" flight of the tiny Diving-Petrels, emerging like little polaris missiles straight from underwater to plop back into the water at some distance, and continue their "flight" after their krill-like food.

There is a panoramic view from a balcony on the north side of the entrance road which takes in the Beagle Channel, Hoste and Navarine Islands and the Murray Narrows which run between them in the general southerly direction of Cape Horn, every one of these names steeped in historical recollections. On the shores of Lapataia Bay as elsewhere too, the tight lawn-like sward covers much greater history, the hummocks of piles of mussel-shells and bones left by the native Yamana or Yaghans, the tribe of canoe-indians which were displaced and eventually inadvertently erradicated by the invasion of the "white" man and his diseases. Their numbers fell alarmingly between the 1830's (HMS Beagle's expeditions) and the 1880's (first governor) as a result of the illnesses brought by early settlers and the navy and in spite of the concern of the missionaries who had been so dedicated to their wellfare for decades.

There is a small waterfall which prevents the salt-water of the higher tides backing up into short Rio Roca and the lake of the same name which lies between the Pirámides and Toro ranges. It is in this valley that the services of a camping area, a small restaurant, and the ranger stations are found, an hostelry to be rebuilt after a fire. Most of the area is closed in winter.

Tierra del Fuego National Park

A view of Roca river which drains Lago Roca, showing the typical landscape with bodies of water, low relief covered with woods of Lenga, and peat bogs developing in certain areas.

Here too one can try one's luck with rod and reel as trout abound in fresh water and Róbalo *(Eleginops maclovianus)* in the estuary and bay.

The road between the bridge over the Roca river and Lapataia Bay skirts peat-bogs. In an area of low temperatures and acid water, decomposition is reduced to an absolute minimum so plant material accumulates on the sides and bottoms of ponds, piling up till they fill. Laguna Negra, close by off the road, is a good example of the early stages of this process. Further along one can see the results of eons of accumulated plant material on which grows a peculiar community of mosses, sedges and lichens which, as time goes on, provide further matter raising the level of the bog well above that of the surrounding area. These hummocks, because of the colour of the fruiting heads, turn orange and golden in summer. Peat-bogs are tremendously treacherous and soft because of the consistency of the "ground".

This is where the small insectivorous Sundew *(Drosera uniflora)* is to be found. As nitrogen in the soil is minimal, to overcome this defficiency the leaves of this plant are equipped with sticky "hairs" on which any unwary insect will be trapped to be digested by excreted enzymes and later absorbed by the leaf. These small plants live among the mosses and consist of a rosette barely an inch or so across, of small roundish leaves.

To the north the park extends some 40 kms or so over the previously mentioned mountain ranges, with valleys of swiftly-running rivers, peat-bogs, "alpine" areas like meadows over the 600 m contour, steeply dropping beyond the Valdivieso range down to lake Kami. This great lake is over 100 kms long, is aligned with the strong westerly winds and is unsuitable for recreational boating because of the waves thus produced. Along the short valley which separates the lake from the marine environment, and into the lake are pushed many sea-birds. Albatrosses, petrels, skuas, diving-petrels and others are trapped in this alien fresh-water environment and eventually perish to be washed up on the eastern beach, dumb evidence of the relative sterility of fresh water versus salt and the unavailability of food for these birds. Along the north shore runs the Marginal Range dropping in elevation eastwards from its highest peak Beauvoir, the northern border of the park. There is a trail which crosses the park from south to north and which starts near the Pipo river's waterfall and ends on the shores of Kami but does not make any circuit. In order to hike this trail one needs excellent equipment for foul weather and overnighting, waterproof clothing and foot-wear as well as special permission obtained from the park authorities.

Forestry was the activity of the prisoners in Ushuaia jail between 1920 and 1940, and they stripped several areas of what is now the park, especially in the Pipo valley, before the park was created. The presence of cattle ever since then has retarded the recuperation of the forests. The exotic Beaver mentioned in the introductory paragraphs of this chapter, as also the rabbit, are being evaluated as the target species of studies to consider their eradication or control. They play no minor role in environmental damage so it is to be hoped that soon some measures will be found to eliminate these anomalies from the national park.

Staten Island National Park Project. As a prolongation of the sinking Andes, at the extreme south-east of the main island of Tierra del Fuego, Staten Island pokes its peaks above the waves. It has been suggested that the otherwise unused, unpopulated island form part of the national park system. Amongst its attributes and worthy of every protection is a population of the little Sea-Otter *(Lutra felina)*, one of the few colonies of Southern Fur-Seals *(Arctocephalus australis)*, rookeries of Rockhopper and Gentoo Penguins *(Eudyptes crestatus* and *Pygoscellis papua)* as well as several breeding petrels.

As to the flora there are species like the Tussock Grass *(Poa flabellata)* which are not found on the mainland. It is a huge grass which covers whole islands in these southern latitudes and provides the breeding habitat for numbers of burrowing sea-birds.

The protection of this island as a breeding ground for so many species amply justifies its creation as a national park, especially if such a park were to include areas of the surrounding sea where several species of cetaceans are to be found.

The Frozen Antarctic

On the basis of its proximity, of Argentine historical presence in the Antarctic region, and for other legal reasons, Argentina claims the sector of that continent between 25°W and 74°W, which would include an area of some 1.4 million square kilometres of ice and rock, and the waters surrounding. One basis for the claim is that Argentina holds the longest unbroken record of occupation by any country; the uninterrupted maintainance of a meteorological station on Laurie Island in the South Orkneys since 1904. This was no obstacle to Argentina's signing the Antarctic Treaty in 1961 for cooperation on a scientific level and the demilitarization of everything south of 60°S, putting all territorial claims on "hold".

The Treaty's scientific coordination and the resultant management parameters are covered in the periodic Consultative Meetings of the Signatory Nations. These are all proof of Argentina's avocation towards the Antarctic Treaty and reason enough to include this sector in the present volume.

With the Antarctic we cover a very singular biogeographic unit - the Antarctic Dominion, part of the Southern Zoogeographic Region - contrasting sharply with the other units of Argentine territory which are all of the Neotropic Region. It is a landscape lacking trees or shrubs, almost devoid of greens or yellows, dominated by the blues, blacks and whites of the sea, the rocks and the ice.

This Antarctic Dominion covers the whole area surrounding the South Pole to a radius of 3,500 to 4,000 kms and it includes the continent called Antarctica, a vast area of marine environment (the Southern Oceans), and all the islands and archipelagos therein, totalling some 45 million square kilometres of very peculiar physiographic, climatological and biological characteristics where the intense cold is the main factor.

The continent acts as a very real refrigerator for the whole region - nay, the southern hemisphere. The low angle of incidence of the sun's rays give the entire area a lessened input of radiated warmth when compared with areas nearer the tropics. But the area is only frigid where permanent ice accumulates. This only happens in places like the continental land mass of Antarctica, not, as is the northern equivalent case, where oceans cover the pole allowing for the transference of warmth by ocean currents to disperse any accumulated winter ice. The fossils of animals and plants which are found in Antarctica - ferns, Araucarias, Southern Beeches (genus *Nothofagus*), and such, even an extinct Patagonian marsupial *Polydolops* - prove that this region has at times had warmer climates. When it split away from the southern super-continent Gondwana (the theory of Continental Drift), and settled in its present position, only then did snow start to accumulate on the land surface over millions of years, and turned to ice under the weight of further snow falling on the top. The average height of the continent - 4,200 m above sea-level on the central plateau - has contributed to this cooling as well as the blinding whiteness of the vast ice-caps which reflects solar radiation instead of absorbing it, so the frozen wastes become permanent.

As a result of all this Antarctica is today covered by a layer of ice which averages over 2,000 m thick and estimated to contain 25,000 billion tons of ice, heavy enough to depress the continent in places to over 1,000 m below sea-level. Barely 2% of the land can be seen as emerging mountains or shorelines.

It is the ice which links the continent (the biogeographical Antarctic Province) to the surrounding seas (the Antarctic Ocean Province) and makes it necessary to deal with both together. In general terms the ice-cap, which is still growing in the interior of the continent, fed by new snow-falls, flows towards the edges at speeds which vary between ten and one thousand metres per year according to topography. At the coast the different glaciers into which the flow is split by the relief, present a snout to the sea from which pieces are continually breaking off to form ice-bergs and drift away on the currents and pushed along by the wind.

Along certain parts of the coast some glaciers appear like shelves, tens or perhaps hundreds of kilometres across. Grounded near the shore under their own weight, they flow out onto the sea, especially in bays, rather like mantles, floating on the surface beyond their point of contact with the bottom. These ice-platforms or ice-shelves can be hundreds of miles long and their fronts cliffs of ice dropping directly into the sea, sometimes 200 m thick, 40 m showing above the water-line. The pieces which break away from these ice-shelves have a peculiar shape and are known a tabular bergs. They tend to be huge and some have been recorded well over 100 miles across.

Contrasting with this continental ice there is also sea-ice; a crust on the freezing sea's surface reaches 100 to 200 Kms from shore, increasing the continent's size by September by some 20 million square kilometres. Wave action, internal pressures and the summer thaw break up this pack-ice into floes which reach three or four metres in thickness if not broken up every year, and, in spring and summer drifting away and dispersing, reduce the surrounding belt to but 3 million square kilometres in March.

The interior of Antarctica is a desert which barely supports any form of life. The climate is cold, dry and windy to the extent that any living thing would be frozen, dessicated and blown away. Where the rock is exposed it is kept free of smaller particles by the action of the wind.

With average temperatures of -60°C at the South Pole, a yearly average precipitation of 150 to 300 mm and a mean wind velocity of 70 km.p.h as at Cape Denison, Wilkes Land, the only plants which can grow are dwarf mosses and lichens, and these restricted to certain areas like rocky promontories which can absorb a bit of warmth in summer.

Animal life on land is limited to minute insects and other arthropods which can shelter under stones or amid the dwarf vegetation.

The Antarctic Peninsula, a substantial part of the Argentine sector, has however much less harsh conditions: the maritime climate is warmer and more humid, its relief more varied with the steep range of mountains - the Antarctandes - ensuring a more frequent breaking up of the otherwise uniform sheets of ice and snow.

On this peninsula and especially on the off-shore islands there appear occasional patches of a tundra-like vegetation: multicolour mosaics of lichens on rock-faces, the crusty species of *Caloplaca, Xanthonia* and *Verrucaria* mixed in with leafy or fruiting-bodied *Usnea* and *Umbilicaria*, carpets of the taller mosses like *Polytrichum* and *Chorisodontium*, or the shorter *Andreaea, Tortula* and *Grimnia*, combined with lichens and algae on stony soils which are clear of snow in summer, even cushion-plants of one of the only species of higher plant in the Antarctic, the grass *Deschampsia antarctica* or *Colobanthus quitensis* the only dicoteledon of the *Cariophillaceae* family.

In summer one can find snow-fields tinged green, red or yellow which indicate the presence of unicellular snow-algae (*Chlamydomonas, Raphidonema* and *Ochromonas* spp., etc.) which confer to the snow the colours of their pigments.

The unimpressive inventory of land animals includes unicellular protozoans, minute rotifera and tardigrades in concentrations of upto 14 million per square metre; successful little nematode worms which are present wherever there is organic material; tiny mites like *Nanorchestes antarcticus* which has been found on nunataks emerging near the South Pole, making it the southernmost being on earth, and *Gamasellus racovitzai* a diminutive predator; and 22 species of free-living insects with collembola (primitive non-flying insects between one and two milimetres long) the most numerous, and a winged but flightless fly the most interesting. There are also fleas, ticks, lice and other parasites on sea-birds and marine mammals, but all these are incapable of independent life.

The only animal life worthy of note are those species of marine vertebrates which come to shore to rest, to reproduce or to moult their fur or feathers, but which depend on the sea for sustenance.

Sharply contrasting with the paucity of life on land, Antarctic waters are biologically very rich.

The Southern Ocean which surrounds the Antarctic is a ring some 200 to 1.000 kms wide, of waters permanently flowing east in a great circumpolar current, pushed along before the dominant winds from the west. It is connected with the warmer Atlantic, Pacific and Indian Oceans, the border being a line or narrow strip where the temperature of the surface water changes sharply from 8° to 4°C in summer and 3° to 1°C in winter. The latitude is variable but is usually around 52°S.

The deep waters of the oceans, on their southward drift, get rich on the nutrients from the plant and animal detritus which "rains" from the upper layers onto the ocean floor. As these waters reach an area known as the Antarctic Divergence they are pushed towards the surface by convection currents of complicated origin. From there part of this water becomes the Antarctic surface water flowing northwards as well as generally east with the circumpolar current. Cooled by the ice-bergs and sea-ice, diluted as summer advances by the melting of snow and ice, it becomes cold and less saline. On meeting the Sub-antarctic surface waters at the Antarctic convergence, it sinks below and becomes the Intermediate Antarctic Water. The difference of temperature at the surface between the Antarctic and the Sub-antarctic waters explains the brusque temperature change at said convergence.

This line also represents a biological frontier since many planktonic organisms cannot survive in the other water and die as they cross the convergence.

These hydrological characteristics explain the high productivity of the Southern Ocean in summer: the high nutrient content of the deep waters brought to the surface, combined with extremely long periods of daylight in summer ensure the enormous production of phytoplankton in that season.

Diatoms are the dominant plant in this soupy sea, the food for the herbivorous zooplankton, particularly rich in crustaceans, mostly copepods (*Calanoides, Calanus* and *Rhincalanus*), and krill *(Euphausia superba)*. This last is a shrimp-like being, some 6 or 7 cms long with 5 pairs of swimming legs and a strong tail with which it propels itself instead of merely drifting with the currents. The population of krill is estimated at 600,000 billion individuals at the peak of the season, which would weigh some 650 million tons - more than twice the weight of all mankind.

Landscape of the Antarctic Peninsula

This peninsula projects northwards from the almost circular antarctic continent, like a finger reaching towards South America. Essentially it is a range of young mountains (the Antarctandes) with steep peaks emerging directly from the sea and abrupt rock-faces peering from between the ice-caps and glaciers which cover most of it.

Lichens and mosses

Exposed rock is the substrate for the meagre and dwarf antarctic vegetation composed mainly of lichens and mosses which include some 400 species of the former, 75 or so of the latter.

Adelie Penguins

These, (Pygoscellis adeliae), are the most typical beings of the antarctic fauna. Their breeding colonies are on fairly level areas and can contain several tens of thousands of pairs.

The Snowy Sheathbill

Looking more like a white pigeon than anything else, Sheathbills (Chionis alba) *are in fact related to gulls and shorebirds and are opportunistic feeders on carrion, waste, eggs and invertebrates.*

The Pintado Petrel

Flocks of this attractively chequered bird (Daption capense) *follow ships in antarctic waters on stiff, longish wings, gliding in a dynamic fashion. It nests in the crevices in rock-faces and cliffs.*

The abundance of zooplankton, including predatory forms like the larvae of certain worms, jelly-fish etc., supports large populations of fish, squid, birds and even mammals.

Krill is the main food of several petrels, three penguins (Chinstrap, Gentoo and Adelie), of the Crabeater Seal *(Lobodon carcinophagus)* and baleen whales *(Mysticeti)*. Krill is so abundant that the populations of just those three species of penguins it feeds, are estimated at between 8 and 20 million, as well as some 15 million Crabeater Seals (between 8 and 40 are the figures given). The population of whales that used to feed on this resource was such that around 1925 whalers killed some 40,000 per year in the region. Today, as a result of such abuse, all are rare except the small Minke *(Balaena acutirostrata)* which usually weighs less than ten tons.

In some bays and straits one can find the Humpback Whale *(Megaptera novaeangliae)* which, like the others, converges on these rich feeding-grounds offered by the Southern Ocean.

Of the toothed whales *(Odontoceti)* the most common, other than the Orca, are Arnoux's Beaked Whale *(Berardius arnuxii)* and the Bottle-nosed Whale *(Hyperodon planifrons)* which feed on squid and fish.

This is also the diet of the rare and barely-known Ross' Seal *(Ommatophoca rossii)* and of the sedentary Weddell Seal *(Leptonychotes weddellii)* which is always near the coasts in summer and on the sea-ice in winter, ice which it chews hole through to dive into the water and find food beneath. In this it differs from the gregarious habits of the Crabeaters which are wanderers and drift with the ice-floes on which they rest and sun, migrating north in winter after the schools of krill.

Other pinnipeds which breed on the off-shore islands are the Elephant Seal *(Mirounga leonina)* and the Antarctic Fur-seal *(Arctocephalus gazella)*.

At the summit of the feeding pyramid in the southern ocean is the usually solitary Leopard Seal *(Hydrurga leptonyx)*, often over three metres long and weighing 350 kgs, a hunter of young seals and penguins, but which also takes fish and krill, and the Orca often referred to as the Killer Whale *(Orcinus orca)* which preys on everything which is large enough to make a mouthful, and even, on occasions, on the great whales which it attacks in a coordinated pack.

The species of penguins already alluded to are the main wildlife attraction of the Antarctic, in their populous breeding colonies on the shores: sheltered low shores provide the nesting habitat for the Gentoo and the Adelie *(Pygoscellis papua* and *P. adeliae)* while steep, rocky coasts exposed to the fury of wind and waves are chosen as the landing-beaches for the colonies of Chinstraps *(P. antarctica)* and the Macaroni *(Eudyptes chrysolophus)*, this last a mainly Sub-antarctic species found in a few places on the periphery.

The giant of the family today is the Emperor Penguin *(Aptenodytes forsteri)* weighing some 30 kgs as against the 5 or less of the others, which differs in its diet (fish and squid) and in its breeding habits: it "nests" in the depths of winter's darkness, over land and beyond the sea-ice. No nest is built, the single egg and later the chick being held on the top of the feet and protected from the extreme cold by an abdominal flap or fold covering all. At nest-relief after about two months of uninterrupted attendance, the relieved bird must cross hundreds of kilometres of sea-ice to reach the feeding-grounds in open water, feed up and head back again.

The other predominant group of ocean birds is the *Procellariiformes*. Most members of this order are masters of the art of dynamic soaring, using the ever-changing air-

The Orca

The Orca (Orcinus orca) *also known as the Killer Whale, is the largest predator of the oceans. This gigantic dolphin, which is found in pods of 6 to 15 individuals, is frequently encountered in antarctic waters and can sometimes be watched in coodinated hunting of seals and of other cetaceans.*

Crab-eater seals

These, (Lobodon carcinophagus), *are the most abundant pinipeds and can be found in numbers resting on the drifting ice-floes and small bergs when not active and feeding on krill.*

currents displaced by advancing waves acting on the wind, to fly for long periods and over vast distances with minimal effort.

The most spectacular of these are the albatrosses; the Black-browed *(Diomedea melanophrys)* is the most common though it does not nest in the Antarctic proper but on cool temperate islands as the Falklands (Malvinas) or around Cape Horn. Others, like the Wandering *(D. exulans)* and the Gray-headed *(D. chrysostoma)* or even the Light-mantled Sooty *(Phoebetria palpebrata)* are birds of the Sub-antarctic islands.

Full citizenship however belongs to several petrels: the Giant *(Macronectes giganteus)*, the Southern or Silvery-gray Fulmar *(Fulmarus glacialoides)*, the Antarctic Petrel *(Thalassoica antarctica)* and the Snow Petrel *(Pagodroma nivea)*, with Wilson's Storm-Petrel *(Oceanites oceanicus)*, all raise their single chick in varying nest-cavities except the first which being six-foot across and armed with a formidable bill, can face any predator on the surface of the ground, like flat gravel areas above the beaches.

The Blue-eyed Cormorant *(Phalacrocorax atriceps)*, the Kelp Gull *(Larus dominicanus)* and the Antarctic Tern *(Sterna vittata)* as well as the three oportunistic scavengers around the colonies of other birds - the Snowy Sheathbill *(Chionis alba)*, and the Antarctic and South Polar Skuas *(Catharacta antarctica* and *C. macckormicki)* complete the list of birds.

Special mention must be made of the fish of the Antarctic because they include a high proportion of endemic forms, both coastal and marine, which possess special metabolic adaptations to be able to survive in icy conditions. They produce their own anti-freeze which allows them to keep active and survive at temperatures of -2°C and they die if it should ever reach 5°C above freezing.

Generally they belong to the sub-order *Nototheniformes* which is named for the Antarctic Cod *(Notothenia rossii)* and includes the surprising family *Chaenichthydae*, the ice-fishes whose blood is colourless lacking haemoglobin, the protein for transporting the oxygen from the lungs (gills in this case) to body tissues.

The whole of the Antarctic has been designated a "nature reserve", dedicated "to peace and to science" by agreement of the parties of the Antarctic Treaty to a Protocol on the Conservation of the Antarctic (Habitat) in 1991. Argentina was of course one of the signatory parties of said protocol though ratification awaits approval by congress.

The wisdom of this agreement lies in the singularity of the biogeographical unit involved and the crucial necessity of a better scientific understanding of the whole planet; in the opportunity to preserve a whole region before it has been totally affected by man, it being the only area in the world which is not permanently inhabited; and in the incalculable danger to the whole planet as the result of changes in its characteristics, bearing in mind its key role in the regulation of atmospheric and ocean currents and its decisive influence on the climate of the whole world.

The treaty is of special importance if one bears in mind the Antarctic's vulnerability in the face of other human activities such as mining which, happily has been forbidden.

Even when there is much to do by many nations for the implementation of this "nature reserve", Argentina is pioneering the management of same by the designation of two qualified Ranger/Wardens per year, resident year-round to effect a certain measure of protection from their own scientific bases.

Antarctic channels and waterways

In the summer the sea-ice which formed here during the winter is broken up and dispersed. These narrow waters which separate the chains of myriad islands, do not impede the passage of ice-bergs, ice-floes and smaller "brash".

New expectations

THE LAST TWELVE YEARS OF CONSERVATION IN ARGENTINA

The first edition of this book was published in 1981 and in spite of the time elapsed since then it can be said to be as pertinent as ever. The subject of conservation and management of natural resources has, however, become an everyday subject of conversation during these last years. The media carry daily reports of the speeches of politicians on this subject. The greatest international meeting in this field - Rio '92 - had as a central theme the conservation and sustainable use of natural resources. The IVth International Congress on National Parks and Protected Areas which also took place in 1992, in Caracas, Venezuela, assembled some 1.200 specialists in the subject from all over the world.

These meetings may seem to promise well for the future but they were the logical outcome of the ever-increasing problems being faced by the whole planet. The pressures on Natural Resources increase daily as a result of the world's growing population and human aspirations for a better life-style. Global warming, the loss of biodiversity, decreasing cultivatable soils, water pollution, air-pollution are all effects which result from human actions of one kind or another, and they affect us all. If we do not react in the face of the seriousness of these threats, and start to search for solutions at once, it may soon be too late. For many species of plants and animals it is already too late - they have become extinct because of man's irrational management. Nothing can bring them back.

Nature in Argentina is still as described in the original version of this book, though perhaps a bit more tattered and deteriorated. At the same time there is a greater public awareness of the problems which affect us all and there are more institutions working towards reverting the situation, searching for solutions. Though there is still a frightening amount to be done and time is running out, the way ahead is clear.

Internationally Argentina has subscribed to many and varied conventions such as RAMSAR, the Convention on Wetlands of International Importance, especially as habitat for wildlife. On signing this accord three areas were set aside which the country guarantees to protect in perpetuity for the international community: Pozuelos Natural Monument in Jujuy, Rio Pilcomayo National Park in Formosa and Laguna Blanca National Park in Neuquén. Argentina also signed the Bonn Convention for the Protection of Migratory Species, and CITES, the Convention on International Trade in Endangered Species. More recently Argentina has become a party to the Convention on Biologigal Diversity at the United Nations Conference on the Environment and Development.

There are also several Biosphere Reserves - a category of protected area bestowed by the Man and the Biosphere (MAB) Programme of the UNESCO. The areas included in this system are the whole of the Pozuelos watershed (Jujuy) including the lake, San Guillermo in San Juan, Laguna Blanca in Catamarca, Ñacuñán in Mendoza and Parque Costero del Sur in the very City of Buenos Aires.

Two National Parks have deserved designation as World Heritage Sites because of the international value of the resources they protect: Los Glaciares in 1981 and Iguazú in 1984.

A programme for the protection of the migratory shorebirds of the Western Hemisphere includes a network of sites. Critical areas are identified so that protection can be given to the special needs and habitats that these birds require. Though there is no formal agreement, it demonstrates willingness on the part of the parties involved. Laguna Mar Chiquita in Córdoba, the foreshore of Tierra del Fuego and the San Antonio Oeste reserve in Rio Negro province are all part of this international system which stretches from end to end of the Americas.

Provincial Departments for the Protection of Natural Areas have also been developing, and though the standards are by no means uniform, several provincial reserves are today adequately managed. There is a National Network of Protected Areas, an agency which endeavours to coordinate the management policies of these areas throughout the country.

An initiative which was barely under way twelve years ago but which is gaining tremendous momentum is that of individuals or associations which keep nature reserves on a private scale. This worthy endeavour is bearing fruit.

Though there has been an important increase of interest in, awareness of and concern for environmental issues, in many cases this has come too late. Unfortunately Argentina has already lost several species of its fauna, and possibly of its flora also. Furthermore, much of the natural habitat outside the protected areas is impoverished compared to what it was in the not-so-distant past. Overgrazing, erosion, indiscriminate forest clearing, pollution and other forms of mismanagement have affected negatively, sometimes irremediably, the natural resources of the country. The Patagonian Steppes can no longer hold the numbers of sheep they once did because overstocking for better short-term yields has affected the vegetation and reduced the soil to a mere vestige of what it was in former years. The Chaco, one of the richest regions in the country, is today alarmingly impoverished, its principal woods cleared, its generous grasslands vanished by overgrazing and fire, replaced with a shrubby secondary growth. Marginalization and poverty close in inexorably.

The exuberant forests of the north-west are still being exploited at an alarming rate to be replaced with soy-beans, other legumes, tobacco, sugar-cane, citrus groves. The soil, lacking its original cover, is eroded and subjected to unsustainable uses which augur further poverty in the future. The very Pampas grasslands which have been the principal natural investment - a sort of blank cheque - have nearly all vanished under the plough and are showing signs of severe over-use and erosion.

In the midst of all this sad reality the National Parks and the other protected areas play a fundamental part in the nation's conservation field. However, only a carefully drawn up management plan for the country's natural resources (soils, water, air, flora and wildlife) can guarantee that Argentina continue to display all its amazing natural diversity in the future. What is even more important is that these very resources are the basis of the country's sustainable development and we will be depending on them for any future welfare and participation in a full and meaningful life.

NATURAL REGIONS - RECENT CHANGES IN PROTECTED AREAS

Subtropical Rainforests. In extreme north-east Misiones, in the departments of General Belgrano and Iguazú, the provincial authorities in 1988, created the Uruguá-í Provincial

Park, covering a surface area of about 84,000 has and connected via the northern border with the extreme south-east corner of Iguazú National Park. In this way, to the coordinated effort of the Argentine and Brazilian National governments is added the initiative and vision of Misiones Province. This has resulted in the protection of by far the largest piece of Paranaense Rainforest, well over 300,000 has total area. Nine other provincial parks and private reserves have been added during recent years to protect parts of this, the richest of Argentine biomes.

At the Earth Summit meeting the president of Argentina announced the start of the Yabotí project whereby a further 223,220 has are protected in the east-centre of Misiones as a Biosphere Reserve.

In 1990, by presidential decree, a new category of protected area is considered - Strict Nature Reserve. This has since been split into two: Educational, and Wilderness. These are all under the National Parks Administration, and one such in Misiones is the San Antonio reserve to preserve an area of the Paraná Araucaria forests in the north-east of the province.

The Chaco and the Espinal. Two contiguous estancias, Santa Teresa and Santa María, cover some 15,000 has in the north of Corrientes Province, near Mburucuyá and are only 150 kms from the capital city, Corrientes.

As if to demonstrate yet again the importance of the private sector in the general conservation effort, the area is being donated to the National Parks Administration by the concerned owner of both estancias, Dr. Troels M. Pedersen.

In 1988 Dr. Pedersen offered them to be conserved just as he had kept them carefully natural or under restoration. For various reasons the project was delayed but in 1991 the generous donation was accepted as a future National Park.

At the time of writing legalities are being arranged for passing this area of Corrientes over to the National Government, naturally a provincial decision.

Meanwhile National Parks, which owns the land, has started effectively managing the area and staffing it while setting up the necessary infrastructure - housing, light, mobility, communications and such.

The same Province of Corrientes, in 1982, had declared the Iberá marshes and lakes a wildlife refuge with an area of 1.2 million has, with the central area around Iberá lake staffed and patrolled by locals as rangers.

The province of Córdoba effectively administers a system of provincial reserves staffed with trained rangers. Among these are Chancaní of 5,000 has at the foot of the westernmost slopes of the Pocho range of mountains. Here there is a relict of the White Quebracho woods mixed with Algarrobo which once covered all western Córdoba. The Wildlife refuge of Laguna La Felipa holds a good sample of the marsh and lake habitat of the ecotone between the Espinal and the Pampas. The birdlife is especially spectacular.

On the south shore of vast Mar Chiquita a Visitors' Centre with Interpretive displays has been set up, and includes simple accomodation for scientists. A population of some 70,000 flamingos (mostly Chilean Flamingos) is an important feature of Mar Chiquita where some half million Wilson's Phalarope "winter". This was reason enough to have Mar Chiquita declared a site of the Western Hemisphere Shorebird Reserve Network.

The Delta. Decree # 2149/90 of the executive included the Otamendi Strict Nature Reserve to be administered by

Diamante National Park

The Riacho de las Mangas is typical of the area protected by this National Park, with dense riverine forests on its banks. This is the first sample of the famous delta of the Paraná river to be effectively protected.

The Curiyú Water-Boa

Chaco wetlands such as the marshes at Mburucuyá, are the habitat for this Water-Boa (Eunectes notaeus), *a smaller version of the Amazon's Anaconda. Though some specimens have measured over four metres, the usual size reaches slightly over three.*

Black Howler Monkeys

Perhaps the most conspicuous mammal of the islands of woods and forest in the eastern Humid Chaco, this Howler (Alouata caraya) *can be heard at a great distance as it roars in chorus to stake out its home range. While the adult male is black, females and young are yellowish brown.*

The Roadside Hawk

*This raptor, (*Buteo magnirostris*) is solitary and perches on exposed look-outs to watch for potential prey - insects, reptiles or rodents.*

The Red Quebracho

*This species of Red Quebracho (*Schinopsis balanzae*) of the Humid Chaco, as also the similar Dry Chaco species which replaces it to the west, was once the dominant species of tree of the Chaco woods. Years of gross over-exploitation for its wood and the tannin it contains have reduced it to a vestigial population.*

The Pampas Deer *The Pampas Deer (*Ozotoceros bezoarticus*) is a smallish deer with antlers which are typically three-tined and a sandy-lion coloured pelage. Once very abundant on the grasslands of the country, it is today one of Argentina's most threatened species.*

National Parks. It covers some 2,600 has and is situated in the north of Buenos Aires province bordering on the Paraná de las Palmas river. It is especially important as it is only 60 kms from the city of Buenos Aires and in an area of great urban and industrial development. The shores of the Paraná have a good sample of the riverine gallery forest with a vegetation which includes Seibos (Coral Trees), Native Willows, Curupís, Anacahuitas and other trees. In this forest there still exists a fair population of the Dusky-legged Guan at the extreme south of its range. Also, but in the quieter and denser parts of this delta vegetation there are still some Paraná Otters, Capybaras and Murine Opossums.

Most of the reserve is a sedimentary flood-plain created by the Paraná itself. On it grow the tall *Typha, Cortadeira,* and *Stipa* grasses and others. The higher areas are only occasionally covered by the great floods and there the Hog's-hair grass dominates as it does on most slightly saline soils, and also Espartillo and the sharp-pointed *Carex* sedge.

In the centre a depression holds several bodies of standing water, Laguna Grande being the largest where a fair sample of the Pampas' waterbirds can be found. The dense, tall grasslands and the riverine forests are the home of the southernmost relict population of the Marsh Deer which, during severe flooding, move to the uplands where they are hunted.

The flood-plain ends at the old river-bank, beyond which the rolling countryside is dotted with the Tala woods characteristic of the Espinal.

Otamendi is a good example of private contributions to the overall conservation effort. Thanks to private foundations it has an administrative centre and information office. From this reception area interpretive trails lead to look-outs on the old river-bank. Further donations provided the means for surveillance by the rangers there operating. A nursery of the local flora, also a private donation, provides plants for recuperation programmes in certain areas and also for the neighbouring municipalities and for environmental extension.

The next step, again from private donations, is the setting-up of an interpretive centre, of great importance because of the reserve's closeness to a huge potential public.

Diamante National Park in Entre Rios province was created in 1992 to protect an area of the Upper Paraná Delta Islands. With a surface area of 2.400 has it is only 6 kms south of Diamante, the town which was decisive in its creation.

The group of islands which make up the park are typically edged with forested levées and low flooded interiors with tall grassland and lakes. The fauna is abundant and varied and includes caymans and otters. There are at present but two rangers in the area.

The Pampas Grasslands. The 3.000 has of Campos del Tuyú loaned to Fundación Vida Silvestre to protect the Pampas Deer were finally purchased in 1986 and 1988 with donations earmarked to this end, so the reserve's future should now be certain. Most of the countryside around acts as a buffer zone thanks to the support of the owners. The population of Pampas Deer has grown to somewhat over 100 specimens which, together with the 250 or so all along the Samborombón Bay's shore still does not guarantee its survival, so further efforts are still necessary.

The Monte. In 1991 the nation passed a law for the creation of Sierra de las Quijadas National Park which had been a long-awaited dream. However the estimated cost of expropriation was extremely high which set back any rapid take-over by the Parks Administration. At present there is an attempt at zoning to reduce the area of the Park which will be subject to expropriation in order to reduce the cost to within the budget, calling the surrounding area a "Reserve".

In San Juan province the local authorities, in conjunction with National Parks, are working towards the concretion of a new protected area at Leoncito, some 70,000 has of the pre-andean Tontal range. Here there is an astronimical observatory under the National Council for Science and Technology (CONICET) which has also to date administered the rest of the area. The area itself contains much of conservation interest, endangered species and endemic taxa.

The Puna and High Andes. Laguna de los Pozuelos in Jujuy is the focal point of the projected National Park of Carahuasi. It has been a National Monument since 1981. The area of some 18,000 has is almost exclusively water most years. On a high plain at about 3,600 m above sea-level, it holds a spectacular bird fauna with some 25,000 flamingos of three species at peak periods. Occasionally the Chilean Flamingo breeds there.

At the mouth of the Cincel river there are great concentrations of ducks - Brown Pintail, the local Sharp-winged Teal (a form of the Speckled), Crested Ducks, Cinnamon Teal and the endemic Puna Teal.

Because of its singular characteristics the lake was declared a RAMSAR site and the whole watershed on which it entirely depends - the projected park of Carahuasi - a Biosphere Reserve in 1991 and 1990 respectively. These international classifications show the value Pozuelos has on a world scale.

With an area of nearly 70,000 has, the future Los Cardones National Park near Cachi in the province of Salta is planned. The area would include the Pre-puna, High Andes and an edge of the Monte at its eastern extremity. The provincial government ceded the area to the national government and it is hoped that not too long will pass before this project comes to fruition.

Tha Patagonian Andes. The existing protected areas in this region have changed but slightly in the time elapsed. They still more than adequately cover all the representative areas. However some endangered species have been the target of some moves to guarantee their survival.

The Andean Deer or Huemul has been the object of much research by the National Parks Administration and CONAF (the Chilean equivalent), the Fundación Vida Silvestre and other organizations. Remnant populations have been located and certain measures taken to ensure successful protection.

As to the Pudu the situation is improving. A greater understanding of its biology has demonstrated that the population within the National Parks is fairly sound and holding its own. The specimens in captivity for breeding on Isla Victoria have been released and monitored on that island and there is a fairly healthy population there today.

The Patagonian Steppe. There have been few advances in the conservation field in this biome. In the south of the province of Mendoza there is a sector of purely patagonian characteristics and influence. In 1980 and 1982 the 200,000 ha La Payunia Reserve and 40,000 ha Llancanelo were created, the last with a lake rich in bird-life.

The Atlantic Seabord. Punta Rasa, the southern headland of the Samborombón Bay in the province of Buenos Aires is an area of critical importance to migratory birds which concentrate there on arrival and before departure each year. Here the Fundación Vida Silvestre has a biological station by agreement with the Navy who own the land. The vast sea-water lagoon at Mar Chiquita further south along the coast of Buenos Aires, which is also important to these birds, has a ranger to protect the populations of the migratory species.

In Rio Negro the sea-lion colony at Punta Bermeja near Viedma has resident wardens, signposting and a visitors' centre.

The Least Cavy

The smallest of this group is Microcavia australis. *These tail-less rodents were the origin of "guinea-pigs" In spite of the "guinea" in the name, they are exclusively South-american.*

Covered with thorns

In the inhospitable Monte plants must adapt to the scarcity of water, so the foliage which grows under such hardship must be protected from herbivores. Many small bushes achieve this with formidable thorns as in this Alpataco (Prosopis alpatacos).

Land Tortoises

This Geochelone chilensis *of the Monte, must face the constant threat to survival from the huge pet-trade for which tortoises are caught in large numbers.*

Leoncito

The new National Park is situated in the pre-andean heights of San Juan province, like the Tontal range which has some endemic plants and animals. It represents an excellent sample of the Monte's shrubland steppe with some puna elements from the higher Andes.

Bibliography

Aves de Argentina y Uruguay.
A. O. P., 345 pp., Bs. As.

Aparicio, F. de y Difieri, H. A.
1958. **La Argentina, Suma de Geografía.**

Archangelski, S.
1970. **Fundamentos de Paleobotánica.**
Serie técnica y didáctica N° 11. Facultad de Ciencias Nat. y Museo Univ. Nac. de la Plata.

Archangelski, S.
1978. **Bosques Petrificados. Áreas naturales y turismo, selección de conferencias presentadas en el I, II, III y IV Seminario Internacional sobre áreas naturales y turismo.**
Rawson, Sec. de Turismo, p. 93-98.

Auer, V.
1951. **Consideraciones científicas sobre la conservación de los recursos naturales de la Patagonia.**
IDIA. N° 40-41, p. 36.

Bellisio, Norberto B., López, Rogelio B., Torno, Aldo.
1979. **Peces Marinos Patagónicos.**
Min. Econ. Sec. Est. Int. Marít. Subsec. Pesca.

Boltovskoy, E.
1968. **Hidrología de las Aguas Superficiales en la parte Occidental del Atlántico Sur.**
Revista Museo Arg. Cien. Nat. "Ber. Riv". Hidrobiología, Tomo II, 6.

Bonaparte, José F.
El Mesozoico de América del Sur y sus tetrapodos.
Op. Lilloana 26, 596 p.

Boschi, E.
1978. **Los Crustáceos Decápodos en las Comunidades Bentónicas del Mar Epicontinental Argentino.**
Ecol. Bentónica y Sed. Plat. Cont. Atlántico Sur. Seminario UNESCO.
1978, Montevideo. Page 279.

Boswall, J., MacIver, D.
1975. **The Magellanic Penguin Spheniscus magellanicus.**
The Biology of Penguins. Ed. Bernard Stonehouse. Univ. Bradford. 271 p.

Boswall, J., Prytherch, R. J.
1972. **Some Notes on the Birds of Point Tombo, Argentina.**
Bull. Brit. Ornith. Club. Vol. 92, N° 5, pág. 118.

Bruning, D.F.
1974. **Social structure and reproductive behaviour in the Greater Rhea.**
The Living Bird. p. 251-294.

Burkart, R. L. Ruiz, Daniele C. A. Maranta y F. Ardura.
1991. **El Sistema Nacional de Áreas Naturales Protegidas de la República Argentina. Diagnóstico de su Desarrollo Institucional y Patrimonio Natural.**
127 p., Inf. Inéd., A.P.N., Bs. As.

Burton, M.
1975. **How Mammals Live.**
Elsevier/Phaidon - London.

Cabrera, Ángel.
1957. **Catálogo de los Mamíferos de América del Sur, I.**
1960. **Catálogo de los Mamíferos de América del Sur, II.**
Revista del Mus. Arg. Cien. Nat., Zool., IV (1).

Cabrera, A. L.
1957. **La Vegetación de la Puna Argentina.**
Rev. Invest. Agric. Bs. As. 11 (4): 317-512.

Cabrera, A. L.
1964. **Las Plantas Acuáticas.**
Ed. Universitaria de Bs. As. Libros del caminante.

Cabrera, A. L.
1968. **Ecología Vegetal de la Puna Argentina.**
Colloquium Geographicum (Bonn) 9: 91-116.

Cabrera, A. L.
1976. **Regiones Fitogeográficas Argentinas.**
Enciclopedia Argentina de Agricultura y Jardinería. 2ª Ed. II (1) Ed. ACME.

Cabrera, A. L. y Dawson, G.
1944. **La Selva Marginal de Punta Lara.**
Rev. Museo de la Plata, 5: 267-382.

Cabrera, A. L. y Willink, A.
1973. **Biogeografía de América Latina.**
Washington, D. C. VI, 117 p.

Cabrera, A. y Yepes, J.
1940. **Mamíferos Sudamericanos.**
Comp. Arg. de Editores. Bs. As.

Cabrera, A. L. y Zardini, E. M.
1979. **Manual de la Flora de los alrededores de Bs. As. 2nd Ed.**
Editorial ACME. Bs. As.

Cajal, J.
1977. **Observaciones sobre hábitos alimenticios de la vicuña (Nota).**
Informe interno Dirección Nac. de Fauna.

Canevari, M., P. Canevari, G. R. Carrizo, G. Harris, J. Rodríguez Mata, R. J. Straneck.
1991. **Nueva Guía de las Aves Argentinas.**
Tomo I: 411 p. y Tomo II: 497 p. Fund. Acindar. Bs. As.

Cano, E. y Movia, C.
1967. **La Vegetación de la República Argentina. VIII. Utilidad de la Fotointerpretación de la Cartografía de Comunidades Vegetales del Bosque de Caldén (Prosopis caldenia Burk)**
INTA. Serie Fitogeográfica N° 8.

Carcelles, A., Pozzi, A.
1933. **Apuntes sobre la Fauna del Golfo San Matías.**
Museo Nac. Hist. Nat. "Ber. Riv". Bs. As.

Castellanos, A.
1975. **Cuenca Potamográfica del Río de la Plata.**
Geog. Rep. Arg. VII (2) Hidrog. Soc. Arg. Est. Geog.

Cei, J. M.
1969. **La Meseta Basáltica de Somuncurá, Río Negro.**
Physis, XXVIII, (77): 257-271.

Cei, J. M.
1977. Los Andes.
Ed. Urbión, Madrid.

Cei, J. M.
1980. **Amphibians of Argentina.**
Monitore Zool. Ital. Monogr. 2. Torino.

Cei J. M.
1986. **Reptiles del Centro, Centro-Oeste y Sur de la Argentina. Herpetofauna de las Zonas Áridas y Semiáridas. Monografie IV.**
Museo Regionale di Scienze Naturali. Torino. 527 p.

Chebez, J. C.
1987. **El Pastizal Pampeano.**
Guía Educativa de Vida Silvestre (2): 30 p., E.V.S.A., Bs. As.

Chevez, J. C.
1987/1990. **La Selva Misionera I y II.**
Guía Educativa de Vida Silvestre (3) y (4): 82 p., F.V.S.A., Bs. As.

Chevez, J. C. y L. H. Rolón.
1989. **Parque Provincial Urugua-í.**
Edic. Montoya, Posadas Provincia de Misiones.

Cloudsley-Thompson, J. L.
1979. **El hombre y la Biología de las Zonas Áridas.**
Ed. Blume, Barcelona.

Correa Luna, H.
1955. **Reserva Nacional Finca El Rey, su valor como representante de una importante región fitogeográfica.**
Natura, I (2): 113-120.

Correa Luna, H.
1960. **Hoyada de Ischigualasto y Valle Fértil. Parque Nacional proyectado.**
Inf. Interno Serv. Nac. Parq. Nac., 8 p.

Correa Luna, H.
1973. **Laguna de Pozuelos (Jujuy), un importante regufio de aves.**
Inf. Interno Serv. Nac. Parq. Nac.

Correa Luna, H.
1977. **La Conservación de la Naturaleza: Parques Nacionales Argentinos.**
Serv. Nac. Parq. Nac. Bs. As., 169 p.

Crespo, J. A.
1950. **Nota sobre Mamíferos de Misiones.**
Inst. Nac. Inv. Ciencias Nat., Cien. Zool. I (14).

Crespo, J. A.
1974. **Comentario sobre Nuevas Localidades para Mamíferos de Argentina y Bolivia.**
Rev. Mus. Arg. Cien. Nat. "Ber. Riv." Zool. XI (1).

Crespo, J. A.
1974. **Incorporación de un Género de Cánidos a la Fauna de Argentina.**
Com. Mus. Arg. Cien. Nat. "Ber. Riv." Zool. IV (6)

Daciuk, J.
1979. **Contribuciones sobre Protección y Conservación de la Vida Silvestre y Áreas Naturales.**
Acta Zool. Lilloana XXXIV. Min. Cult. y Educ. Bs. As.

Digilio, A. P. L. y Legname, P. R.
1966. **Los Árboles Indígenas de la Provincia de Tucumán.**
Opera Lilloana XV. Tucumán.

Dimitri, M. J.
1972. **La Región de los Bosques Andino-Patagónicos.**
Col. Cient. INTA, Bs. As., 381 p.

Dimitri, M. J.
1977. **Pequeña Flora Ilustrada de los Parques Nacionales Andino-Patagónicos.**
(2nd ed.) Separata Anales Parq. Nac. XIII, p. 1-122.

Dimitri, M. J. (Dirección), Santos Biloni, J. (Textos)
1973. **Libro del Árbol.**
T. I. Celulosa Argentina.

Dorst, J.
1967. **South America and Central America. A Natural History.**
Chanticleer Press Ed.

Erize, F.
1978. **El Parque Marino Golfo San José.**
Inf. Fund. Vida Silvestre Arg. N° 3/4. Bs. As.

Erize, F.
1979. **El Mayor Espectáculo de Fauna Argentina.**
Fund. Vida Silvestre Argentina.

Erize, F.
1993. **El Gran Libro de la Naturaleza Argentina.**
Revista Gente, 336 p. y 21 Afiches. Edit. Atlántida, Bs. As.

Eskuche, U.
1968. **Fisonomía y Sociología de los Bosques de** Nothofagus dombeyi **en la Región de Nahuel Huapi.**
Vegetatio. 16: 192-204.

Eskuche, U.
1973. **Estudios fitosociológicos en el norte de Patagonia. I. Investigación de algunos factores de ambiente en comunidades de bosque y de chaparral.**
Phytocoenología 1 (1): 64-113.

Franceschi, E. A. y Lewis, J. P.
1979. **Nota sobre la Vegetación del Valle Santafecino del Río Paraná (Rep. Arg.).**
Ecosur, Arg. 6 (11): 55-82

Frangi, J. L.
1975. **Sinopsis de las Comunidades vegetales y el Medio de la Sierra de Tadil (Pcia. Bs. As.).**
Bol. Soc. Arg. Bot. 16 (4): 293-319.

Frangi, J. L.
1976. **Descripción florística-estructural de un "stand" de bosque de** Nothofagus dombeyi **en Lago Gutíerrez (Prov. de Río Negro).**
Darwiniana. 20: 577-585.

Gallardo, J. M.
1966. **Zoogeografía de los anfibios chaqueños.**
Physis. XXVI (71): 67-81.

Gallardo, J. M.
1974. **Anfibios de los alrededores de Buenos Aires.**
Ed. Univ. Bs. As., 231 p.

Gallardo, J. M.
1976. **Estudio Ecológico sobre los Anfibios y Reptiles de la Depresión del Salado, Prov. de Bs. As., Argentina.**
Museo Arg. Cien. Nat., Ecología, II, (1): 26 p.

Gallardo, J. M.
1977. **Reptiles de los alrededores de Buenos Aires.**
Ed. Univ. Bs. As., 213 p.

Gallardo, J. M.
1980. **Observaciones Ecológicas sobre los Anfibios y Reptiles del Parque Nacional de Lihué Calel, La Pampa, Rep. Arg. Resumen del trabajo presentado en la II Reunión Iberoamericana de Conservación y Zoología de Vertebrados.**
15 al 20 de junio, 1980. Cáceres (España).

Gallardo, J. M.
1987. **Anfibios Argentinos, Guía para su identificación.**
Bibl. Mosaico. 98 p.

Grasse, P. P.
1970. **La Vida de los Animales.**
3 T., Ed. Planeta, Barcelona.

Haene, E. y G. Gil.
1991. **El Proyectado Parque Nacional Sierra de las Quijadas. Provincia de San Luis. República Argentina.**
102 p., Inf. Inéd. A.P.N., Bs. As.

Haene, E., S. Heinonen, J. C. Chebez.
1993. **Proyecto de Parque Nacional El Leoncito (Departamento Caligasta. Provincia de San Juan).**
47 p., A.P.N., Bs. As.

Halloy, S.
1978. **Un Parque Nacional en el Aconquija.**
Bol. Fund. Vida Silvestre Arg., 5, Abril 1978.

Hamilton, J. E.
1934. **The Southern Sea Lion** Otaria byronia.
De Blainville Discovery Reports 8: 269: 218.

Hueck, Kurt.
1978. **Los Bosques de Sudamérica, Ecología, Composición e Importancia Económica.**
Soc. Alemana Coop. Técnica.

Humphrey, P. S., Bridge, D., Reynolds, P. y Peterson, R. T.
1970. **Birds of Isla Grande (Tierra de Fuego).**
Univ. Kansas Mus. Nat. Hist. 411 p.

Jackson, J.
1980. **Campos del Tuyú Reserve; an Ecological Panorama.**
Informe para la Fundación Vida Silvestre Argentina (no publicado).

King, J. E.
1964. **Seals of the World.**
Trustees of Brit. Museum (Nat. Hist.) London.

Koford, Carl
1957. **The Vicuña and the Puna.**
Ecol. Monog. 27 (2): 152-219.

Kühnemann, O.
1978. **Megafitobentos del Litoral Argentino. Ecología Bentónica y Sedimentación de la Plataforma Continental del Atlántico Sur.**
Unesco. Seminario 9-12 mayo 1978. Montevideo. p. 94.

Laurent, Raymond F.
1967. **Descubrimiento del Género Gastrotheca Fitzinger en Argentina.**
Acta Zoológica Lilloana XXII, 1967.

Laws, R. M.
1953. **The Elephant Seal (Mirounga leonina, Linn.) I. Growth and Age.**
Falkland Islands Dependencies Survey Sci. Rep. Nº 8. 62 p.

Laws, R. M.
1956. **The Elephant Seal (Mirounga leonina, Linn.) II. General social and reproductive behaviour.**
Falkland Islands Dep. Survey. Sci. Rep. Nº 13, 88 p.

Laws, R. M.
1956. **The Elephant Seal (Mirounga leonina, Linn.) III. The physiology of reproduction.**
Falkland Islands Dep. Survery. Sci. Rep. Nº 15.

Leonardis, Rosario F. J. (Dirección y textos)
1975. **Libro del Árbol. II.**
Celulosa Argentina.

Lewis, J. P. y Collantes, M. B.
1973. **El Espinal Periestépico.**
Ciencia e Investigación. 29: 345-408

Llanos, A. C. y Crespo, J.
1954. **Ecología de la Vizcacha** (Lagostomus maximus maximus Blainv.) **en el Nordeste de la Provincia de Entre Ríos.**
Nueva Serie, Nº 10, Rev. Inv. Agric. VI, (3-4).

Mann Fischer, Guillermo.
1978. **Los pequeños mamíferos de Chile (marsupiales quirópteros, edentados y roedores).**
Goyana, Zoología Nº 40. Ed. Univ. Concepción.

Marconi, P., Lisi, J. y Canevari, P.
1980. **Proyecto de Parque Nacional Sierra de las Quijadas.**
Informe interno Serv. Nac. Par. Nac., 29 p.

Marconi, P., Lisi, J. y Canevari, P.
1980. **Proyecto de Parque Nac. Comechingones.**
Informe interno Serv. Nac. Parq. Nac. 27 p.

Márquez, J., J. C. Chebez, E. Haene, A. Flores y E. Sánchez.
1991. **Sistema Provincial de Áreas Naturales Protegidas. Provincia de San Juan. República Argentina.**
65 p., Inf., Inéd., Gobierno de la Provincia de San Juan, A.P.N. y Fund. Amb. Sanjuanina.

Martínez-Crovetto, R y Piccimini, B. C.
1950. **La vegetación de la República Argentina. I. Los Palmares de Butia yatay.**
Rev. Invest. Agric. Buenos Aires, 5 (2): 153-242.

Massoia, Elio (Director R. A. Rinquelet).
1976. **Fauna de Agua Dulce de la República Argentina.**
Vol. XLIV. Mammalia. FECIC.

Massoia, Elio.
1970. **Contribución al conocimiento de los mamíferos de Formosa con noticias de los que habitan en zonas viñateras.**
IDIA. Dic. 1970: 55-63.

Massoia, Elio.
1972. **La presencia de Marmosa cinerea paraguayana en la Rep. Arg., Prov. Misiones.**
Rev. Invest. Agrop. INTA, Bs. As. Serie 1, IX (2).

Massoia, E. y J. C. Chebez.
1993. **Mamíferos Silvestres del Archipiélago Fueguino.**
261 p., Edic. Lola, Bs. As.

Meyer, T.
1944. **Un viaje Botánico al Departamento de Orán (Prov. de Salta).**
Rev. Geográfica Americana, XXI (128).

Meyer, T.
1959. **Características de la Flora Salteña, Estudios realizados sobre ella y orientación sobre trabajos futuros.**
Rev. Fac. Cienc. Nat. Salta, 1: 15-33.

Meyer, T.
1963. **Estudios sobre la selva Tucumana. La Selva de Mirtáceas de "Las Pavas".**
Opera Lilloana X. Tucumán.

Meyer, T. y Weyrauch, W.
1966. **Guía para dos excursiones biológicas en la provincia de Tucumán.**
Miscelánea N° 23. Tucumán.

Meyer de Schauensee, R.
1970. **A guide to the Birds of South America.**
Acad. Nat. Scien. Philadelphia Livingston Pub. Co. 470 p.

Minoprio, J. D. L.
1945. **Sobre el** *Chlamyphorus truncatus* **Harlan.**
Acta Zool. Lilloana, 3: 14-25.

Morello, J.
1958. **La provincia fitogeográfica del Monte.**
Opera Lilloana, 2: 11-155.

Morello, J. y Adamoli, J.
1968. **La vegetación de la Rep. Arg. Las Grandes unidades de vegetación y ambiente del Chaco Arg. 1ª parte: Objetivos y metodología.**
INTA. Serie fitogeográfica. N° 10.

Morello, J. y Adamoli, J.
1974. **Las grandes unidades de vegetación y ambiente del Chaco Argentino. 2ª parte: Vegetación y ambiente de la Prov. del Chaco.**
INTA, Serie fitogeográfica, n° 13.

Morello, J. H., Crudeli, Nedo E. y Saraceno, M.
1971. **La vegetación de la Rep. Arg. Los vinalares de Formosa (R. A.). La colonización leñosa** *Prosopis ruscifolia*.
INTA. Serie fitogeográfica. N° 11.

Navas, J. R.
1977. **Fauna de agua dulce de la Rep. Arg.**
Vol. XLIII (Aves); 2 (Anseriformes): 14 p. (Director R. A. Ringuelet).

Navas, J. y Bo, N.
1977. **Ensayo de tipificación de nombres comunes de las aves argentinas.**
Rev. Mus. Arg. Cien. Nat., Zool., XII (7): 69-111.

Olivier, S. R., Paternoster, I. K. de, Bartida, R.
1966. **Estudios Biocenóticos en las costas de Chubut (Arg.) I. Zonación Biocenológica de Puerto Pardelas (Golfo Nuevo).**
Min. Econ. Prov. Chubut. Dir. Asun. Agr. Rawson. Apart. del Bol. Inst. Biol. Marina, N° 10.

Olivier, S. R., Torti, M. R., Bastida, R.
1968. **Ecosistema de las aguas litorales.**
Armada Argentina. Serv. Hidrog. Naval.

Olrog, C. C.
1959. **Las Aves Argentinas, una guía de campo.**
Instituto "Miguel Lillo", Tucumán.

Olrog, C. C.
1968. **Las aves sudamericanas. Una guía de campo.**
Fundación Inst. Miguel Lillo, Tucumán.

Olrog, C. C.
1969. **Birds of South America.**
Fittkau et al. (editores), Biogeography and Ecology in Touth America, Hillary, Nueva York, N. Y., Vol. 2: 849-878.

Olrog, C. C.
1979. **Los Mamíferos de la Selva Húmeda, Cerro Calilegua, Jujuy.**
Acta Zool. Lilloana XXXIII (2): 9-14.

Olrog, C. C., Ojeda, R. A. y Bárquez, R. M.
1976. **Catagonus wagneri (Rusconi) en el Noroeste Argentino.**
Neotropica, 22 (67).

Olrog C. Ch y M. Lucero.
1981. **Guía de Mamíferos Argentinos.**
151 p., Fund. M. Lillo, Tucumán.

Olrog C. Ch.
1984. **Las Aves Argentinas. Una Nueva Guía de Campo.**
352 p., A.P.N. Edit. INCAFO, Madrid.

Orians, G. H. y Solbrig, O. T. (Editores)
1977. **Convergent evolution in warm deserts.**
US/IBP Synthesis series - 3. Dowden, Hutchinson & Ross, Inc., Stroudsburg, Pennsylvania.

Otte, K. y Hofmann, R. K.
1977. **Utilización racional de la vicuña silvestre en el Perú.**
28 p. mimeograf. Congreso Latinoamericano de Zool. Tucumán.

Parodi, L. R.
1940. **La distribución geográfica de los talares en la Prov. de Buenos Aires.**
Darwiniana, 4 (1): 33-56.

Partridge, W. H.
1956. **Notes on the Brazilian Merganser in Argentina.**
The Auk, Journ, Ornith. 73 (4) American Ornith. Union. p. 473.

Payne, Roger.
1976. **At Home with Right Whales.**
Nat. Geog. Vol. 149 (3).

Peña de la, M. R.
1985-9. **Guía de Aves Argentinas.**
Tomo I-IV. Sta. Fe. Bs. As.

Peters, J. A. y Orejas Miranda, B.
1970. **Catalogue of the Neotropical Squamata.**
Part. I. Snakes. Smithsonian Instituion. United States National Museum.

Peters, J. A. y Donoso Barros, R.
1970. **Catalogue of the Neotropical Squamata: Part. II: Lizards and Amphisbaenians.**
Smithsonian Institution. United States National Museum.

Povilitis, A.
1978. **The Chilean Huemul Project-A Case History (1975-76).**
Threatened Deer. IUCN - 1978.

Prichard, H. H.
1902. **Field notes upon some of the larger mammals of Patagonia between September 1900 and June 1901.**
Proc. Zool. Soc., London. 1: 272-277.

Ragonese, A.
1970. **La vegetación del Parque Chaqueño.**
Bol. Soc. Arg. Bot. 11 (Supl.) 133-160.

Redford, K. H. y J. F. Eisenberg.
1992. **Mammals of the Neotropics. The Southern cone. Volume II: Chile, Argentina, Uruguay, Paraguay.**
430 p., Univ of Chicago Press, Chicago y London.

Reig, O.
1960. **La región de la hoyada de Ischigualasto (Dpto. Valle Fértil, San Juan) como Yacimiento Paleontológico.**
Informe interno Serv. Nac. Parq. Nac. 5 p.

Ringuelet, R. A.
1955. **Panorama Zoogeográfico de la Provincia de Buenos Aires.**
Univ. Nac. de la Plata. Notas del Museo. XVIII, Zool. N° 156, 15 p.

Ringuelet, R. A.
1961. **Rasgos Fundamentales de la Zoogeografía de la República Argentina.**
Physis, 22, 151-170.

Ringuelet, R. A.
1962. **Ecología Acuática Continental.**
Bs. As. Manuales de Eudeba, Cien. Nat.: VII (XI): 1-138.

Ringuelet, R. A., Aramburu, R. H. y Aramburu, A. A. de.
1967. **Los Peces Argentinos de Agua Dulce.**
Comisión Invest. Cient., La Plata, 601 p.

Roig, F. A.
1971. **Flora y vegetación de la Reserva Forestal de Ñacunán.**
Deserta. 1: 25-232.

Roquero, M. J.
1969. **La vegetación del Parque Nacional Laguna Blanca.**
Anal. Parq. Nac., 11 (2): 129-207.

Ruthsatz, B.
1974. **Los arbustos de las estepas andinas del noroeste argentino y su uso actual.**
Bol. Soc. Arg. Bot. XVI (1-2): 27-45.

Ruthsatz, B. y Movia, C.
1975. **Relevamiento de las estepas andinas del noroeste de la provincia de Jujuy.**
F.E.C.I.C. Bs. As. 131 p.

Schulz, A. G.
1961. **Nota sobre la vegetación acuática chaqueña. Esteros y embalsados.**
Bol. Soc. Arg. Bot. 9: 141-150.

Scolaro, A.
1979. **Punta Tombo. La vida en una pingüinera.**
Rev. Periplo. Año V, N° 28, Madrid.

Soriano, A.
1956. **Los distritos florísticos de la Prov. Patagónica.**
Rev. Inv. Agríc. X (4).

Sota de la, E.
1977. **Flora de la Provincia de Jujuy.**
Parte II (XIII), Col. Cient. INTA, Director Ángel Cabrera. Bs. As. 277 p.

Tarak, A.
1978. **Proyecto de Parque Nacional Carahuasi –Circuito oeste– Desarrollo de Circuitos turísticos multinacionales.**
Resumen - BID - INTA.

Thomasson, K.
1959. **Nahuel Huapi. Plankton of some lakes in an Argentine National Park, with notes on terrestrial vegetation.**
Acta Phytogeogr. Suec., 42: 1-83.

Vellard, J.
1948. **Batracios del Chaco Argentino.**
Acta Zool. Lilloana. Instituto Miguel Lillo. V: 137-174.

Vervoorst, F.
1967. **Las comunidades vegetales de la depresión del Salado (Prov. de Bs. As.)**
INTA. La vegetación de la República Argentina, 7: 1-262.

Villa-R. B. y Villa Cornejo, M.
Algunos murciélagos del norte de Argentina.
Misc. Publ. 51, Univ. Kansas Mus. Nat. Hist.

Walker, E. P. y Paradiso, J. L.
1968. **Mammals of the World. I and II.**
John Hopkins Press. Baltimore.

Weller, M. W.
1967. **Notes on some marsh birds of Cape San Antonio, Argentina.**
Ibis, 109: 391-411.

Wetzel, R. M., Dubos, R. E., Martín, R. L. y Myers, P.
1975. **Catagonus, an "Extinct" Peccary, Alive in Paraguay.**
Science. 189: 379-381.

Wetzel, R. M. y Crespo, J. A.
1975. **Existencia de una tercera especie de Pecarí Fam. "Tayassuidae, Mammalia" en Argentina.**
Rev. Museo Arg. Cien. Nat., Zool., XII (3).

Index of common and scientific names

A

Aa paludosa 156
ABADEJO - 137
Abrocoma cinerea - 159
Abrothrix - 176
Aburria jacutinga - 33
ACACIA - 87
Acacia caven - 56, 68, 80, 87, 120
Acacia furcatispina - 120
Acacia macracantha - 83
Acanthistius brasilianus - 137
Acanthochelys spixi - 73
Acarosphora - 163
Accipiter bicolor - 55, 78, 180
Accipiter striatus - 78
ACHICORIA - 155
Acrocinus longimanus - 34
Acrocomia chunta - 59
Acromyrmex - 63
Acromyrmex lundi - 102
Adenanthera macrocarpa - 39, 49
Adenocalymma marginatum - 22
Adesmia horridiuscula - 156
Adesmia incana - 104
Adesmia lihuelensis - 119
Adesmia nanolignea - 163
Adesmia oborata - 163
Adesmia pinnifolia - 163
Adesmia schickendantzii - 155
Adesmia spinossisima - 156
Adesmia uspallatensis - 163
Adiantopsis radiata - 22
Aechmea bromeliaefolia - 22
Aechmea calyculata - 22
Aechmea distichantha - 22, 50
Aeghus spp. - 83
Aequidens portalegrensis - 72
Aextoxicon punctatum - 175, 194
AFATA BLANCA - 49
Agelaius flavus - 76
Agelaius ruficapillus - 107
Agelaius thilius - 76, 107
AGOUTI - 27, 44, 55
Agouti paca - 27, **32**
Agriornis montana - 160
AGUAÍ - 21, 39
Ajaia ajaja - 75, **85**, 105
Akodon sp - 55, 98
ALBATROSS, Black browed - 137, 207, 209, 216
ALBATROSS, Gray headed - 216
ALBATROSS, Light mantled sooty - 216
ALBATROSS, Wandering - 216
Alchornea iricurana - 43
ALECRON - 21, 43
ALDER - 50, 56
Alectrurus risorius - 76
Alectrurus tricolor - 76
ALERCE - 194, 195, 197
ALGARROBO - 87, 89, 92
ALGARROBO, Black - 63, 87
ALGARROBO DULCE - 113
ALGARROBO, Patagonian - 124
ALGARROBO, White - 63, 87, 113
ALISO DEL RIO - 50
Allophylus edulis - 49, 92
Alnus jorullensis - 50
Alouata caraya - 29, 79, 83, **220**
Alouata guariba - 29
ALPACA - 156
ALPAMATO - 49
ALPATACO - 87, 128
Alstromeria spp - 50
Alstromeria aurantiaca - 173
AMANCAY - 50, 114, 173
Amaryllidaciae - 114
Amaryllis tucumana - 114
Amazona aestiva - 67

Amazona tucumana - 50
Amazonetta brasiliensis - 92
AMBA1 - 26
Amblyramphus holosericeus - 76
Ambarana cearensis - 49
ANACAHUITA - 223
Anaea morvus - 36
AÑAGUA - 156
ANAGGILLA, 156
Anairetes parulus - 117, 180
Anas cyanoptera - 104
Anas flavirostris - 104, 131
Anas georgica - 104, 131
Anas leucophrys - 67
Anas platalea - 104, 131
Anas puna - 160
Anas sibilatrix - 130, 131
Anas specularis - 179
Anas versicolor - 104, **111**
Anatidae - 178
ANCHICO COLORADO - 21
ANCHOÍTA - 137, 148
Andreaea - 211
Anhimidae - 111
ANHINGA - 75
Anhinga anhinga - 75
ANI, Greater - 43
ANI, Smooth-billed - 43, 80
ANT, Aztec - 26
ANT, Fire - 102
ANT, hunting - 167
ANT, Leaf-cutting - 63
ANT, Seed-gathering - 63
ANTS, Predatory - 63
ANTBIRDS - 53, 67
ANTEATER, Collared - 29, 40, 79
ANTEATER, Giant - 64, **69**, 71
Anthaxia concinna - 182
Anthus sp. - 101
ANTPITTA, Variegated - 33
ANTS, ARMY - 30
ANTS, ARMY - 30
ANTSHRIKE, Giant - 53
ANTSHRIKE, Great - 53
ANTSHRIKE, Large-tailed - 30
ANTSHRIKE, Spot-backed - 30
ANTSHRIKE, Tufted - 30
ANTSHRIKE, Variable - 53
ANTTHRUSH, Short-Tailed - 33
Annumbius annumbi - 64
Aotus azarae - 79
Aphrastura spinicauda - 179
Aphyllocladus spartioides - 163
Aphyocharax rubripinnis - 72
Apocinaceae - 50
Aptenodytes forsteri - 215
Apuleia leiocarpa - 21
Ara auricollis - 67
Ara militaris - **54**
Arabidea chica - 22
ARACARI, Chestnut-eared - 33
Aracea - 50
Aracnites uniflora - 175
Aramides saracura - 30
Aramides ypecaha - 76
Aratinga acuticaudata - 67
Aratinga leucophthalma - 33, 67
Aratinga mitrata - 50
ARAUCARIA - 131, 170, 175, **197**, 211
Araucaria - 170, 186, 188
Araucaria araucana - 170
Araucaria mirabilis - 131
Arbacia dufresnei - 137
Arctocephalus australis - 137, **141**, 151, 209
Arctocephalus gazella - 215
Ardea cocoi - 75, 105
ARDEGRAS - 113
Arilus carinatus - **38**
Aristida - 97
Aristida humilis - 156
Aristofelia - 169

ARMADILLO - 114, 127, 159
ARMADILLO, Bare-tailed - 64
ARMADILLO, Fairy - 64, 114
ARMADILLO, Giant - 64, 68, 71
ARMADILLO, Hairy - 83, 101, 127
ARMADILLO, Nine-banded - **42**, 43
ARMADILLO, Seven-banded - 101
ARMADILLO, Six-banded - 55, 64, **78**
ARMADILLO, Three-banded - 64, 114
ARMADILLO, Wailing Pichi - 64, 114
ARRAYAN - 176, **190**
Arremon flavirostris - 53
Artemesia longinaris - 137
Artibeus lituratus - 29
Artibeus planirostris - 55
Asio flammenus - 79, 101, 127
Asio stygius - 55
Aspidosperma polyneuron - 39
Aspidosperma quebracho-blanco - 63, 87, 120
Asterostigma vermicida - 150
Asthenes baeri - 64
Astragallus garbancillo - 156
Astronium balansae - 71
Atelognathus patagonicus - 134
Atelognathus reverberii - 128
Athene cunicularia - 89, 101, 160
Atlapetes citrinellus - 53
Atlapetes torquatus - 53
Atriplex lampa - 114, 124
Atriplex madariagae - 156
Atriplex microphylla - 156
Atriplex sagittifolium - 124
Atriplex vulgarissima - 151
Atta - 63
Attagis malouinus - 127, 186
Aulacomia magellanica - 137
Aurancomyrmex tewer - 167
Austrocedrus - 170
Austrocedrus chilensis - 173
AVELLANO SILVESTRE - 194
AVOCET, Andean - 160
Axis axis - 184
Azara lauceolata - 173
Azolla spp. - 104
Azolla filiculoides - 72
Azorella - 170, 185
Azorella compacta - 156
Azorella trifurcata - 123
AZOTA - CABALLO - 21, 92
Azteca - 26

B

Baccharis boliviensis - 156
Baccharis magellanica - 185
Baccharis salicifolia - 83
Bahuinia candicans - 80
Baillonus bailloni - 33
Balaena acutirostrata - 215
Balanus psittacus - 137
Balfourodendron reidelianum - 21
BARNACLES - 137
Baryphthengus ruficapillus - 33
Basileuterus bivittatus - 53
Bastardiopsis densiflora - 21
BAT, Big-eared Brown - 127, 159, 176
BAT, Fishing - 30, 79, 83
BAT, Free-tailed - 55, 159
BAT, Long-tongned - 29
BAT, Mouse-eared - 176
BAT, Palm - 79
BAT, Peters Wooly False Vampire - 30, 55
BAT, Red - 55, 176

BAT, Velvety free-tailed - 79
BAT, Yellow - 55
BAT, Yellow-shouldered - 29, 55
Batara cinerea - 53
Bathyraja brachyurops - 138
Batrachyla leptopus - 180
BATRACIANS - 117
BEAR, Brown - 182
BEAR, Spectaded - 59
BEAVER - 182, 209
BECARD, Crested - 53
BECARD, Green-bached - 53
BECARD, White-winged - 53
BEECH - 169, 211
BEECH, Coihue Southern - 173
BEECH, Deciduous - 189
BEECH, Low Deciduous - 173
BEECH, Tall Deciduous - 173
BEETLE, Rhinoceros - 34, **38**
BEETLE, Stag - 182
Begonia cucullata - 50
Begonia micrantha - 50
Benettitales - 131
Benthoctopus tehuelchus - 137
Berardius arnouxii - 215
BERBERIS - 173
Berberis cuneata - 124
Berberis darwinii - 173
Berberis empetrifolia - 185
Berberis ilicifolia - 176
Berberis microphylla - 176
Betulaceae - 50
Bignonia ungiscati - 50
Billbergia nutans - 22
BITTERN, Strie-bached - 105
BIVALVES - 137
BLACK EAR TREE - 21, 49, 56, 80
BLACKBIRD, Austral - 180
BLACKBIRD, Saffron-cowled - 76
BLACKBIRD, Scarlet-headed - 76
BLACKBIRD, White-browed - 101
BLACKBIRD, Yellow-winged - 76, 107
BLACK-TYRANT, White-winged - 117
Blastocerus dichotomus - 76, **81**
Blechnum chilensis - 175
Blechnum penna-marina - 176
Blepharocalyx gigantea - 49
Blepharocalyx tweediei - 92
BOA - 67, 71, 119, 120
Boa constrictor - 67, 71, 119
BOA, Water - 72, 73
Boehmeria caudata - 49
BOAR, Wild European - 89, 95, 119, 182, 197
BOG, Balsam - 176
BOGA - 83
BOGAVANTE - 137
Bolax gummifera - 176
Bombacaceae - 63
Bombus dahlbomi - 182
Bosquila trifoliolata - 194
Bothrops alternata - 34, 68, **92**
Bothrops ammodytoides - 128
Bothrops jararaca - 34
Bothrops jararacussu - 34
Bothrops neuwiedii - 34, 68, 119
Bougainvillea spinosa - 113-128
BONITO - 137
Bouteloua aritidoides - 113
Bouteloua barbata - 113
Bouteloua simplex - 156
Brachyclados caespitosus - 123
Brachyodontes purpuratus - 137
Bradypus variegatus - 59
BRAMA - 155, 156

Brassolis sophora vulp - **28**
Brassolidae - 36
Brassavola perrini - 25
BREA - 63, 119
BROCKET, Brown - 27, 44, 55, 64, 95
BROCKET, Dwarf - 27, 44
BROCKET, Red - 27, 44, 55
Bromelia hieronymii - 63
Bromelia serra - 63
BROMELIAD - 120
BRUSH-FINCH, Fulvous-headed - 53
BRUSH-FINCH, Stripe-headed - 53
BRUSH-FINCH, Yellow-striped - 53
BRUSQUILLA - 102
Bubo virginianus - 127
Bucconidae - 33
Bufo spp. - 55
Bufo arenarum - 102, **102**, 117
Bufo granulosus - 63, 102
Bufo paracnemis - 63, 73
Bufo spinolosus - 160, 180
Bufo variegatus - 180
Bullia spp. - 137
Bulnesia sarmientoi - 71
Buprestidae - 34, 182
Burmaniaceae - 175
Busarellus nigricollis - 78
BUSH DOG - 29
BUSH TANAGER, Common - 50
Buteo poecilochrous - 160
Buteo polysoma - 114, **125**, 127, 134
Buteo swainsoni - 101
Buteogallus urubitinga - **74**, 78
Butia yatay - 80, 87
BUZZARD-EAGLE, Black-chested - 114, 127, 160, **175**, 180, 204

C

Cabassous chacoensis - 64
Cabralea canjerana - 21, 43
CABRILLA - 138
CACHIAL - 156
CACHIYUYO - 156
CACHOLOTE, Brown - 90, 92, 119
CACHOLOTE, White-throated - 90, 117
Cacicus haemorrhous - 34
CACIQUE, Red-rumped - 34, 43
CACTUS, Treacherous - 119
Caenopetta - 182
Caesalpinia gilliesi - 119
Caesalpinia paraguariensis - 63
Caiman latirostris - 41, 44, **45**, 75, 82
Caiman yacare - 75, **82**
Cairina moschata - 44, 75
Cajophora - 156
CALAFATE - 124
Calandria punae - 156
Calanoides - 212
Calanus - 212
Calceolaria spp. - 173, 175, 176
CALDEN - 87, 89, **118**, 119
Calidris alba - 138
Calidris bairdii - 138
Calidris canutus - 138
Calidris fuscicollis - 138
Calliandra - 41
Callichthys callichthys - 73
Callicore hydaspes - 36
Callicore hystaspes - 36
Calligo memnon - 36
Callorhymus callorhymus - 138
Callomys - 98
Caloplaca - 24
Caltha sagittata - 185

231

Calycophyllum multiflorum - 49
Calydon submetallicum - 182
CAMELIDS - 163, 167
Campephilus leucopogon - 64
Campephilus magellanicus - 175, 179
Campephilus robustus - 44
Campomanesia xanthocarpa - 27
Camponotus - 63
Camponetus mus - 102
Campsidium valdivianum - 175
Campylocentrum - 25
CANASTERO, Short-tailed - 64
CANCHARANA - 21, 43
CANE, Colihue - 170, **172, 177**, 195
CANELON GUAZU - CANJIA - 155
Capparis - 63
Caprimulgus rufus - 55
CAPYBARA - 27, 44, **67**, 75, 83, 92, 98, 111, 223
Carabidae - 34, 182
CARACARA, Chimango - 78, 101, 104, 114, 127
CARACARA, Crested - **74**, 78, 114, 127
CARACARA, Mountain - 160
CARACARA, White-throated - 180
CARANDAY - 72, 79
CARDINAL, Red-crested - 67, 92, 109
CARDINAL, Yellow - 117
Cardiospermum - 22
CARDOON - 63, 156, **167**
CARDOON, Poco - 156
Carduelis barbatus - 180
Carex - 223
Cariamidae - 64
Cariama cristata - 56, 64, **81**
Caryophyllaceae - 211
CASCARUDOS - 73
Casearia sylvestris - 26
Cassia aphylla - 113
Cassia crassiramea - 163
Cassia hookeriana - 156
Castor canadensis - 182
CAT, Andean - 163
CAT, Geoffroys - 64, 89, 98, 111, 114, 127
CAT, Guigna - 195
CAT, Little-Spotted - 29
CAT, Pampas - 55, 64, 89, 98, 114, 127, 159
CAT, Spotted - 176
Catagonus wagneri - 64
CATFISH, Armoured - 83
CATFISH, Surubı́ - 83
CATFISH, Velvety - 181
CATFISH, White - 83
CATFISH, Yellow-Bagre - 83
Catharacta spp. - 138
Catharacta antarctica - 216
Catharacta macckormickii - 216
Catharacta skua - 148, **151**
Cathartes aura - 53, 78, 114
Cathartes burrovianus - 78
CATTAILS - 72, 104
Cavia aparea - 55
Cavia tschudi - 55
CAVY - 55, 98, 114
CAVY, Least - 127
CAVY, Pampas - 98
CAVY, Patagonian - 89, 114, 120, 124
CAYMAN, Black - 75, 80, **82**
CAYMAN, Broad-snouted - 41, 44, **45**, 75, 80, 82
CEBIL COLORADO - 49
CEBIL, Red - 83
Cebus apella - 29, 55
Cecropia pelfata - 26, 43
Cedrella - 56
Cedrella angustifolia - 49
Cedrella fissilis - 21
Cedrella lilloi - 49

CEDRILLO - 49
CEDRO MISIONERO - 21
CEDRO SALTEÑO - 49
Celtis - 26
Celtis spinosa - 87, 113
Celtis triflora - 50
CENTOLLA - 137
CENTOLLON - 137
Cephalopods - 138
Cephalorhynchus commersoni - 137
Ceramaster patagonica - 137
CERAMBICIDS - 68, 182
Ceratophrys ornata - **102**, 102
Cercidium australe - 63, 119
Cerdocyon thous - 29, 55, 64, 92
CERELLA - 27
Cereus validus - 63
Ceroglossus - 182
Ceroglossus bugueti - 182
Cervus elaphus - 89, 182
Ceryle torquata - 43
Cestrum parqui - 87
CHACAY - 173
Chacaya trinervis - 173
CHACHALACA, Chaco - 56, 64
CHACHI - 21
CHACO-FINCH, Many-coloured - 117
Chacophrys cranwelli - 63
Chacophrys pierotti - 63
Chaenichthydae - 216
Chaetophractus nationi - 159
Chaetophractus vellerosus - 64, **89**, 114
Chaetophractus villosus - 83, 101, 127
Chaetopleura isabelli - 137
Chaetopleura tehuelche - 137
CHAGUARES - 63
CHAL-CHAL - 49, 92
Chamaeza campanisoma - 33
CHANCHITAS - 72
CHAÑAR - 63, 87, 95, 113, 120
Characidae - 131
Charadrius alticola - 160
Charadrius falklandicus - 131, **134**, 138
CHARAO - 132
Chauna torquata - 80, 104, **111**
Chelemys macronyx - 176
Chelonoidis chilensis - 119
Chenopodiaceae - 156
Chenopodium scabricaule - 151
Chiacognathus - 182
CHIJUA - 156
CHILCA - 83, 102
CHILLHAUA - 159
CHINCHILLA - 163
Chinchilla brevicaudata - 163
Chinchilla sahamae - 159
Chione antigua - 137
Chionis alba - 138, **215**, 216
Chirodecton sanguineum - 33
Chironectes minimus - 29
Chiroxiphia caudata - 33
Chlamydomonas - 211
Chlamyphorus retusa - 64
Chlamyphorus truncatus - 114
Chloephaga spp. - 102
Chloephaga hybrida - 178, 207, **207**
Chloephaga melanoptera - 160
Chloephaga picta - 128, **170**, 178
Chloephaga poliocephala - **170**, 178
Chloraea membranacea - 87
Chloroceryle amazona - 43
Chloroceryle americana - 43
Chlorophora tinctoria - 49
Chlorospingus ophthalmicus - 50
Chlorostilbon aureoventris - **38**

Chorizia - 43
Chorisia insignis - 63
Chorisia speciosa - 71
Chorisodontium - 211
CHOZCHORIS - 159
Chrysocyon brachyurus - **70**, 76
Chrotopterus auritus - 30, 55
Chrysanthemum leucanthemum - 182
Chrysomelidae - 34
Chrysophyllum gonocarpum - 21
Chrysoptilus melanochlorus - 64
Chunga burmeisteri - 56, 64
CHUNTA - 59
Chuquiraga avellanedae - 123
Chuquiraga erinacea - 113
CHURQUI - 68, 155
Chusquea - 170
Chusquea culeou - 170
Chusquea lorentziana - 50
Chusquea ramosissima - 22, 43
Cicadidae - 34
Cichlasoma facetum - 72
CICONIIFORMS - 85
CINCLODES, Dark-bellied - 179
Cinclodes patagonicus
CINCLODES, White-winged - 160
Cinclodes atacamensis - 160
CIPRES DE LAS GUAITECAS - 175, 204
Circus buffoni - 78, **98**
Circus cinereus - 78, 160
Cissopis leveriana - 33
Cissus striatus - 50
CLAMS - 137
Classis - 182
Clupeiform - 137
Clystostoma callistegioides - 22
Cnemidophorus leachi - 67
Cnesterodon decemmaculatus - 107
COATI - 29, 43, 55, 79, 83
COCK-FISH - 138
COCO - 68, 95
COD, Antarctic - 216
Codium decorticatum - 137
Codium fragile - 137
Codonorchis lessoni - 175
Coendou prehensilis - **51**
COENDU, Dark - 29, **51**
COIHUE - 172, 175, **180**
COIRON - 123, **134**, 173, 199
COIRON, Bitter - 123, 132, 163
COIRON, White - 124
COLA DE LEON - 156
COLA DE ZORRO - 156
COLAPICHE - 123, 124, 132
Colaptes campestris - 90, 119
Colaptes melanolaimus - 119
Colaptes pitius - 179
Colaptes rupicola - 160
Coleoptera - 182
Colletia - 63
Colletia paradoxa - 102
Colletia spinosissima - 87, 173
Colliguaya intergerrima - 124
Colobanthus quitensis - 211
Colonia colonus - 33
Colossoma mitrei - 83
Columba picazuro - 67
COMET, Red-tailed - 95
COMPOSITAE - 104, 156, 163
CONCHS - 137
Condalia microphylla - 87, 119, 128
CONDOR, Andean - 71, **157**, 160, 180, 193, 199
Conepatus chinga - 101, 114
Conepatus humboldtii - 127, 176, **192**

Conepatus rex - 159
Congiopodus peruvianus - 138
Conopophaga lineata - 33
Contopus cinereus - 53
COOTS - **100**, 105, 204
COOT, Andean - 160
COOT, Giant - 160
COOT, Horned - 160, 163
COOT, Red-fronted - **100**, 105
COOT, Red-gartered - **100**, 105, 131, 183
COOT, White-winged - **100**, 105
Copaitera langsdorfi - 39
COPEPODS - 212
Copernicia alba - 72, 77
Coragyps atratus - 53, 78, 114, 180
CORAL SNAKE - 34
CORAL TREE - 39, 80, 223
Cordia trichotoma - 21
Coriaria - 170
Coriaria ruscifolia - 194
CORMORANTS - 137, 138, 147, 148
CORMORANT, Blue-eyed - 153, 193, 216
CORMORANT, Guanay - 147, 148
CORMORANT, King - 148, **149**, 153
CORMORANT, Olivaceus - 145, 179, 182
CORMORANT, Red-legged - 147, 151, 153
CORMORANT, Rock - 147, 148, 151
CORONILLO - 87, 92, 109
CORPUS - 213
Cortadeira spp - 109, 223
Cortadeira pilosa - 132
Cortadeira selloana - 95, 104
Cortadeira speciosa - 156
Coryphospingus cucullatus - 53
COSCOROBA - 71, 104, 131
Coscoroba coscoroba - 71, 104, 131
Cosmaterias lurida - 137
COTINGAS - 53, 67
COWBIRD, Giant - 43
COYPU - 76, 92, 107, 184, 195, 197
CRABS - 83, 109
CRAB, Hermit - 137
CRAB, Patagonian - 137
CRAKE, Ash-throated - 44
CRAKE, Rufous-sided - 44
Cranioleuca pyrrhophia - 90
Crax fasciolata - 79
Crematogaster - 28
CREOSOTE BUSH - 87, 113, 120
Cricetidae - 159, 176
Crinodendrum tucumanum - 50
Crotalus - 34
Crotalus durisus - 68, 119
Croton urucurana - 26, 80
Crotophaga ani - 43, 80
Crotophaga major - 43
CROWN OF THORNS - 71, 92
CRUSTACEANS - 212
Crypturellus obsoletus - 30
Ctenomys spp. - 98, 114, 127
Ctenomys frater - 159
Ctenomys mendocinus - 89
Ctenomys opimus - 159
CUCKOO, Guira - 80, 101
CUCKOO, Squirrel - 44, 92
Cucurbitaceae - 50
CUPA-I - 39, 43
Cupania vernalis - 49
Cupriguanus achalensis - 68
Curaeus curaeus - 180
CURASSOW, Bare-faced - 79
Curculionidae - 34, 182
CURIYU WATER BOA - **220**
CURIEW, Eskimo - 107
CURROMAMUEL - 102

CURUPAY - 39, 43
CURUPI - 80, 223
Cuspidaria pterocarpa - 22
Cyanocorax chrysops - 44, 53, 92
Cyanoliseus patagonus - 115, 117
Cyathea o'donelliana - 59
Cyclagras gigas - 73
Cyclanthera thamnifolia - 50
Cyclopogon elatus - 87
Cygnus melancoryphus - 71, 104, 131, **133**, 178
Cynolebias spp. - 73
Cyeraceae - 104
Cyperus giganteus - 72
CYPRESS, Andean - 173
Cypseloides semex - **42**, 43
Cyrtograpus angulatus - 137
Cyttaria darwinii - 173

D

Dacrydium foonckii - 176
DAISY - 182
Dama dama - 182
Dasypodidae - 114
Dasyprocta azarae - 27
Dasyprocta punctata - 55
Dasypus hybridus - 101
Dasypus mazzai - 55
Dasypus novemcinctus - **42**, 43
DEER, Andean - 176, 186, 188, 193, 195, **197**, 223
DEER, Axis - 182
DEER, Fallow - 182
DEER, Marsh - 76, 80, **81**, 83, 85, 223
DEER, Northern Andean - 163
DEER, Pampas - 76, 80, **95**, 98, 109, 111, 223
DEER, Red - 89, 182, 188
DEER, Spotted - 184
Dendrocincla fuliginosa - 33
Dendrocolaptidae - 64
Dendrocolaptes platyrostris - 33
Dendrocygna autumnalis - 75
Dendrocygna bicolor - 75, 104
Dendrocygna viudata - 75, 104
Dermatonotos mulleri - 63
Deschampsia antarctica - 211
Desfontainiaceae - 170
Desmodus rotundus - 29, **51**, 55
DEU - 194
Diaethria candrena - 36
Diaethria clymena - 36
Diatenopteryx sorbifolia - 21
Dicotyles tajacu - 27, 55, 64, **67**, 120
Dicranopteryx cuadriparfila - 175
DIDDLE-DEE - 186
Didelphidae - 29
Didelphis albiventris - 29, 79, 101, **103**, 114
Didelphis marsupialis - 29
Didonis biblis - 36
Dinoponera australis - 34
Diomedea chrysostoma - 216
Diomedea exulans - 216
Diomedea melanophrys - 137, 216
Diplolaemus bibronii - 128
Dilomystes viedmensis - 181
Discaria longispina - 102
Discopyga tschudii - 138
Dissostichus eleginoides - 138
Distichlis - 104
Distichlis humilis - 155
Distichlis scoparia - 124
Distichlis spicata - 124
Diuca diuca - 117, 128
DIUCA FINCH, Common - 117, 128
DIUCON, Fire-eyed - 119, 180, **189**

DIVING PETREL - 207, 209
Dodonaea viscosa - 102
Dolichandra cynanchoides - 22
Dolichotis patagonum - 89, 114, 120, 124, **125**
DOLPHIN, Bottlenosed - 137, 144
DOLPHIN, Commersonis - 137, 151
DOLPHIN, Dusky - 137, 144
DOLPHIN, Peal's - 137
DORADO - 83
Dorymyrme - 63
DOTTEREL, Rufous-chested - 131
DOTTEREL, Tauiny-throated - 102, 128
DOVE, Eared - 67, 92
DOVE, White-tipped - 67, 92
Doxantha unguiscati - 22
Dromiciops australis - 170
Dromiciops gliroides - 176
Drosera uniflora - 176, 209
Drymis - 170
Drymis winteri - 176
Drymornis bridgesii - 90
Dryocopus galeatus - 44
Dryocopus lineatus - 44
DUCK, Andean Ruddy - **130**, 132, 204
DUCK, Black-headed - 104
DUCK, Brazilian - 92
DUCK, Comb - 75
DUCK, Crested - 131, 223
DUCK, Lake - 179
DUCK, Muscovy - 44, 75
DUCK, Spectacleed - 179
DUCK, Torrent - **177**, 178, 193
DURAZNILLO - 124, 132
DURAZNILLO, White - 104
Duruillea spp. - 137
Dusicyon culpaeus - 55, 95, 114, 128, 159, 176
Dusicyon griseus - 114, 176
Dusicyon gymnocercus - 89, 92, 98, **99**
Dyckia - 120

E

EAGLE, Black and chestnut - 53
EAGLE, Crested - 53
EAGLE, Crowned - 78
EAGLE, Solitary - 53
EARTHCREEPER, Band-tailed - 128
EARTHCREEPER, Rock - 160
EARTHCREEPER, Rock - 160
EARTH-CREEPER, Scale-throated - 117, 128
Eciton spp. - 30
Ecphysis - 182
Ectatomma - 63
EDIBLE - FROG - 34
EEL, Local - 107
EGRET, Common - 105
EGRET, Great - 75, 105
EGRET, Snowy - 75, 105
Egretta alba - 75, 105
Egretta thula - 75, 105
Eichornia spp. - 83
Eichornia azurea - 72
Eichornia crassipes - 72
Eira barbara - 29, 55
Elachistochleis bicolor - 63
Elanoides forficatus - 33, 53
Elaenia albiceps - 179
ELAENIA, Gray - 53
ELAENIA, White-crested - 179
Elanus leucurus - 78
Elaphroptera - 182
ELDER - 50, 87
Eleginops maclovinus - 138, 209
Eleocharis - 156

Eleutherodactylus discoidalis - 55
Eligmodontia hirtipes - 159
Eligmodontia typus - 114
Elionurus muticus - 72
Embernagra platensis - 76
Embothrium coccineum - 173, **177**
EMERALD, Glittering-bellied - **38**
Empetrum rubrum - 186
Enema pau - 34, **38**
Engraulis anchoita - 137
Enicognathus ferruginea - 179
Enterolobium contortisiliquum - 21, 49, 56, 80
Ephedra breana - 155, 156
Ephedra frustillata - 124
Ephedra ochreata - 124
Epiphylum - 22
Eragrostis argentina - 113
Eragrostis migricans - 156
Eremobius phoenicurus - 128
Eryphanes polyxena - **28**
Erythrina cristagalli - 39, 80
Erythrina falcata - 27
Erythrolamprus aesculapii - 34
ESCUERZO, Common Lesser - 107
ESCUERZO, Lesser - 117
ESPADAÑA - 109
ESPINILLO - 80, 87, 95, 120
ESPINO NEGRO - 87, 173
ESPORAL - 155
Eubalaena australis - 137, 141, **143**
Euryphia cordifolia - 194
Eudromia elegans - 89, 114, 124
Eudromia formosa - 64
Eudyptes chrysolophus - 215
Eudyptes crestatus - 209
Eugenia involucrata - 27
Eugenia mato - 49
Eugenia pseudo-mato - 49
Eugenia pungens - 49
Eugenia uniflora - 27, 92
Eunectes notaeus - 73, **220**
Euneomys chinchilloides - 176
Euphausia spp. - 137
Euphausia superba - 212
Euphonia spp. - 33
Euphractus sexcinctus - 55, **78**
Eurydites stenodesmus - 36
Eusophus - 180
Eusophus taeniatus - 193
Eustephiopsis speciosa - 156
Euterpe edulis - 39
Euxenura maguari - 105
Exacretodon frenguelli - 120

F

Fabiana densa - 156
Fabiana peckii - 123
Fabiana - 155
Fagara coco - 68, 95
Falco femoralis - 78, 101, 114, 160
Falco rufigularis - 78
Falco sparverius - 114, 180
FALCON, Aplomado - 76, 101, 114, 127, 160
FALCON, Bat - 78
FALCON, Peregrine - 127, 134
FALCONET, Spot-winged - 67, 117
Faramea syanea - 41
FERNIS, Floating - 104
Felis colocolo - 55, 64, 89, 98, 114, 159
Felis concolor - 29, 64, 87, 114
Felis geoffroyi - 64, 89, 98, 114

Felis guigna - 176
Felis jacobita - 163
Felis pardalis - 29, **52**
Felis tigrina - 29
Felis wiedii - 29, 55
Felis yaguaroundii - 29, 89, 114
FERN, Adder's-tongue - 175
Festuca - 123, **161**, 185
Festuca chrysophylla - 160
Festuca ortophulla - 156, 160
Festuca pallescens - 124
Festuca pampeana - 102
Festuca scirpifolia - 156
Festuca ventanicola - 102
Ficus enormis - 25
Ficus maroma - 59
FIG, Giant - 59
FIG, Strangler - 25, 43, 113
FINCH, Black-thorated - 128
FINCH, Great Pampas - 76
FINCH, Red-crested - 53
FINCH, Yellow-bridled - 128, 186
FIRE-BUSH, Chilean - 173, **177**, 207
FIRECROWN, Green-backed - 179, 204
FIREWOOD GATHERER - 64
FISH - 107, 160
FISH, Characiform - 72
FISH, Cichlid - 72
FISH, Elephant - 138
FISH, Float - 138
FISH, Ice - 216
FISH, Loricarid - 83
FISH, Pig - 138
Fitzroya - 170, 184, 193, **196**
Fitzroya cupressoides - 173
FLAMINGO - 163, 223
FLAMINGO, Andean - 71, 160
FLAMINGO, Chilean - 71, 105, 160, **165**, 220, 223
FLAMINGO, James - 163
FLICKER, Andean - 160
FLICKER, Chilean - 179, 204
FLICKER, Field - 90, 119
FLYCATCHER, Dusky-capped - 53
FLYCATCHER, Fork-tailed - 67, 102, 109
FLYCATCHER, Piratic - 43
FLYCATCHER, Vermilion - 67, 102, 109
FOREST-FALCON, Barred - 34
FOREST-FALCON, Collared - 34
Formicariidae - 53
Forpus xanthopterygius - 33
FOX, Crab-eating - 29, 55, 64
FOX, Pampas-grey - 89, 98, **99**
FOX, Red - 55, 95, 114, 127, 128, 159, 176
FOX, Small-gray - 114, 127, 176
FRANCISCO ALVAREZ - 92
Frankenia patagonica - 124
FRINGILLIDS - 53, 90, 117, 128, 180
FROG - 107, 128, 134
FROGS, Burrowing - 102
FROG, Coraline - 63
FROG, Darwin's - 180
FROG, Four-eyed - 180
FROG, Green Leaf-folding - 67
FROG, Horned - 63, **102**, 102
FROG, Marsupial - 55
FROG, Monsctached - 73
FROG, Weeping - 73
FRUIT-BAT, Neotropical - 29, 55
FRUIT-CROW, Red-ruffed - 33
Fuchsia - 170
Fuchsia boliviana - 50
Fuchsia magellanica - 175
FUINQUE - 175

Fulica ardesiaca - 160
Fulica armillata - **100**, 105, 131
Fulica cornuta - 160
Fulica gigantea - 160
Fulica leucoptera - **100**, 105
Fulica rufifrons - **100**, 105
FULMAR, Southern or Silvery-gray - 216
Fulmarus glacialoides - 216
FUMO BRAVO - 26
FUNGUS, Darwin's - 173
FURNARIIDS - 64, 89, 90, 160, 179, 180
Furnarius cristatus - 64
Furnarius rufus - 64, 92, 101
FUR-SEAL, Antarctic - 215
FUR-SEAL, Southern - 137, 138, **141**, 151, 209
FUSCHIA - 50

G

Gaillardia cabrerae - 119
Galaxia - 169, 181
Galaxia attenuatus - 181
Galaxia variegatus - 181
Galaxidae - 181
Galbula ruficauda - 33
Galea musteloides - 98, 114
Galictis cuja - 98, 114, 159, 176
Galictis furax - 92
Gallinago andina - 160
Gallinago gallinago - 76
Gallinago paraguaiae - **170**
GALLITO, Crested - 117
GALLITO, Sandy - 117
Gamaselly racovitzai - 212
GARABATO - 120
GARBANCILLO - 156
GARRAPATA YUYO - 49
Gastrotheca christiani - 55
Gastrotheca gracilis - 55
Gavilea lutea - 175
GEKKO - 67, 119
Genypterus blacodes - 137
Geoffroea decorticans - 63, 87, 113, 120
Geositta cunicularia - 90, 128
Geositta punensis - 160
Geositta tenuirostris - 160
Geoxus valdivianus - 176
Geoxus valdivianus - 176
Geranoaetus melanoleucus - 114, 160, **175**, 180
GINKO - 120
Glandularia peruviana - 97
Glaucidium nanum - 180
Gledistia amorphooides - 71, 92
Glossophaga soricina - 29
GLYPTODON - 64
GNATEATER, Rufous - 33
Gnetalae - 155
GOAT'S BEARD - 119
GOAT'S HORN - 163
Gochnatia glutinosa - 163
GODWIT, Hudsonian - **111**, 138
Gomortegacae - 170
GOOSE, Andean - 160
GOOSE, Ashy-headed - **170**, 178, 192, 204
GOOSE, Kelp - 178, 192, **207**, 207
GOOSE, Upland - **170**, 178, **192**, 202, 204
Govenia tinguens - 50
Gracilinanus agilis - 29
Grallaria varia - 33
Graomys griseoflavus - 89
GRASS, Arrow - 97, 101, 109
GRASS, Hog's hair - 223
GRASS, Pampas - 95, 104, 109, 132, 156
GRASS, Roofing - 72, 83
GRASS, Salf - 104, 155
GRASS, Tussock - 209
GRASS, Yellow - 72
GREBE - 204

GREBE, Great - 131, 178, 193
GREBE, Hooded - 131, **134**
GREBE, Least - 44, 75
GREBE, Pied-billed - 75, 178
GREBE, Silvery - 131, 132
GREBE, White-tuffed - 75,**130**, 178
Grilina - 170
Grimmia - 211
GRISON - 92, 98, 114, 127, 159, 176
GRISON, Dwart - 114
GROUND-DOVE, Bare-eyed - 163
GROUND-LIZARD, Large - 92
GROUND-LIZARD, Red - 67, 119
GROUND-TYRANTS - 128
GROUND-TYRANTS, Dark-faced - **189**
GROUND-TYRANTS, Puna - 160
GUABIROBA - 27
Guadua trinii - 22, 43
Guadua angustifolia - 22
GUAN, Dusky-legged - 50, 85, 223
GUAN, Red-faced - 50
GUANACO - 71, 89,98, 114, **118**, 120, 124, **199**, 199, 204
GUANAY - 147
GUAPO-Í - 113
GUATAMBU BLANCO - 21
GUAVA - 27
GUAYACAN - 63
GUAYAIBI -21, 49
GUAYCA - 21
Gubernatrix cristata - 117
Gubernetes yetapa - 76
GUEMBE - 25
GUEVIN - 194
Guevina avellana - 194
GUILI - 49
GUILI BLANCO - 49
GUINDO - 176, 199, 204, 207
Guira guira - 80, 101
GULL - 137, 138
GULL, Brown-hooded - 138
GULL, Dolphin - 138, 148, 151
GULL, Franklins - 71
GULL, Kelp - 138, 145, 148, 179, 216
GULL, Olrog's - 138
Gunnera - 170
Gunnera chilensis - 175
Gymnocharacinus bergi - 131
Gymnodactylus - 55
Gyrophora - 163

H

Habranthus spp. - 114
Habroncus - 55
Haematopus ater - 138, 207
Haematopus leucopodus - 131, 138, **170**, 207
Haematopus palliatus - 138, 170
HAKE - 137
HAKE, Black - 138
HAKE, Long-tailed - 137
Halaelurus bivus - 138
Halicarcinus planatus - 137
Hamadryas - 36
Haplochiton - 181
Haplochiton taeniatus - 181
Haplopappus pectinatus - 124, 132
HARE, Eurpean - 89, 184, 197, 204
Harpagifer spp. - 138
Harpia harpyja - 34, 53
Harpiprion caerulescens - 80
HARPY EAGLE - 34, 53
Harpyhaliaetus coronatus - 53, 78
Harpyhaliaetus solitarius - 53

233

HARRIER, Cinereus - 78, 127, 160
HARRIER, Long-winged - 78, **98**
Hatcheria - 181
HAWK, Bay-winged - 78
HAWK, Bicoloured - 55, 78, 180
HAWK, Black-collared - 78
HAWK, Great-black - **74**, 78
HAWK, Harris' - 78
HAWK, Puna - 160
HAWK, Red-backed - 114, **125**, 127, 134
HAWK, Savannah - 80
HAWK, Sharp-shinned - 78
HAWK, Swainson's - 102
HAWK-EAGLE, Black - 34
HAWK-EAGLE, Black and White - 41
HAWK-EAGLE, Ornate - 34
HAWK-MOTHS - 37
HAZEL, Native - 194
Hebe - 169
Heliocarpus popayanensis - 49
Hemigrammus caudovittatus - 72
HERMIT, Scale-Throated - 33
HERONS - 85
HERON, Cocoi - 75, 105
HERON, White-necked - 75, 105
HERON, Whisthing - 92
Heteronetta atricapilla - 104
Heterospizias meridionalis - 80
HILLSTAR, Andean - 160, **161**
HILLSTAR, White-sided - 160, 186
Himantopus mexicanus - 107
Hippocamelus antisensis - 163
Hippocamelus bisulcus - 176, **197**
Histiotus - 159
Histiotus montanus - 127, 176
Historis orion - 36
Hoffmanseguia gracilis - 156
Holocalyx balansae - 21, 43
Holochilus brasiliensis - 107
Holochilus chacarius - 76
Homonota borelli - 67
Homonota horrida - 67, 119
Homonota darwinii - 128
Hoplias malabaricus - 72, 107
Hopolosternum thoracatum - 73
HORCO CEBIL - 49, 83
HORCO MATO - 49
HORCO MOLLE - 49
HORNERO, Crested - 64
HORNERO, Rufous - 64, 92, 101
HOWLER MONKEY - See Monkey
HUEMUL - 163, 176, 186, 193, 195, **197**, 204, 223
HUET-HUET - 179
HUMMINGBIRD, White-throated - 33
Hyalis argentea - 104
Hydrangea intergerrima - 175
Hydrochaeris hydrochaeris - 44, **67**, 75, 92
Hydropsalis brasilianum - 79
Hydrurga leptonyx - 215
Hyla faber - 34
Hyla minuta - 34
Hyla phrynoderma - 73
Hyla pulchella - 73, 107, 160
Hylorina sylvatica - 180
Hymenophyllum - 175
Hymenops perspicillata - 107
HYMENOPTERA - 182
Hyperodon planifrons - 215
Hypochoeris meyeniana - 155
Hypoedaleus guttatus - 30

I

IBIS, Bare-faced - 75, **85**
IBIS, Black-faced - 180, **183**, 204
IBIS, Baff-necked - 80
IBIS, Plumbeous - 80
IBIS, Puna - 160
IBIS, White-faced - 75, **107**, 107
ICHNEUMONIDIS - 182
Ictinia plumbea - 33, 53
Iguanidae - 67
Ilex argentina - 49
Ilex paraguariensis - 22
Illex argentinus - 138
ILUCA - 156
INCIENSO - 21, 87
INGA - 41, 80, 92
Inga marginata - 41
Inga uruguensis - 41, 80, 92
IRARA - 29
Irenomys tarsalis - 176
IRIS - 175
IROS - 150, 160, **162**, 163
ITIN - 63
Ixobrychus involucris - 105

J

Jabiru mycteria - 75, **85**
JACAMAR, Rufous-tailed - 33
Jacana jacana - **54**, 75
JACANA, Wattled - **54**, 75
JACARANDA - 49, 83
Jacaranda mimosifolia - 49, 83
Jacaratia dodecaphylla - 27, 43
JAGUAR - **24**, 29, 44, 55,56, 64, 71, 83, 98
JAGUARUNDI - 29, 89, 114, 119
JARILLA MACHO - 113, 120
JARILLAL - 113
JASMIN DE CHILE - 50
JAY, Plush-crested - 44, 53, 92
Jochina rhombifolia - 63, 87, 113, 119
Juglaus australis - 49, 50
JUME - 114, 124
Juncus - 156
Juncus acutus - 109
Juncus lesearii - 124
Junellia ligustrina - 124
Junellia seriophioides - 156
Junellia tridens - 123, 124

K

KELP - 137
KESTREL, American - 114, 127, 180
KINGBIRD, Tropical - 102
KING-CRAB, Fuegan - 137
KINGFISHER, Amazon - 43
KINGFISHER, Green - 43
KINGFISHER, Ringed- 43, **81**, 195
KITE, Plumteous - 33, 53
KITE, Snail - 78, 107
KITE, Swallow-tailed - 33, 53
KITE, White-tailed - 78
Knipolegus aterrimus - 117
KNOT, Red - 138
KONGOY 0 176
Krameria iluca - 156
KRILL, Lobster - 137

L

LADY'S SLIPPER - 173
Lagenorhynchus australis - 137
Lagenorhynchus obscurus - 137
Lagidium viscacia - 127, 131

Lagidium wolffsohni - 199
Lagostomus maximus - 64, 89, **95**, 98, 114
LAHUAN - 194
Lama glama- 156
Lama guanicoe - 71, 89, 124
Lama pacos - 156
LAMPAYA - 156
Lampaya schickendantzii - 156
LAPACHO, Black - 71
LAPACHO NEGRO - 26
LAPACHO, Pink - 49
LAPACHO, Yellow - 26, 49
LAPWING, Andean - 160
LAPWING, Southern - 101, 180, **192**
LARREA - 114
Larrea cuneifolia - 113, 120
Larrea divaricata - 87, 113, 120
Larrea nitida - 113
Larrea tridentata - 113
Larus atlanticus - 138
Larus dominicanus - 138, 148, 179, 216
Larus maculipennis - 138
Larus pipixcan - 71
Lasiurus borealis - 55
Lasiurus varius - 176
Lasiurus ega - 55, 79
LATA - 113
LATA DE POBRE - 49
Laterallus melanophaius - 44
Laurantaceae - 63
LAUREL - 175
LAUREL AMARILLO - 21
LAUREL DE LA FALDA - 49, 50
LAUREL NEGRO - 21
LAUREL, River - 39, 80
LAUREL, White - 39, 80
Laurelia - 169
Laurelia philippiana - 175
LECHERON - 80
Leguminosae - 87, 163
Leiosaurus - 104
Leiosaurus belli - 128
Leiosaurus paronae - 67
Lemna spp. - 104
LEÑA AMARILLA - 163
LEÑA DURA - 207
LENGA - 170, 173, 175, 176, **178**, **189**, 199, 204, 207
Lepechinia graveolens - 50
Lepidobatrachus laevis - 63
Lepidobatrachus llanensis - 63
Lepidosiren paradoxa - 73
Leporinus obtusidens - 83
Leptasthenura aegithaloides - 90, 180
Leptodactylidae - 169, 180
Leptodactylus bufonius - 63
Leptodactylus gracilis - 102
Leptodactylus laberinthicus - 34
Leptodactylus laticeps - 63
Leptodactylus latinasus - 102
Leptodactylus mystacinus - 73, 102
Leptonychotes weddelli - 215
Leptotila verreauxi - 67, 92
Lepus c.europaeus - 89, 184
Lessonia spp. - 137
Lessonia rufa - 117
Lestodephis halli - 127
Leucippa - 137
Leucochloris albicollis - 33
Leucophaeus scoresbii - 138, 148
Leurocyclus - 137
Lebinia - 137
LICHENS - 163, **205**, 211, **212**
Limosa haemastica - 111, 138
LIMPET - 137
LINGUE - 175
Liolaemus spp. - 119, 128, 160, 180
Liolaemus chacoensis - 67
Liolaemus huacahuasicus - 164
Liolaemus multiformis - 160

Liolaemus multimaculatus - 104
Liolaemus wiegmanni - 104
LION-LIZARD - 119, 128
Liophis anomalus - 102
Liophis miliaris- 55
Liophis occipitalis - 55
Lithodes antarcticus - 137
Lithraea molleoides - 95
Lithraea ternifolia - 68
LIZARDS - 44, 89, 102, 104, 160
LIZARD, Legless - **98**
LLAMA - 156, 163
LLAO-LLAO - 173, **207**
Lobelia tupa - 175
Lobodon carcinophagus - 215, **216**
Loligo spp. - 138
Lomatia - 169
Lomatia ferruginea - 175
Lomatia hinsuta - 173
Lonchocarpus leucanthus - 21
Lonchocarpus muehlbergianus - 21
Lohonetta specularioides - 131
Lophortyx californicus - 184
Lophosoria quadripartitum - 175
Lophotus vitulus - 182
Loricaria spp. - 83
LORO BLANCO - 21
Lucilia araucana - 185
Luciopimelodus pati - 83
Luehea divaricata - 21, 92
Luma apiculata - 176, **190**
LUNG, FISH - 73
Lutra felina - 209
Lutra longicaudis - 29, 55, 83, 92
Lutra provocax - 176
Lutreolina crassicaudata - 29, 55, 79, 101, 107
Luzuriaga radicans - 175
Lycium ameghinoi - 124
Lyncodon patagonicus - 114
Lysapsus mantidactylus - 107
Lystrophis dorbignyi - 98, 102

M

MACAW, Golden-collared - 50
MACAW, Military - **54**
Mackenziaena leachii - 30
Mackenziaena severa - 30
MACKERELL - 137
Macrobrachium borelli - 83
Macrocystis pyrifera - 137
Macronectes giganteus - 137, 216
Macruronus magellanicus - 137
MACRECITA - 107
MAILLICO - 185
MAITEN - 173, 194
MAITIN - 49
MALASPINA - 124
Malaxis padilliana - 50
MANAKIN, Baud-tailed - 33
MANAKIN, Swallow-tailed - 33
Mandevilla laxa - 50
MANDIOCA BRAVA - 26
MANED WOLF - **70**
Manettia cordifolia - **42**
MANGURUYU - 83
Manihot flavellifolia - 26
MANIU HEMBRA - 175
MANIU MACHO - 175
MARA - 89, 98, 114, 120, 124, **125**
MARA, Woodland - 64, 71, 120
Marcia exalbida - 137
MARGAY - 29, 55
MARIA PRETA - 21
MARMELERO - 71
Marmosa spp. - **51**
Marmosa cinerea - 29
Marmosa constantiae - 55

MARSHBIRD, Brown and yellow - 76, 107
MARSHBIRD, Brown-headed - 107
MARSHBIRD, Yellow rumped - 76
MARSUPIAL, Patagonian - 127, **211**
MATA LAGUNA - 124
MATA NEGRA - 123, 124
MATA SEBO - 113
MATAOJO - 80, **91**
MATO - 49
Maytenus - 63
Maytenus boaria - 173
Maytenus magellanica - 207
Mazama americana - 27, 55
Mazama gouazoubira - 27, 55, 64, 95
Mazama rufina - 27
MBOREVI-CAA - 41
MEADOWLARK, Long-tailed - 180
Megaptera novaeangliae - 215
Melanerpes cactorum - 64
Melanerpes candidus - 90
Melanerpes flavirostris - 44
Melanodera melanodera - 128
Melanodera xanthogramma - 128, 186
Melanophryniscus rubriventris - 55
Melanophryniscus stelzneri - 102
Membracidae - 34
Merganetta armata - **177**, 178
MERGANSER, Brazilian - 41
Mergus octosetaceus - 41
Merluccius australis - 137
Merluccius merluccius - 137
MERO - 137
Merostachys clausseni - 22, 43
Mesoplodon sp. - 144
Metachirus nudicaudata - 29
Metamorpha stelenes - **28**
Meteoropsis onusta - 50
Metriopelia morenoi - 163
MICHAY - 173
Micrastur ruficollis - 34
Micrastur semitorquatus - 34
Microcavia australis - 114, 127
MICROHYLIDS - 63
Micromesisitius australis - 137
Micropalama himantopus - 107
Micrurus sp. - **32**
Micrurus corallinus - 34
Micrurus frontalis - 34, 119
Migidae - 25
Miltonia flavescens - 25
Milvago chimango - 78, 101, 114
Mimodromius crinum - 182
Mimulus luteus - 175
Mimus dorsalis - 163
Mimus saturninus - 92
MINER, Common - 90, 128
MINER, Puna - 160
MINER, Slender-billed - 160
MINK - 184, 197
Mirounga leonina - 137, 143, 215
Misochocyttarus lules - 55
Misochocyttarus lilae- 119
Misodendrum - 173
MISTOL - 63
MITES - 212
MOCKINGBIRD, Brown-backed - 163
MOCKINGBIRD, Chalk-biowed - 92
MOCORACA - 156
Moenkhansia sanctae-filomenae - 72
MOJARRA BRONCEADA - 128
MOJARRA DESNUDA - 131
MOLLE - 87, 92, **115**, 124, 132

MOLLE DE BEBER - 95
Molossus ater - 798
MONCHOLO - 83
MONITO DE MONTE - 170
MONJITA, Dominican - 67
MONJITA, Gray - 67
MONJITA, White - 67, 119
MONKEY, Black Howler - 29, 79, 83, **220**
MONKEY, Brown Capuchin - **22**, 29, 55
MONKEY, Night - 79
MONKEY, Red- Howler - 29
MONKEY, Puzzle - 170
MONKEY'S, Comb- 72
Monodelphis dimidiata - 101
Monodelphis henseli - 29
MONTE NEGRO - 113, 128
Monttea aphylla - 113
MORA AMARILLA - 49
MORA BLANCA - 43
Morphidae - 34
Morpho achilles - 34
Morpho aega - 34
Morpho anaxibia - **28**, 34
Morpho catenarius - 34, 41
MOSSES - 211, **212**
MOTMOT, Blue-crowned - 53
MOTMOT, Rufous - 33
MOUSE, Leaf-eared - 55, 89, 127
MOUSE, Paint-nosed - 167
MOUSE, Puna - 159
Mulinum spinosum - 123
MUÑA-MUÑA - 156
Munida gregaria - 137
Muscisaxicola - 128
Muscisaxicola juninensis - 160
Muscisaxicola macloviana - **189**
MUSKRAT - 182
MUSSEL - 137
Mustela vison - 184
MUTISIA - 173
Mutisia campanulata - 22
Mutisia decurrens - 173
Mutisia retusa - 173
Mycteria americana - 75, **85**
Myiarchus tuberculifer - 53
Myioborus brunniceps - 53
Myiopagis caniceps - 53
Myopsitta monachus - 67, 90
Myocastor coypus - 76, 92, 107
Myotis spp. - 114
Myotis chiloensis - 176
Myrceugenia exsucca - 176
Myrcia ramulosa - 92
Myriophyllum - 132, **134**
Myriophyllum elatinoides - 131, **134**
Myrmecophaga tridactila - 64
Myrocarpus frondosus - 21
Myroxolon peruiferum - 49
Myrrhinium rubiflorum- 49
Myrtacere - 92, 170
Mystaceti - 215
Mytilus magellanicus - 137

N

Nanorchestes antarticus - 212
Nardophyllum armatum - 156
Nassauvia axillaris - 132, 163
Nassauvia glomerulosa - 123
NASTURTIUM, Ground-creeping - 163
Nasua nasua - 29, 55
Nectandra falcifolia - 39, 80, 92
Nectandra lanceolata - 21
Nectandra saligna - 21
NEGRITO, Rufous-backed - 117
Nemosia pileata - 50
NENEO - 123, 124,132, 173,185
Neotomys ebriosus - 167

Neoxolmis rufiventris - 128
Netta peposaca - 104
NETTLE - 156
NETTLE, Giant - 22, 50
NIGHT-HERON, Black-crowned - 105, 178, **183**
NIGHTJAR, Rufous - 55
NIGHTJAR, Scissov-tailed - 79
Noctilio labialis - 30, 79
Noctilio leporinus - 30, 79, 83
Notharchus macrorhynchus - 33
Nothofagus spp. - 169, 170, 182, **207**, 211
Nothofagus alpina - 175, 188
Nothofagus antarctica - 173, **175, 189**
Nothofagus betuloides - 176
Nothofagus dombeyi - 173
Nothofagus obliqua - 175, 188
Nothofagus pumilio - 170, 173, **189**
Nothoprocta cinnerascens - 64, 114
Nothoprocta ornata - 160
Nothura darwinii - 114
NOTHURA, Darwin's - 114
Nothura maculosa- 101
Notiochelidon cyanoleuca - 43
Notothenia - 138
Notothenia rossii - 216
Nototheniidae - 138
Nototheniformes - 216
NOTRO - 173
NUDIBRANCHS - 137
Numenius borealis - 107
Nuncia - 169
Nyctibeus griseus - 55
Nycticorax nycticorax - 105, 179, **183**
Nymphalidae - 36
Nystalus chacuru - 33

Ñ

ÑANDUBAY - 80, 87, 92
ÑIRE - 173, **175**, 176, **178, 189**, 199, 207

O

Oceanites oceanicus - 216
OCELOT - 29, **52**
Ochromonas - 211
Ochthoeca parvirostris - 180
Ocotea acutifolia - 39, 80
Ocotea puberula - 21
Octodontomys gliroides - 159
Octomys mimax - 163
OCTOPUS - 137
Odontesthes bonariensis - 107
Odontesthes microlepidotus - 181
Odontoceti - 215
Odontophorus capueira - 30
Odontophrynus americanus - 107
Odontophrynus occidentalis - 117
OLD MAN'S BEARD - 50, 173, **190**
Oligorizomys - 176
Oligorizomys longicaudatus - 176
Olyra latifolia - 22
OMBU - 87
Ommatophoca rossii - 215
Oncidium - 25
Oncidium viperinum - 50
Ondatra zybethica - 182
Ophioglossum vulgatum - 175
Ophioides spp. - **98**
OPOSSUM, Bare-tailed - 29
OPOSSUM, Black-eared - 29
OPOSSUM, Four-eyed - 29, **70**, 79
OPOSSUM, Mouse - 55
OPOSSUM, Murine - 29, **51**, 55, 79, 115, 223

OPOSSUM, Rat-tailed - 29
OPOSSUM, Short-tailed - 101
OPOSSUM, White-bellied - 29
OPOSSUM, White-eared - 79, 101, **103**, 114
Opuntia spp. - 113
Opuntia puelchiana - 119
Opuntia quimilo - 63, 64
ORCA - 137, **141**, 144, 215, **216**
ORCHIDS - 25, 175
ORCHIDS, Ground - 87
ORCHID, Puna - 156
Orcinas orca - 137, 215, **216**
OREGANO - 124
Oreopholus ruficollis - 102, 128
Oreopolus glacialis - 185
Oreotrochilus estella - 160, **161**
Oreotrochilus leucopleurus - 160, 186
Ortalis canicollis - 56, 64
Oryctolagus cuniculus - 182
Oryzomys longicaudatus - 55
Ostrea puelchana - 137
Otaria byronia - 137, 138, **141**
OTTER, Andean - 182
OTTER, Giant - 29, 41, 83
OTTER, Paran - 29, 55, 56, 80, 83, 92, 223
OTTER, Sea - 209
Ourisia - 170
Ourisia alpina - 185
Ovalipes punctatus - 137
Ovidia pillopillo - 175
OWL, Barn - 127, 180
OWL, Burrowing - **89**, 101, 127, 160
OWL, Great Horned - 127, 199
OWL, Rufous-legged - 79
OWL, Short-eared - 79, 101, 127
OWL, Spectacled - 55
OWL, Stygian - 55
Oxalis - 97
Oxalis valdiviensis - 173
Oxidores kneri - 83
Oxypeltus quadrispinosus - 182
Oxyura j.ferruginea - **130**, 132
Oyura vittata - 179
OYSTER - 137
OYSTERCATCHER, Blackish - 138, 148, 207
OYSTERCATCHER, Common - 138, 148, 170
OYSTERCATCHER, Magellanic - 131, 138, **170**, 207
Ozotocerus bezoarticus - 768, 95, 98

P

PACA - 27, **32**, 44
PACARA - 49
Pachyramphus polychropterus - 53
Pachyramphus viridis - 53
Pacysiphonaria lessoni - 137
PACU - 83
PACURI - 27
Pagodroma nivea - 216
Pagurus comptus - 137
PAHUELDIN - 175
PAJA BOBA - 72
PAJA COLORADA - 102
Palaemonetes argentinus - 83
Palaina - 55
PALM, Caranday - 79, 87, **92, 95**
PALM, Pind; - 21, **26**, **43**, 80
PALM, Spiny Palmate - 63
PALM, White - 72, **77**, 79
PALM, Yatay - 80, 87, 90
PALMITO EDIBLE PALM - 39, 44

PALO AMARILLO - 49
PALO BARROSO - 49
PALO BLANCO - 49
PALO BOBO - 50, 80
PALO BORRACHO - 43
PALO BORRACHO, Pink - 71
PALO BORRACHO, Yellow - 63
PALO BRILLADOR - 49
PALO LATA - 49
PALO LUZ - 50
PALO POLVORA - 26
PALO ROSA - 39, 44
PALO SAN ANTONIO - 49
PALO SANTO - 71
PALOMETA - 83, 137
PAMPANITO - 137
PANGUE - 175
Panicum elephantipes - 83
Panicum fasciculatum - 83
Panicum prionitis - 72, 83
Panicum racemosum - 104
Panicum urvilleanum - 104
Panthera onca - 29, 55, 64
Pantodactylus schreibersi - 55
Papilio anchisiades - 36
Papilio hectorides - 36
Papilio lycophron - 36
Papilio scamander - 55
Papilio thoas - 36
Papilionidae - 36
Parabuteo unicinctus - 78
PARAKEET, Austral - 179, 204
PARAKEET, Blue-Crowned - 67
PARAKEET, Green-Cheeked - 50
PARAKEET, Mitred - 50
PARAKEET, Monk - 67, 90, 104, 109
PARAKEET, Reddish-bellied - 33, 43
PARAKEET, White-eyed - 33, 67
Paralomis granulosa - 137
Parapiptadenia excelsa - 49, 83
Parapiptadenia rigida - 21
Parastrephia lepidophylla - 156
Parastrephia phylicaefornis - 156
Parmelia cirrhata - 50
Paroaria coronata - 67, 92
PARROT, Alder - 50
PARROT, Burrowing - **115**, 117, 120
PARROT, Red-capped - 30
PARROT, Scaly-headed - 33
PARROT, Turquoise-fronted - 67
PARROTLET, Blue-winged - 33
Paspalum intermedius - 72
Paspalum lilloi - 43
Paspalum quadrifarium - 102
Paspalum repens - 83
Pastrania - 119
Patagonula americana - 21, 49
PATAGUA - 176, 194
PATI - 83
Paulicea lutkeni -83
PAWPAW (native) - 27, 43
PECCARY, Collared - 27, 44, 55, 64, **67**, 120
PECCARY, White-lipped - 27, 29, 44, 55, 64, 67
Pectinibruchus - 119
Pectis sessiliflora - 113
Pediolagus salinicola - 64, 71
PEHUAJO - 72
PEJE - 63
PEHUEN - **197**
PEJERREY - 107
PEJERREY Patagonian - 181
PELADILLAS - 181
Peltarion spinulosum - 137
Peltophorum dubium - 21, 26, 80
Penelope dabbenei - 50

Penelope obscura- 35, 50, 85
Penelope superciliaris - 33
PENGUIN, Adelie - 215
PENGUIN, Chinstrap - 215
PENGUIN, Emperor - 215
PENGUIN, Gentoo - 209, 215
PENGUIN, Macaroni - 215
PENGUIN, Magellanic - 138, 147, **151**
PENGUIN, Rockhopper - 209
Pennistum chilense - 155
Peperomia - 22
PERCH, Native - 181
Percychthys spp. - 181
Percychthys colhuapiensis - 181
Percychthys trucha - 181
Percychthys vinciguerrai - 181
Pernethya - 170
Pernethya mucronata - 173
Pernethya pumila - 185
Perona signata - 137
Persea lingne - 175
PETERIBI - 21
PETREL - 209, 216
PETREL, Antarctic - 216
PETREL, Giant - 137, 216
PETREL, Snow - 216
PETREL, White-chinned - 137
PEWEE, Tropical - 53
Phacellodomus ruber- 64
Phacellodomus striaticollis - 64
Phaetornis eurynome - 33
Phalacrocorax albiventer - 138, 153
Phalacrocorax atriceps - 138, 153, 193, 216
Phalacrocorax bongaivilli - 138, **147**
Phalacrocorax gaimardi - 138, **147**, 151
Phalacrocorax magellanicus - 138, **147**, 151
Phalacrocorax olivaceus - 138, 178
PHALAROPE, Wilson's - 216
Phalcobaenus albogullaris - 180
Phalcobaenus megalopterus - 160
Pharus glober - 22
Phegornis mitchelli - 160
Philander opossum - 29, **70**, 79
PHILODENDRON - **23**, 25
Philodendron bipinnatifidum - **23**, 25
Philodryas baroni- 67
Phimosus infuscatus - 75, **85**
Phioceramis jannarii - 137
Phlebodium aureum - 50
Phleocryptes melanops - 107
Phoebe porphyria - 49
Phoebetria palpebrata - 216
Phoebis cipris - 36
Phoenicoparrus andinus - 71, 160
Phoenicoparrus jamesi - 160
Phoenicopterus chilensis - 71, 105, 160, **165**
Phoneutria nigriventer - 34
Phrygilanthus acutifolius - 113
Phrygilus alaudinus - 160
Phrygilus atriceps - 160
Phrygilus dorsalis - 160
Phrygilus fruticeti - 117, 128
Phrygilus gayi - 128
Phrygilus patagonicus - 180
Phrynohyas venulosa - 34
Phrynops hilarii - 73
Phycoides claudina - 55
Phycoides telutosa - 55
Phyllomedusa hypocondrialis - 63
Phyllomedusa sauvagii - 63
Phylloscartes ventralis - 53
Phyllostylon ramnoides - 49
Phyllotis darwinii - 55, 127

235

Phyllotis osilae - 159
Phyllotis sublimis - 159
Phymaturus flagellifer - 163
Physalaemus biligonigerus - 73
Phytolacca dioica - 87
Phytotoma rara - 180
Phytotoma rutila - 90
Piaya cayana - 44, 92
Pieridae - 36
PICHANA - 113
PICHI, Patagonian - 114, 127
PICHI, Wailing - **89**
Picidae - 64
Picoides lignarius - 179
PIGEONS - 90
PIGEON, Picazuro - 67
Pilgerodendron uviferum - 175
PIL PIL VOQUI - 175
Piltotrichella versicolor - 50
Pimelodus albicans - 83
Pimelodus clarias - 83
PINDO-PALM - 21, **26**, 43, 80
PINGO-PINGO - 155, 156
Pinguicula antarctica - 185
Pinguipes fasciatus - 138
Pinguipes somnambula - 138
PINO DEL CERRO - 50
PINTAIL, Brown - 104, 131, 223
Pinopsitta pileata - 33
Pionus maximilaini - 33
Piper hieronymi - 49
Piper tucumanum - 49
PIPITS - 101
Pipra fasciicauda - 33
Piptadenia macrocarpa - 83
Piptocarpha sellowi - 22
Piptochaetium - 97
PIQUILLIN - 87, 119, 128
PIRAÑAS - 83
Piranga flava - 50
PIRAYU - 83
Pisonia aculeata - 22
Pisonia zapallo - 71
Pistia stratiotes - 72
PIT-VIPERS - 34, 68, **92**, 119
Pithecellobium scalare - 71
Pithecoctenium cynanchoides - 72
Plagiostelum - 182
Plantago bismarckii - 102
PLANTCUTTER, Rufous-tailed - 180
PLANTCUTTER, White-tipped - 90
Plataxanthus patagonicus - 137
Plataypsaris rufus - 53
Plecostomus spp. - 83
Plectocarpa rougessi - 113
Plectocarpa tetracantha - 113
Plegadis chihi - 75, **107**, 107
Plegadis ridgwayi - 160
Pleoticus muelleri - 137
Pleurodema - 180
Pleurodema bibroni - 180
Pleurodema bufonina - 180
Pleurodema kriegi - 68
Pleurodema nebulosa - 117
PLOVER, Golden - 107
PLOVER, Magellanic - 131
PLOVER, Puna - 160
PLOVER, Two-banded - 131, **134**, 138
PLOVERCREST, Black-breasted - 33
Pluvialis dominicus - 107
Pluvianellus socialis - 131
Pneumatophorus japonicus - 137
Poa - 123, 163, 185
Poa flabellata - 209
Poa gymnantha - 163
Poa holciformes - 163
Poa lanuginosa - 104
Poa lilloi - 163
Poa muñozensis - 163
POCHARD, Rosy-billed - 104
Podager nacunda - 79
Podiceps dominicus - 44, 75

Podiceps gallardoi - 131, **134**
Podiceps major - 131, 178
Podiceps occipitalis - 131
Podiceps rolland - 75, **130**, 178
Podilymbus podiceps - 75, 178
PODOCARPS - 56
Podocarpus nubigenus - 175
Podocarpus parlatorei - 50
Pogonomyrmex - 63
POKEWEED - 87
Polipodiaceae - 50
POLLOCK - 137
Polybia ruficeps - 63
Polyborus plancus - **74**, 78, 114
Polychrus acutirostris - 67
Polydolops - 211
Polylepis australis - 50
Polylepis tomentella - 156, **161**
Polypodium - 22
Polystes buyssoni - 119
Polystes flavogullatus - 55
Polythysana rubrescens - 182
Polytrichum - 211
Pomacea spp. - 78
Pompilidae - 34
POMPILIDS - 182
Poospiza torquata - 117
PORCUPINE, Tree - **51**
Porzana albicollis - 44
Potamogeton pectinatus - 132
Potamotrygon brachyurus - 83
Potamotrygon motoro - 83
POTATO, Native - 155
POTATO, Common - 55
Potostema comata - 43
Potostemaceae - 43
Pouteria gardneriana - 39
Pouteria salicifolia - 80, **91**
PRAWNS - 137
Prepona pheridamas - 36
PRIMAVERA - 185
Primula farinosa - 185
Priodontes maximus - 64
Pristidactylus casuhatiensis - 102
Procellaria aequinoctialis - 137
Procellariiformes - 215
Prochilodus platensis - 83
Proctotretus pectinatus - 104
Procyon cancrivorus - 29, 55
Prosopis - 87, 120
Prosopis affinis - 80, 87
Prosopis alba - 63, 87, 113
Prosopis alpataco - 87, 128
Prosopis caldenia - 87, **118**, 119
Prosopis denudans - 124
Prosopis ferox - 155
Prosopis flexuosa - 87, 113
Prosopis kuntzei - 63
Prosopis nigra - 63, 87
Prosopis ruscifolia - 68
Prosopis torquata - 113
Proteaceae - 173
Prunus tucumanensis - 50
Psamnobatis scobina - 138
Pseudochinus magellanicus - 137
Pseudocaryophyllus guili - 49
Pseudoleistes guirahuro - 76
Pseudoleistes virescens - 76, 107
Pseudopanax laetevirens - 176
Pseudoplatystoma coruscans - 83
Pseudoplatystoma fasciatum - 83
Pseudorhombus isosceles - 138
Pseudoseisura gutturalils - 90, 117
Pseudoseisura lophotes - 90, 92, 119
Psidium guajava - 27
Pteris deflexa - 49
Pterocactus tuberosus - 114

Pteroclidae - 124
Pterocnemia pennata - 124, **129**, 160, 167, 200
Pterotocos tarwii - 179
Pteroglossus castanotis - 33
Pteronura brasiliensis - 29, 41
PUDU - 176, **184**, 188, 193, 197, 223
Pudu pudu - 176
PUFFBIRD, Eared - 33
PUFFBIRD, White-necked - 33
PUFFING PID - 137
Puffinus spp. 137
Pulsatrix perspicillata - 55
PUMA - 29, 64, 87, 98, 114, 120, 127, 159, 176
Puya spp. - 120
PUYA-PUYA - 156
Pygarrhichas albogularis - 175, 179
Pygidium alterum - 160
Pygidium boylei - 160
Pygidium spegazzini - 160
PYGMY-OWL, Austral - 180, 204
Pygoscelis adeliae - 215
Pygoscelis papua - 209
Pyrocephalus rubinus - 67, 102
Pyroderus scutatus - 33
Pyrope pyrope - 119, 180, **189**
Pyrostegia venusta - 22
Pyrrulina australis - 72
Pyrrhura frontalis - 33, 43
Pyrrhura molinae - 50

Q

QUAIL, Californian - 184, 193
QUEBRACHO, Chaco Red - 71, 79, 87
QUEBRACHO, Horco - 68
QUEBRACHO, Santiago Red - 61
QUEBRACHO, White - 63, 87, 92, 120
QUEÑOA - 50, 156, **161**, 164
QUILENBAI - 123, 124
QUIMIL - 63
QUIMILERO - 64
QUINA - 49
QUINTRAL - 173, 179

R

RABBIT - 55, 79, 182, 209
RABO-ITA - 21
RABO-MACACO - 21
RACCOON, Crab-eating - 29, 55, 79
RADAL - 173
Radiodiscus - 55
RAIL, Blackish - 44
Rallus nigricans - 44
Rallus sanguinolentus - 76, **89**
RAMIO TUCUMANO - 49
RAMO - 49
Ramphastos dicolorus - **26**, 33, 44, 80
Ramphastos toco - 33, 44, 53, 80
Rangifer tarandus - 182
Rapanea ferruginea - 49
Rapanea laetevirens - 49
Rapanea lorentziana - 21
RAT, Chinchilla - 159
RAT, Coney - 127
RAT, Large-eared - 159
RAT, Rock - 159
RATTLESNAKE - 34, 119
RAULI - 175, **170**, 188
RAY, Electric - 138
RAYADITO, Thron-tailed - 179, 204
Recurvirostra andina - 160

REDSTART, Brown-capped - 53
REINDEER - 182
Reithrodon - 98, 176
Reithrodon auritus - 89, 127
Reussia subovata - 72
Rhaphidonema - 211
Rhea americana - 64, 90, 98, **108**, 114
RHEA Greater - 64, 89, 90, 98, **108**, 111, 114, 120
RHEA Lesser or Darwin's - 124, **129**, 160, 167, 200
RHEA Puna - 160
Rheidae - 64
Rheedia brasiliensis - 27
Rhincalanus - 212
Rhinocrypta lanceolata - 117
Rhinocriptids - 179
Rhinoderma darwinii - 180
Rhipidomys leucodactylus - 55
Rhipsalis - 22
Rhipsalis aculeata - 63
Rhipsalis lorentziana - 50
Rhodoficeae - 137
Rhyephens maillei - 182
Rhynchotus rufescens - 80, 90, 101
Rigalites ischigualastianus - 120
RIGHTWHALE, Southern - 137, 141, **143**
ROBALO - 138, 209
ROBLE - 49
ROBLE PELLIN - 175, **178**, 188
RODAJILLO - 113
Rosa moschata - 182
Rosaceae - 155
ROSE, Briar - 182, 194, 197
ROSITA - 156
Rosthramus sociabilis - 78, 107
Rothschildia jacobae - 37
Roupala cataractarum - 39
Rubiaceae - **42**
Ruprechtia laxiflora - 71
RUSHBIRD, Wren-like - 107
RUSHTYRANT, Many-coloured - 107
Rynchops nigra - 83
Rynchosaurus - 120

S

SABALO - 83
SACHA UVA - 50
Salicornia - 109, 114
Salix humboldtiana - 50, 80, 92, 114
Salminus maxillosus - 83
Salmo fario - 182
Salmo gardneri - 182
Salmo salar sebago - 182
SALMON, Laud-locked - 182, 193
SALMON, False Sea - 138
Saltator aurantiirostris - 67, 117
Saltator caerulescens - 67
SALTATOR, Golden-billed - 67, 117
SALTATOR, Grayish - 67
Saltatricula multicolor - 117
Salvelinus fontinalis - 182
SALVIA, White - 50
Salvinia spp. - 104
Salvinia auriculata - 72
Sambucus australis - 87
Sambucus peruviana - 50
SANDERLING - 138
SANDPIPER, Band's - 138
SANDPIPER, Buff-breasted - 107
SANDPIPER, Solitary - 80
SANDPIPER, Spotted - 80
SANDPIPER, Stilt - 107
SANDPIPER, White-rumped - 138
SANDPIPER-PLOVER, Diademed - 160

SANGRE DE DRAGO - 26, 80
Sapium haematospermum - 80
Sappho sparganura - 95
Sarcoramphus papa - 53, **54**
Sarda sarda - 137
SARDINE, Fuegan - 137
Sarkidiornis melanotos - 75
Satureja parviflora - 15
Saturnidae - 37
SATURNIDS - 182
Saxgothaea conspicua - 175
Scaphidura oryzivora - 43
Scaphonyx sanjuanensis - 120
Scapteromys aquaticus - 76
Scarabeidae - 34
Scelorchilus rubecula - 179
Schinopsis balansae - 71
Schinopsis haenkeana - 68
Schinopsis quebracho-colorado - 61
Schinus fasciculatus - 87
Schinus longifolia - 87
Schinus polygamus - **115**, 124
Scirpus - 156
Scirpus californicus - 104
Sciurus ignitus - 55
Scopaneidae - 138
SCORPIONFISH - 138
SCREAMER, Southern - 80, 104, **111**
Scutalus - 55
Scutia buxifolia - 87
Scytalopus magellanicus - 179
Synax fuscovaria - 34
Synax nasica - 34, 73
SEAL, Gabeater - 215, **216**
SEAL, Leopard - 215
SEAL, Ross - 215
SEAL, Southern Elephant - 137, 138, 143, 153, 215
SEAL, Weddell - 215
SEA-LION, South American - 137, 138, **141**, 141 153
SEA-URCHINS - 137
Sebastes aculeatus - 138
SEDGE - 109, 223
SEDGE, Giant - 72
SEEDEATERS - 76
SEEDEATER, Double-collared - 101
SEEDSNIPE - 124, 199
SEEDSNIPE, Gray-breasted - 124, 200
SEEDSNIPE, Least - 102, 124, 199
SEEDSNIPE, White-bellied - 127, 186
SEIBO - 80
Selenidera maculirostris - 33
SENECIO - 132, 185
Senecio crassiflorus - 104
Senecio filaginoides - 124
Senecio julietti - 185
Senecio poepiggi - 185
Senecio portalessianus - 185
Senecio ventanensis - 102
Sephanoides sephanoides - 179
SERIEMAS - 64, **81**
SERIEMA, Black legged - 56, 64
SERIEMA, Red legged - 56, 64, **81**
Serjania - 22
Serpyllopsis caespitosa - 175
Serrasalmus spp. - 83
Serrasalmus nattereri - 83
SHARK - 138
SHEARWATERS - 137
SHEATHBILL, Snowy - 138, 148, **215**, **216**
SHELD-GOOSE - 102, 195
SHELD-GOOSE, Upland - 128
SHOVELLER, Red - 104, 131
SHRIKE-TYRANT, Black-billed - 160
SHRIMPS - 137
SHRIMPS, Fresh-water - 83

Sicalis lebruni - 128
Sicalis luteola - 101
Sicalis lutea - 160
Sicyos odonelli - 50
Sicyos polyacanthus - 50
SIERRA-FINCH, Band-tailed - 160
SIERRA-FINCH, Black-hooded - 160
SIERRA-FINCH, Mourning - 117, 128
SIERRA-FINCH, Patagonian - 128,180
SIERRA-FINCH, Red-bached - 160
SILK-MOTHS - 37
SISKIN, Black-chinned - 180
Sittasomus griseicaillus - 33
SKIMMER, Black - 83
SKUA - 138, 148, **151**, 151, 209
SKUA, Antarctic - 216
SKUA, South Polar - 216
SKUNK - 83
SKUNK, Hog nosed - 101, 114, 159
SKUNK, Patagonian Hog-nosed - 176, **192**
SLOTH - 59
SLOTH, Giant ground - 64
SLUG - 55
Smilax campestris - 72
SNAILS - 55
SNAIL, Apple - 78
SNAKE, Colubrid - 160
SNAKE, Coral - 34, 119
SNAKE, Grass - 102
SNAKE, Hog-nosed - **98**
SNAKE, Water - 73
SNIPE, Common - 76, **170**
SNIPE, Puna - 100
Solanaceae - 123
Solanum acauk - 155
Solanum glaucophyllum - 104
Solanum granuloso-leprosum - 26
Solanum megistracolobum - 155
Solenopsis richteri - 102
Solenopsis saevissima - 55
SOLUPE - 124
SOMBRA DE TORO - 63, 87, 95, 113, 119
Somuncuria somuncurensis - 128
Sophronitis coccinea - 25
Sorghastrum agrostoides - 72
SPARROW, Rufous - collared - 128
SPARROW, Saffron-billed - 53
Spartanoica maluroides - 101
Spartina ciliata - 104
Speothos venaticus - 29
Sphagnum - **205**
Spheniscus magellanicus - 151
Sphiggurus spinosus - 29
Sphingidae - 37
SPINETAIL, Stripe-crowned - 90
Spizaetus isidori - 53
Spizaetus ornatus - 34
Spizaetus tyrannus - 34
Spizaetus melanoleucus - 41
Spiziapteryx circumcinctus - 67, 117
SPOONBILL, Roseate - 75, **85**,105, **107**
Sporophila spp. - 76
Sporophila caerulescens - 101
Sprattus fueguinsis - 137
SQUID - 138
SQUIRREL - 55
STAR, Serpent - 137
STARFISH, Giant Magellanic - 137
STEAMER DUCK, Chubut Flightless - 148
STEAMER DUCK, Flightless - 178, 207

STEAMER DUCK, Flying - 131, 178, **199**, 200
Steirastoma marmoratum - **32**
Stephanoxis lalandi - 33
Sterna eurygnatha - 138
Sterna hirundinacea - 138
Sterna maxima - 138
Sterna trudeaui - 138
Sterna vittata - 216
Stetsonia coryne - 63
Stigmatura budytoides - 117
STILT, Black-necked - 107
STING-RAYS - 83, 138
STING-RAY, Boba - 83
STING-RAY, Overa - 83
Stipa spp. - 97, 123, **134**, 156, 163, 199, 223
Stipa caespitosa - 163
Stipa chrysophilla - 163
Stipa humilis - 123
Stipa frigida - 163
Stipa juncoides - 102
Stipa neai - 123
Stipa pampeana - 102
Stipa scirpea - 163
Stipa speciosa - 123, 163
Stipa tenuissima - 163
Stipa vaginata - 163
STORK, Jabiru -75, **85**
STORK, Mageari - 105
STORK, Wood - 75, **85**
STORM-PETREL, Wilson's - 216
STRANGER-FIG - 25, 43
Strix rufipes - 79
Stromateus brasiliensis - 137
Sturnella loyca - 180
Sturnella supucilliaris - 101
Sturnira lilium - 29, 55
Styrax subargenteus - 49
Suaeda divaricata -114, 124
SUNDEW - 176, 209
SURUBI, Spotted - 83
SURUBI, Striped - 83
Sus scrofa - 89, 95, 119, 182
SWALLOW, Blue and white - 43
SWALLOW, Chilean - 180
SWALLOW, White winged - 43
SWALLOW TANAGER - 33
SWAN, Black necked - 71, 104, 131, 132, **133**, 178, 204
SWIFT, Great Dasky - **42**, 43
Syagrus romanzoffianum - 21, **26**, 80
Sybinomorphus turgidus - 68
Sybinomorphus ventrimaculatus - 68
Sylvilagus brasiliensis - 55
Sylviorthorhynchus desmursii - 179
Synbranchus marmoratus - 73, 107
Syrigma sibilatrix -92

T

Tabebuia alba - 26
Tabebuia avellanedae - 49
Tabebuia impetiginosa - 26
Tabebuia lapacho - 49
Tachuris rubrigastra - 107
Tachycineta albiventer -43
Tachycineta leucopyga -180
Tachyeres brachypterus - 148
Tachyeres patachonicus - 131, 178, **199**, 200
Tachyeres pteneres - 178
Tachymenis scripta - 92
Tachymenis chilensis - 181
Tachymenis peruviana - 160
Tachyphonus coronatus - 33
TACUAPI - 22
TACUAREMBO - 22
TACUARUZU - 22, 43
Tadarida spp. - 114, 159
Tadarida brasiliensis - 55

TAGUA - 64
TALA -26 ,87, 92, 95, 109, 111, 113, 223
TALILLA - 50
TAMANDUA - 29, 40, 79
Tamandua tetradactyla - 29, 40, 79
TAMBOATA - 73
TANAGER, Black-goggled - 33
TANAGER, Blue and yellow - 109
TANAGER, Blue hoaded - 50
TANAGER, Green-headed - 33
TANAGER, Hepatic - 50
TANAGER, Hooded - 50
TANAGER, Magpie - 33
TANAGER, Orange-headed - 50
TANAGER, Ruby-crowned - 33
Tangara seledon - 33
TAPACULO, Andean - 179
TAPACULO, Chucao - 179, 195
TAPACULOS - 67
TAPIR - **23**, 27, 44, 55, 56, 64, **70**, 71
Tapirus terrestris - 27, 55, 64, **70**
Taraba major - 53
TARARIRA - 72, 107
TARCO - 49
TARUCA - 163, 164
Tasmacetus shepherdi - 144
TATARE - 71
Tayassu pecari - 29, 55, 64
TAYRA - 29, 55
TEAL, Cinnamon - 114, 223
TEAL, Puna - 160,223
TEAL, Ringed - 67
TEAL, Sharp-winged - 223
TEAL, Silver - 104, **111**
TEAL, Spechled - 104, 223
TEAL, Yellow billed - 131
Teidae - 67
Teius cyanogaster - 67
Teledromas fuscus - 117
Telmatobius - 160
Telmatobius barrioi - 55
Telmatobius ceirum - 55
TENTREDINIDS - 182
Tepualia stipularis - 176
TERN - 137
TERN, Antarctic - 216
TERN, Cayenne - 138
TERN, Royal - 138
TERN, Snowy-crowned - 138
TERN, South American - 138
Tersinia viridis - 33
Tessaria integrifolia - 50, 80
Tetraglochin cristatum - 155
Thalassoica antarctica - 216
Thalia geniculata - 72
Thalurania glaucopis - 33
Thamnophilus caerulescens - 53
Theristicus caudatus - 80
Theristicus melanopis - 180
Thinocoridae - 199
Thinocorus orbignyianus - 124, 199
Thinocorus rumicivorus - 102, 124, 199
Thlypopsis sordida - 50
THORNBIRD, Freckle-breasted - 64
THORNBIRD, Greater - 64
Threskiornithidae - **85**
THRUSH, Austral - 180
THRUSH, Chignanco - 95
THRUSH, Creamy-bellied - 92
THRUSH, Rufous-bellied - 92
Thylamys spp. -79
Thylamys elegans - 55
Thylamys pusilla - 114
Thylamys venusta - 55
Thyniae - 169
THYNNIDS - 182
TIGER-ANT - 34

TIGER-HERON, Rufescent - 80
Tigrisoma lineatum - 80
Tillandsia spp. - 50, 63
Tillandsia maxima - 50
Tillandsia meridionalis - 22
Tillandsia usneoides - 50
TIMBO - 21, 49
Tinamidae - 64
Tinamotis ingoufi - 124
Tinamotis pentlandii - 163
TINAMOU, Brown - 30
TINAMOU, Brushland -64, 114
TINAMOU, Elegant-crested - 89, 114, 120, 124, **125**
TINAMOU, Lillo -64
TINAMOU, Ornate - 160
TINAMOU, Pale-Spotted - 114
TINAMOU, Patagonian - 124
TINAMOU, Puna - 163
TINAMOU, Red-winged - 80,90, 101, 111
TINAMOU, Solitary - 30
TINAMOU, Spotted - 101
Tinamus solitarius - 30
TIPA - 83
TIPA BLANCA - 49
Tipuana tipa - 49, 83
TIQUE - 175, 194
TIT-SPINETAIL, Plain-mantled - 90, 180
TIT-TYRANT, Tufted - 117, 180
TITYRAS - 33
Tityra cayana - 33
Tityra inquisitor - 33
TOADS - 55, 102
TOAD, Common - **102**
TOAD, Cururú - 73
TOAD, Giant - 63
TOAD, Gray-green woodland - 180
TOAD, Sand - 117
TOLA - 156, 159
TOLA DEL RIO - 156
TOLA VACA - 156
TOLILLA - 156
Tolypeutes matacos - 64, 114
Tomodon ocellatus - 102
TORPEDO -138
TORTOISE - 89, 119, 120
Tortula - 211
TOUCAN, Red-breasted - **26**, 33, 44, 80
TOUCAN, Toco - 33, 44, 53, 80
TOUCANET, Saffron - 33
TOUCANET, Spot-billed - 33
TRAMONTANA - 156
TREE, Black Ear - 21, 49, 56, 80
TREE, Orchid - 80
TREE-DUCK, Black-bellied - 753
TREE-DUCK, Fulvous - 75, 104
TREE-DUCK, White-Faced - 75, 104
TREE-FERNS - 59
TREE-FROGS - 34, 160
TREE-FROG, Hylid - 73, 107
TREE-PORCUPINE - 29
TREERUNNER, White-throated - 175, 179
Trema micrantha - 26
Tremarctos ornatus - 59
Trevoa patagonica - 124
Trichocereus spp. - **167**
Trichocereus pascana - 156, 163
Trichocereus poco - 156
Trichocereus terscheckii - 163
Trichodactylus sp. - 83
Trichopteris atravirens - 21
Trichotrauphis melanops - 33
Tringa flavipes - 107
Tringa macularia - 80
Tringa melanoleuca - 107
Tringa solitaria - 80
Tristerix tetraudrus - 173

Trithrinax campestris - 63, 87, **92**, 95
Trochilidae - 38
TROGON, Blue-crowned - 53
Trogon curucui - 53
Trogon rufus - 33
Trogon surucura - 33
Tropedaceae - 50
Tropaeolum pentaphyllum - 87
Tropaeolum polyphyllum - 163
Tropidurus - 55
Tropidurus catyalanensis - 55
Tropidurus spinulosus - 67
TROUT, Brook - 182
TROUT, Brown - 182
TROUT, Rainbow - 182
Tryngites subruficollis - 107
TUCOTUCO - 89, 98, 114, 127, 159, 167
TULE - 164
Tupinambis rufescens - 67, 119
Tupinambis teguixin - 44, 92
TURCO - 138
Turdus amaurochalinus - 92
Turdus chiguanco - 95
Turdus falcklandii - 180
Turdus rufiventris - 92
Tursiops truncatus - 137
TURTLE, Painted - 92
TURTLE, Water - 73
TUSCA - 68, 83
Typha spp. - 104, 223
Typha dominguensis - 72
Tyrannidae - 102, 128
TYRANNULET, Mottle-cheeked -53
Tyrannus melancholicus - 102
Tyrannus savana - 67, 102
TYRANTS - 53, 67, 90
TYRANT, Chocolate-vented - 128
TYRANT, Cock-tailed - 76
TYRANT, Long-tailed - 33
TYRANT, Patagonian - 180
TYRANT, Spectacled - 107
TYRANT, Strange-tailed - 76
TYRANT, Streana-tailed - 76
TYRANT, Yellow-browed - 109
Tyto alba - 127, 180

U

UCLE - 63
UKUMARI - 59
Ulmaceae - 50, 87
ULMO - 194
Umbellifera - 123
Umbilicaria - 211
Upucerthia andaecola - 160
Upucerthia dumetaria - 117, 128
Urera caracascana - 50
URMO - 194
Ursus arctus - 182
Urtica baccifera - 22
Urtica chamaedryoides - 156
URUNDAY - 71
Urvillea - 22
Usnea - 50, 173, 176, **190**, 211
Utricularia platensis - 104

V

Vaginulus borellianus - 55
VAMPIRE (Bat) - 29, **51**, 55
Vanellus chilensis - 101, 180, **192**
Vanellus resplendens - 160
Vanessa - 182
VERBENA - 124
Verbena seriphioides - 156
Verrucaria - 211
Veurice - 119
Victoria cruziana - 72

Vicugna vicugna - 156, **158**
VICUÑA - 156, **158**, 159, 163, 167
VINAL - 68
Viola - 185
Viola maculata - 173
VIOLET, Yellow - 173
VIRARO - 71
VISCACHA, Mountain - 127, 131, 163, 199
VISCACHA, Plains - 63, 64, 67, 89,92, **95**, 98,101, 114
VISCACHERA - 163
Vitaceae - 50
Voluta spp. - 137
VOQUI BLANCO - 194
Vultur gryphus - 71, **157**, 180
VULTURE, Black - 43, 53, 78, 180
VULTURE, King - 53, **54**, 114
VULTURE, Lesser Yellow-headed - 78
VULTURE, Trukey - 53, 78, 114

W

WAGTAIL-TYRANT, Greater - 117
WALLNUT, Southern - 49, 50, 56
WARBLERS - 67
WARBLER, Two-banded - 53
WARBLING FINCH, Ringed - 117
WASP - 63
WASPS, Social - 55
WATER BOA - 72, 73
WATER HYACINTH - 72, 83
WATER LETTUCE - 72
WATERFALL FLOWER - 185
WATER-POSSUM, Red - 29, 55, 79, 101, 107
WATER-RAT, Red - 76, 107
WATER-WEED - 131
WEEVILS - 68
Weinmannia - 170
WHALES - 215
WHALE, Arnoux's Beaked - 215
WHALE, Beaked - 144
WHALE, Bottle-nosed - 215
WHALE, Humpback - 215
WHALE, Killer - 215, **216**
WHALE, Minke - 215
WHALE, Tasman-Beacked - 144
WIDGEON, Chiloe - 130, 131
WILLOW, Native - 50, 80, 92, 95, 115, 223
WINTER'S BARK - 176, 207
WIRETAIL, DesMurs' - 179
WOLF, Maned - **70**, 76, 80
Wolfia spp. -104
WOODCREEPER, Great Rufous - 64
WOODCREEPER, Narrow-billed - 64
WOODCREEPER, Olivaceous - 33
WOODCREEPER, Plain-brown - 33
WOODCREEPER, Planalto - 33
WOODCREEPER, Scynitar-billed - 90
WOODCREEPER, White-throated - 33
WOODNYMPPH, Violet-capped - 33
WOODPECKER, Cream-backed - 64
WOODPECKER, Golden-brested - 64, 119
WOODPECKER, Helmeted - 44
WOODPECKER, Lineated - 44
WOODPECKER, Magellanic-175, 179, 195, 204
WOODPECKER, Robust - 44
WOODPECKER, Striped - 179
WOODPECKER, White - 90
WOODPECKER, White-fronted - 64
WOODPECKER, Yellow-fronted - 44
WOOD-QUAIL, Spot-winged - 30
WOOD-RAIL, Giant - 76
WOOD-RAIL, Slaty-breasted - 30
WREN-SPINETAIL, Bay-capped - 101

X

Xanthornia - 211
Xilophanes tersa - 34
Xiphocolaptes albicollis - 33
Xiphocolaptes major - 64
Xolmis cinerea - 67
Xolmis dominicana - 67
Xolmis irupero - 67, 119
Xystreuris rasile -138

Y

YAGUAPINDA - 22
YAPOK - 29
YARETTA - 123, 156
YATEVO - 22
YBIRA PERE - 21
YBIRA PYTA - 21, 26, 80
YELLOWFINCH, Grassland - 101
YELLOWFINCH, Patagonian - 128
YELLOWFINCH, Puna - 160
YELLOWLEGS, Greater - 107
YELLOWLEGS, Lesser - 107
YERBA MATE - 22
Yramea - 182
YUCHAN - 63

Z

Zaedyus pichyi - 114, 127
ZAMPA - 114, 124
ZAPALLO CASPI - 71
ZARZAPARRILLA, White - 72
Zenaida auriculata - 67, 92
Zidama angulata - 137
Zizaniopsis bonariensis - 109
Zizyphus mistol - 63
Zonibyx modestus - 31
Zonotrichia capensis - 128
ZOOPLANKTON - 212, 215
Zuccagnia punctata - 113, 120
Zygophyllaceae - 113

Photography:

Argenpress: 54 - 126; Estudio Bechis: 175 inf.; J. y J. Blassi/Incafo. 48 - 57 - 58 - 60 - 73 - 77 - 86 - 91 - 168 - 181 - 187 - 189 sup. - 191 - 195; Andrés Bosso: 221 super.; Roberto Bunge: 8; Roberto Bunge/Argenpress: 202; A. Camoyán/Incafo: 62 -201 - 202- 206 - 208; Marcelo Canevari: 23 - 31 - 45 - 51 - 67 - 89 - 92 - 100 - 106 - 110 - 118 - 192 - 198; Pablo Canevari: 32 - 92 - 102 - 106 - 118 - 158 - 161 - 165 - 197; Francisco Erize: 23 - 24 - 26 - 32 - 35 - 38 - 40 - 42 - 46 - 51 - 52 - 54 - 67 - 69 - 70 - 78 - 81 - 82 - 84 - 85 - 88 - 89 - 94 - 96 - 98 - 99 - 102 - 103 - 105 - 108 - 110 - 111 - 115 - 125 - 129 - 130 - 135 - 136 - 139 - 140 - 142 - 145 - 146 - 149 - 150 - 152 - 157 - 175 sup. - 166 - 171 - 174 - 175 - 177 - 178 - 183 - 185 - 189 inf. - 191 - 192 - 196 - 197 izda. - 198 - 205 - 206 - 210 - 212 - 213 - 214 - 216 - 217 - 218 - 221 - 222 - 224 - 225 super.; Guillermo Gil: 116; Mario Gustavo Costa; 154; Stephan Halloy: 161 - 162; Sofía Heinonmu: 225 inf.; Hunters: 54; Gerhard Jurzitza: 28; Mandojana: 118 - 121; Loren A. McIntyre: 20 - 36 - 202; M. Olano-J. Echevarri/Incafo: 150 - 177; H. Rivarola: 28 - 32 - 38 - 42; Mauricio Rumboll: 140; Arturo Tarak: 167; Juan Carlos Vázquez: 216; Michel Thibaud: 65 - 81; Günter Ziesler: 62 - 66 - 70 - 74 - 93 - 122 - 133.

Cover: Francisco Erize